Cretaceous–Tertiary High-Latitude Palaeoenvironments: James Ross Basin, Antarctica

Geological Society books refereeing procedures

The Society makes every effort to ensure that the scientific and production quality of its books matches that of its journals. Since 1997, all book proposals have been refereed by specialist reviewers as well as by the Society's Books Editorial Committee. If the referees identify weaknesses in the proposal, these must be addressed before the proposal is accepted.

Once the book is accepted, the Society Book Editors ensure that the volume editors follow strict guidelines on refereeing and quality control. We insist that individual papers can only be accepted after satisfactory review by two independent referees. The questions on the review forms are similar to those for *Journal of the Geological Society.* The referees' forms and comments must be available to the Society's Book Editors on request.

Although many of the books result from meetings, the editors are expected to commission papers that were not presented at the meeting to ensure that the book provides a balanced coverage of the subject. Being accepted for presentation at the meeting does not guarantee inclusion in the book.

More information about submitting a proposal and producing a book for the Society can be found on its web site: www.geolsoc.org.uk.

It is recommended that reference to all or part of this book should be made in one of the following ways:

Francis, J. E., Pirrie, D. & Crame, J. A. (eds) 2006. *Cretaceous–Tertiary High-Latitude Palaeoenvironments: James Ross Basin, Antarctica.* Geological Society, London, Special Publications, **258.**

Crame, J. A., Pirrie, D. & Riding, J. B. 2006. Mid-Cretaceous stratigraphy of the James Ross Basin, Antarctica. *In*: Francis, J. E., Pirrie, D. & Crame, J. A. (eds) *Cretaceous–Tertiary High-Latitude Palaeoenvironments: James Ross Basin, Antarctica.* Geological Society, London, Special Publications, **258**, 7–19.

GEOLOGICAL SOCIETY SPECIAL PUBLICATION NO. 258

Cretaceous–Tertiary High-Latitude Palaeoenvironments: James Ross Basin, Antarctica

EDITED BY

J. E. FRANCIS
University of Leeds, UK

D. PIRRIE
University of Exeter in Cornwall, UK

and

J. A. CRAME
British Antarctic Survey, UK

2006
Published by
The Geological Society
London

THE GEOLOGICAL SOCIETY

The Geological Society of London (GSL) was founded in 1807. It is the oldest national geological society in the world and the largest in Europe. It was incorporated under Royal Charter in 1825 and is Registered Charity 210161.

The Society is the UK national learned and professional society for geology with a worldwide Fellowship (FGS) of 9000. The Society has the power to confer Chartered status on suitably qualified Fellows, and about 2000 of the Fellowship carry the title (CGeol). Chartered Geologists may also obtain the equivalent European title, European Geologist (EurGeol). One fifth of the Society's fellowship resides outside the UK. To find out more about the Society, log on to www.geolsoc.org.uk.

The Geological Society Publishing House (Bath, UK) produces the Society's international journals and books, and acts as European distributor for selected publications of the American Association of Petroleum Geologists (AAPG), the American Geological Institute (AGI), the Indonesian Petroleum Association (IPA), the Geological Society of America (GSA), the Society for Sedimentary Geology (SEPM) and the Geologists' Association (GA). Joint marketing agreements ensure that GSL Fellows may purchase these societies' publications at a discount. The Society's online bookshop (accessible from) offers secure book purchasing with your credit or debit card.

To find out about joining the Society and benefiting from substantial discounts on publications of GSL and other societies worldwide, consult www.geolsoc.org.uk, or contact the Fellowship Department at: The Geological Society, Burlington House, Piccadilly, London W1J 0BG: Tel. +44 (0)20 7434 9944; Fax +44 (0)20 7439 8975; E-mail: enquiries@geolsoc.org.uk.

For information about the Society's meetings, consult *Events* on www.geolsoc.org.uk. To find out more about the Society's Corporate Affiliates Scheme, write to enquiries@geolsoc.org.uk.

Published by The Geological Society from:
The Geological Society Publishing House, Unit 7, Brassmill Enterprise Centre, Brassmill Lane, Bath BA1 3JN, UK

(*Orders*: Tel. +44 (0)1225 445046, Fax +44 (0)1225 442836)
Online bookshop: www.geolsoc.org.uk/bookshop

British Library Cataloguing in Publication Data

A catalogue record for this book is available from the British Library.

ISBN 10: 1-86239-197-1
ISBN 13: 978-1-86239-197-0

Typeset by Type Study, Scarborough, UK

Printed by Antony Rowe, Chippenham, UK

Distributors

USA

AAPG Bookstore, PO Box 979, Tulsa, OK 74101-0979, USA
Orders: Tel. + 1 918 584-2555
Fax +1 918 560-2652
E-mail bookstore@aapg.org

India

Affiliated East-West Press PVT Ltd, G-1/16 Ansari Road, Darya Ganj, New Delhi 110 002, India
Orders: Tel. +91 11 2327-9113/2326-4180
Fax +91 11 2326-0538
E-mail affiliat@vsnl.com

Japan

Kanda Book Trading Company, Cityhouse Tama 204, Tsurumaki 1-3-10, Tama-shi, Tokyo 206-0034, Japan
Orders: Tel. +81 (0)423 57-7650
Fax +81 (0)423 57-7651
Email geokanda@ma.kcom.ne.jp

Contents

Cretaceous–Tertiary high-latitude palaeoenvironments, James Ross Basin, Antarctica: introduction

J. E. FRANCIS[1], J. A. CRAME[2] & D. PIRRIE[3]

[1]*School of Earth and Environment, University of Leeds, Leeds LS2 9JT, UK (e-mail: j.e.francis@leeds.ac.uk)*

[2]*British Antarctic Survey, Natural Environment Research Council, High Cross, Madingley Road, Cambridge CB3 0ET, UK (e-mail: JACR@bas.ac.uk)*

[3]*Camborne School of Mines, School of Geography, Archaeology and Earth Resources, University of Exeter in Cornwall, Tremough Campus, Penryn, Cornwall TR10 9EZ, UK (e-mail: dpirrie@csm.ex.ac.uk)*

The James Ross Basin, at the northern tip of the Antarctic Peninsula, provides the thickest and best-exposed onshore Cretaceous and Early Tertiary sedimentary succession in Antarctica. When compared with other onshore sections, it is clear that the area has a much broader significance as a key reference section for the Cretaceous and Early Tertiary throughout the Southern Hemisphere. The sedimentary record exposed within the basin also provides an unrivalled opportunity to unlock the record of climate change and biotic response within a high-palaeolatitude setting.

James Ross Island was first visited during the heroic age of polar exploration at the start of the 20th century. Swedish geologist Otto Nordenskjöld sailed into the region in 1901 in his ship *Antarctic*, captained by explorer and sealer Carl Larsen. Plans to spend a year in the region for scientific exploration went disastrously wrong when his ship sank near Paulet Island, forcing Nordenskjöld to spend over 2 years in a small hut on Snow Hill Island. Members of his ship-wrecked party survived in horrific conditions, with only penguins for food and small stone huts for shelter at Hope Bay, at the tip of Trinity Peninsula, and also on Paulet Island.

Nordenskjöld's enforced stay in the area was, however, not unprofitable. In 1902 he and his five companions made trips over the sea ice to Seymour Island, where they made the first important fossil discoveries, including the bones of giant penguins (now known to be from the Eocene La Meseta Formation). They also made the first collections of fossil plants from the region, including leaves of flowering plants (from the Palaeocene Cross Valley Formation). This was well before Scott's collection of Permian *Glossopteris* plant fossils from the Transantarctic Mountains. The plants recovered by Nordenskjöld provided the first signal that Antarctica once had a warm climate, setting the agenda for future research.

Since the 1940s, the rocks and fossils of the James Ross region have been the target of serious scientific investigations. One of the first year-round Antarctic bases was established at Hope Bay in 1945, as part of Operation Tabarin, a UK armed forces expedition organized during the Second World War to protect Antarctic waters. Since then, the region has been extensively studied by scientists from many nations, most notably from Argentina, the USA and the UK, and systematic studies have led to comprehensive mapping and stratigraphic research. As such, the geology of the region around James Ross Island is now increasingly well known (Fig. 1).

Subduction of the Pacific Plate under the Antarctic Plate led to the evolution of an extensive magmatic arc, the remnants of which now form the Antarctic Peninsula. A large sedimentary basin, referred to as the Larsen Basin (see fig. 1 of Whitham *et al.*), was initiated in Jurassic times in the early stages of continental rifting to the east of the peninsula. The James Ross Basin, the subject of this collection of papers, is a small sub-basin located at the northern end of the Larsen Basin.

The James Ross Basin initially formed as a back-arc basin, next to the intermittently active volcanic arc. It was continuously subsiding, providing the accommodation space for the deposition of in excess of 5 km of marine sedimentary rocks for over 115 Ma, through the Late Jurassic to the Late Eocene. The evolution of the basin fill includes a regressive megasequence, beginning with Jurassic strata formed in deep-water anoxic conditions, followed by the

From: FRANCIS, J. E., PIRRIE, D. & CRAME, J. A. (eds) 2006. *Cretaceous–Tertiary High-Latitude Palaeoenvironments, James Ross Basin, Antarctica.* Geological Society, London, Special Publications, **258**, 1–5. 0305-8719/06/$15

Fig. 1. Satellite map showing the principal localities in the James Ross Basin. LANDSAT ETM Path 216 Row 105, 21 February 2000.

coarse clastic sediments of the Early Cretaceous Gustav Group that record erosion of the tectonically active arc. The basin continued to fill with finer grained shallow-marine sediments of the Marambio Group, in turn overlain by the shallow-water sediments of the Seymour Island Group.

The result of years of geological research now confirms the James Ross Basin fill as one of the most important Early Cretaceous–early Palaeogene sedimentary sequences in the Southern Hemisphere. It has prolific marine invertebrate faunas, an important vertebrate record, the highest latitude section in the Southern Hemisphere across the Cretaceous–Tertiary boundary, and fossil floras that are key to understanding past climate change and biogeographic evolution. In addition, exceptional fossil preservation has enabled the use of strontium isotopes to provide a more robust chronostratigraphic framework for palaeoclimatic and evolutionary histories, and has highlighted the importance of the basin within a global context. It, undoubtedly, has many more geological secrets to yield.

The papers in this volume represent recent research on various aspects of the James Ross Basin geology. The extensive Cretaceous sedimentary sequence exposed within the James Ross Basin is discussed by **Crame *et al.*** Their studies indicate that this sequence is critical for regional stratigraphic correlations in the Southern Hemisphere, and also our understanding of the radiation and extinction of a range of fossil groups. Their work identifies facies variations and local unconformities that were the result of fault-controlled deep-marine sedimentation along the basin margin, including an unconformity in the Cenomanian–late Turonian sequence. The Turonian–Coniacian boundary is provisionally placed at the junction between the Whisky Bay and Hidden Lake formations. The revised stratigraphic ages for this section indicate that the Late Cretaceous radiations of a number of major plant and animal groups can be traced back to at least the Turonian stage. This raises the possibility that their dissemination might be linked to the global Cretaceous thermal maximum.

Sediments of the Cretaceous basin fill are also discussed in a paper by **Whitham *et al.*** The Coniacian Hidden Lake Formation, exposed on James Ross Island, is a 300–400 m-thick succession of marine volcaniclastic conglomerates, sandstones and mudstones. It occurs at a point of transition in the evolution of the James Ross Basin, as it is underlain by deep-marine strata and overlain by shallow-marine strata. The succession reflects the two main factors controlling the deposition of the formation: (1) the influx of large quantities of volcaniclastic sediment; and (2) a pronounced inversion event in the early Coniacian that heralds the cessation of transpressive tectonic activity in the basin. The succession is dominated by a range of sediment density-flow deposits that, combined with the limited faunas and the lack of wave-induced structures, suggest deposition in a relatively deep marine environment below storm-wave base. The infilling of this basin topography by sediment and waning intrabasinal tectonism during the Coniacian resulted in the progressive elimination of the basin-floor topography and the onset of shallow-marine shelf sedimentation.

James Ross Island has provided some of the earliest angiosperm macrofossils from Antarctica, described here by **Hayes *et al.*** The fossilized remains of Cretaceous angiosperm leaves are preserved within sandstones and siltstones of the Coniacian Hidden Lake Formation (Gustav Group) and the late Coniacian–Campanian Santa Marta Formation (Marambio Group). The leaves represent the remains of vegetation that grew at approximately 65°S on the emergent volcanic arc, and was subsequently transported and buried in marine sediments in the adjacent back-arc basin. On the basis of their morphology, some of the angiosperm leaf morphotypes can be tentatively compared to those of living families such as Sterculiaceae, Lauraceae, Winteraceae, Cunoniaceae and Myrtaceae. Palaeoclimate analysis based on physiognomic aspects of the leaves, such as leaf-margin analysis, indicates that the mean annual temperatures for the Hidden Lake and Santa Marta formations were 12–21 °C (mean 17 °C) and 14–23 °C (mean 19 °C), respectively. The fossil plants are indicative of warm terrestrial climates without extended periods of winter temperatures below freezing and with adequate moisture for growth. This period of Cretaceous warmth in Antarctica corresponds with the mid-Cretaceous thermal maximum, an interval of peak global warmth from the Turonian to the early Campanian.

The past vegetation of the James Ross region is also discussed in a paper by **Poole & Cantrill**. A compilation of data for Cretaceous and Cenozoic Antarctic fossil wood floras, predominantly from the James Ross Basin, provides a different perspective on floristic and vegetation change when compared with previous studies that have focused on leaf macrofossils or palynology. Four phases of vegetation development in the overstorey are recognized based on the distribution and taxonomic composition of wood floras: Aptian–Albian coniferous forests; ?Cenomanian–Santonian mixed angiosperm forests; Campanian–Maastrichtian southern temperate forests; and Palaeocene–Eocene reduced diversity *Nothofagus* forests. Climate change during the Cretaceous and Tertiary influenced the composition of the vegetation, but evolving palaeoenvironments in the Antarctic Peninsula region may have been of equal, if not greater, importance.

It is becoming clear that the Late Cretaceous strata in the James Ross Basin also contain a rich vertebrate fauna. There are fossils of sharks, teleost and chimeroid (shell-crushing) fish, plesiosaurs and mosasaurs, both as juvenile and adult forms. **Kriwet *et al.*** discuss the record of fish diversity. **Martin & Crame** examine the palaeobiological significance of these high-latitude faunas, in particular the difference between abundance and species richness. For example, plesiosaurs are abundant but taxonomically limited, whereas mosasaurs are not abundant but are as taxonomically diverse as elsewhere in the world. Although such patterns may be due to collecting bias, there may be an environmental signal here. It is intriguing that both sea turtles and rays are absent from the fossil record in the James Ross Basin; these are warm water creatures and their absence may be a signal of cooler conditions in high latitudes in the Late Cretaceous. The biostratigraphy of the marine lizards, the mosasaurs, are specifically discussed in a paper by **Martin**. They are present in Late Campanian–Late Maastrichtian deposits, as in the rest of the world, and include some spectacular specimens with jaws and cranial material. Study of the taxa suggests that a mix of both cosmopolitan and endemic genera are present.

During the earliest Cenozoic the James Ross Basin continued to fill, but basin uplift or decreased basin subsidence was outpaced by the sedimentation rate and led to the development of a broad shallow shelf, sporadically emergent during the Palaeogene. **Marenssi** provides a new analysis of sedimentation in the Eocene La Meseta Formation, a composite incised-valley system developed on the emergent marine shelf. Stratigraphic, sedimentological, palaeontological and geochemical data all indicate that Eocene sedimentation in the James Ross Basin was mainly controlled by eustatic sea-level changes. The sedimentary succession features six erosionally based members, defined by Marenssi, that were probably caused by lowstands of sea-level. These episodes also match sea-level changes observed in other parts of the world, taking the James Ross Basin into a global context.

As Nordenskjöld discovered in 1901–1903, Seymour Island, on which the Eocene La Meseta Formation crops out, is richly fossiliferous. A rich vertebrate fauna has been described after many expeditions by Argentinean and American palaeontologists and others, and the emerging information is contributing to our understanding of the evolution of the mammals. **Goin *et al.*** describe an enigmatic group of extinct non-therian mammals, the gondwanatherians. This group developed rodent-like incisors and the earliest known hypsodont cheek-teeth among mammals. A rodent-like dentary fossil from La Meseta Formation suggests that this is the youngest occurrence of a gondwanathere, adding significant direct and indirect evidence to the already documented cosmopolitanism of gondwanatheres among Gondwanan mammals, and the crucial biogeographical role of Antarctica during the Cretaceous–Tertiary mammalian transition.

During his unexpected stay on Seymour Island, Otto Nordenskjöld discovered bones of giant penguins. Penguins are by far the dominant group of marine vertebrates in the Eocene La Meseta Formation, and **Tambussi *et al.*** describe two new species of fossil penguins to add to the known fauna, some small and some large, that lived together. They present a wonderful new pictorial reconstruction of these famous penguins. We are beginning to appreciate that the initial radiation of this key Antarctic group was more complex than originally anticipated.

A new mammal, *Notolophus arquinotiensis*, is described by **Bond *et al.*** Several isolated teeth from the La Meseta Formation on Seymour Island show that this is a new type of ungulate (hoofed animal such as camels and horses) of the extinct South American ungulate order Litopterna. This new taxon shows close affinities to taxa from South America and once again illustrates the importance of Antarctica in the evolution of certain biotas, in this case the ungulates.

The terrestrial vertebrates from the Eocene of the James Ross Basin provide intriguing insights into life at high latitudes. **Case** has analysed the body sizes of the vertebrates that are preserved on Seymour Island. They range from small insectivorous, omnivorous and granivorous marsupials, plus the rodent-like non-therian gondwanathere, to large-sized ungulates, a sloth and cursorial birds (a ratite and a phororachoid). However, he notes that medium-sized, homeothermic animals in the size range represented by rabbits to small ungulates are missing. In comparable faunas in Patagonia the whole range of body sizes are present. Case indicates that the bimodal body size pattern is not unlike that seen in higher latitude mammalian faunas of North America today, which may be a response to cold winter temperatures in these higher latitudes. The smaller mammals have adapted to the cold winter temperatures through physiological means, and the larger animals have adapted to the cold winter conditions by conserving heat through small surface-area-to-volume ratios as a result of their greater bulk. Medium-sized animals would have lacked these thermal strategies and thus be at a selective disadvantage.

The youngest sediments in the James Ross Basin are unconformable over the Cretaceous–Eocene basin fill. These are the Neogene deposits within the James Ross Island Volcanic Group that provide important information about the nature and dynamics of late Cenozoic glaciation. Extensive glacigenic sediments, comprising the Hobbs Glacier Formation, discussed here by **Hambrey & Smellie**, occur extensively within and at the base of this group. The principal facies are diamictite and mudstone, overlain by a variety of volcanic rocks (tuff, lava, breccia). The diamictite is interpreted as remobilized proximal glaciomarine sediment and, in one place at least, as a basal till. Within the volcanic sequence itself, there is further evidence of repeated glacial activity in the form of striated pavements and thin diamictite units, which separate all of the volcanic units mapped so far. Volcanism was contemporaneous with glaciation, as in places the diamictite has been contact-metamorphosed by basaltic lava.

The James Ross Basin is now one of the key sections in the world for high-latitude global-change studies. We now have further evidence of the mid-Cretaceous phase of global warming, cooling in the latest Cretaceous, climatic fluctuations across the Cretaceous–Tertiary (K/T) boundary, a Palaeocene–Eocene thermal maximum and then progressive cooling through the Cenozoic. There are some indications of temperate conditions from the La Meseta Formation, but the onset of glacial conditions is not recorded. The James Ross Island Volcanic Group has considerable potential for refining the Neogene climate record at the northern tip of the Antarctic Peninsula.

Both the Cretaceous faunas and floras from the James Ross Basin have a largely temperate aspect to them, with perhaps a hint of warmer conditions in the late Turonian–Coniacian. However, perhaps the greatest significance of the marine faunas is their record of various biotic events, both leading up to and across the K/T boundary. These events include both phased extinctions and a series of originations of key austral groups. The plants and vertebrates also illustrate the origination of key austral taxa (e.g. mammals, penguins, angiosperms), but there is still much to do to refine this record. Many molecular phylogenetic models are now predicting Cretaceous or earlier origins for modern taxa and it is more than likely that some of these will have been in the southern high latitudes.

It is likely that future studies will need to concentrate on further refinement of both the litho- and biostratigraphies. Strontium isotope dating will be crucial for providing both absolute ages and unequivocal correlations with the Northern Hemisphere. We are also beginning to appreciate that much of this key reference section can now be placed within a sequence stratigraphic framework to detect possible glacio-eustatic sea-level changes at a number of different stratigraphic levels. We suggest that the Coniacian–Campanian Santa Marta Formation may be a key stratigraphic unit for further intense investigation.

It is also likely that many future palaeoenvironmental breakthroughs in Antarctica will be made offshore through drilling programmes such as ANDRILL, Shaldrill and IODP. However, effective correlation of these cores will only be possible with the use of standard onshore reference sections. The most important of these is undoubtedly that exposed within the James Ross Basin.

Mid-Cretaceous stratigraphy of the James Ross Basin, Antarctica

J. A. CRAME[1], D. PIRRIE[2] & J. B. RIDING[3]

[1]*British Antarctic Survey, Natural Environment Research Council, High Cross, Madingley Road, Cambridge CB3 OET, UK (e-mail: JACR@bas.ac.uk)*

[2]*Camborne School of Mines, School of Geography, Archaeology and Earth Resources, University of Exeter, Cornwall Campus, Penryn, Cornwall TR10 9EZ, UK*

[3]*British Geological Survey, Keyworth, Nottingham NG12 5GG, UK*

Abstract: The extensive Cretaceous sedimentary sequence exposed within the James Ross Basin, Antarctica, is critical for regional stratigraphic correlations in the Southern Hemisphere, and also for our understanding of the radiation and extinction of a range of taxonomic groups. However, the nature and definition of Cenomanian–Turonian strata on the NW margins of James Ross Island has previously been difficult, due both to marked lateral facies changes and to stratigraphical discontinuities within the extensive Whisky Bay Formation. Facies variation and local unconformities were the result of fault-controlled deep-marine sedimentation along the basin margin. In this study the Albian–Cenomanian boundary is defined for the first time in the upper levels of the Lewis Hill Member of the Whisky Bay Formation. However, there is a Cenomanian–late Turonian unconformity between the Lewis Hill and Brandy Bay members of the Whisky Bay Formation. Equivalent lithostratigraphical units exposed further to the SW on James Ross Island appear to be more complete with the early Cenomanian–late Turonian interval represented by the upper parts of the Tumbledown Cliffs and the lower part of the Rum Cove members of the Whisky Bay Formation. The Turonian–Coniacian boundary is provisionally placed at the junction between the Whisky Bay and Hidden Lake formations. The revised stratigraphic ages for this section show that the Late Cretaceous radiations of a number of major plant and animal groups can be traced back to at least the Turonian stage. This raises the possibility that their dissemination might be linked to the global Cretaceous thermal maximum.

The extensive Cretaceous sedimentary sequence exposed within the James Ross Basin, NE Antarctic Peninsula (Fig. 1) is assuming an ever greater significance for regional stratigraphical correlations in the Southern Hemisphere (Feldmann & Woodburne 1988; Rinaldi 1992; Crame *et al.* 2004). Totalling more than 5 km in thickness, this succession is now known to range in age from earliest Aptian to the latest Maastrichtian (Riding & Crame 2002), and contains a range of both marine and terrestrial fossil taxa suitable for biostratigraphy. The importance of this locality has been heightened by the successful introduction of a partial strontium isotope stratigraphy (SIS) (McArthur *et al.* 1998, 2000; Crame *et al.* 1999).

As studies have intensified in recent years it has become apparent that two particularly thick and continuous sequences are present within the James Ross Basin: an Aptian–Albian one (which is approximately 1750 m thick and assigned to the Gustav Group); and a Campanian–Maastrichtian one (2150 m thick and assigned to the Marambio Group) (Figs 1 & 2) (Feldmann & Woodburne 1988; Pirrie *et al.* 1991*a*, 1997; Rinaldi 1992; Riding & Crame 2002; Crame *et al.* 2004). Strata representing the intervening Cenomanian–Santonian stages have traditionally been less well defined, due partly to a lack of diagnostic index fossils and partly to pronounced lateral facies changes at key localities along the NW coast of James Ross Island (Ineson *et al.* 1986; Olivero *et al.* 1986). One particularly striking anomaly has been the presence of Cenomanian strata in the Tumbledown Cliffs–Rum Cove region (Fig. 1) but their apparent absence some 25 km to the NE in the Whisky Bay–Brandy Bay region (Fig. 2). Is there a stratigraphical hiatus here of local or even regional extent?

This study examines the mid-Cretaceous (defined here as Cenomanian–Coniacian) stratigraphy of the James Ross Basin. A series of new field observations from the Brandy Bay reference section are used to recalibrate litho-, bio- and chronostratigraphies for the area, and suggest correlations with localities further SW along the coast of James Ross Island (Figs 1 & 2). It is anticipated that this revised stratigraphy will in turn facilitate the investigation of a series of major Cenomanian–Coniacian palaeoclimatic

From: Francis, J. E., Pirrie, D. & Crame, J. A. (eds) 2006. *Cretaceous–Tertiary High-Latitude Palaeoenvironments, James Ross Basin, Antarctica.* Geological Society, London, Special Publications, **258**, 7–19. 0305–8719/06/$15

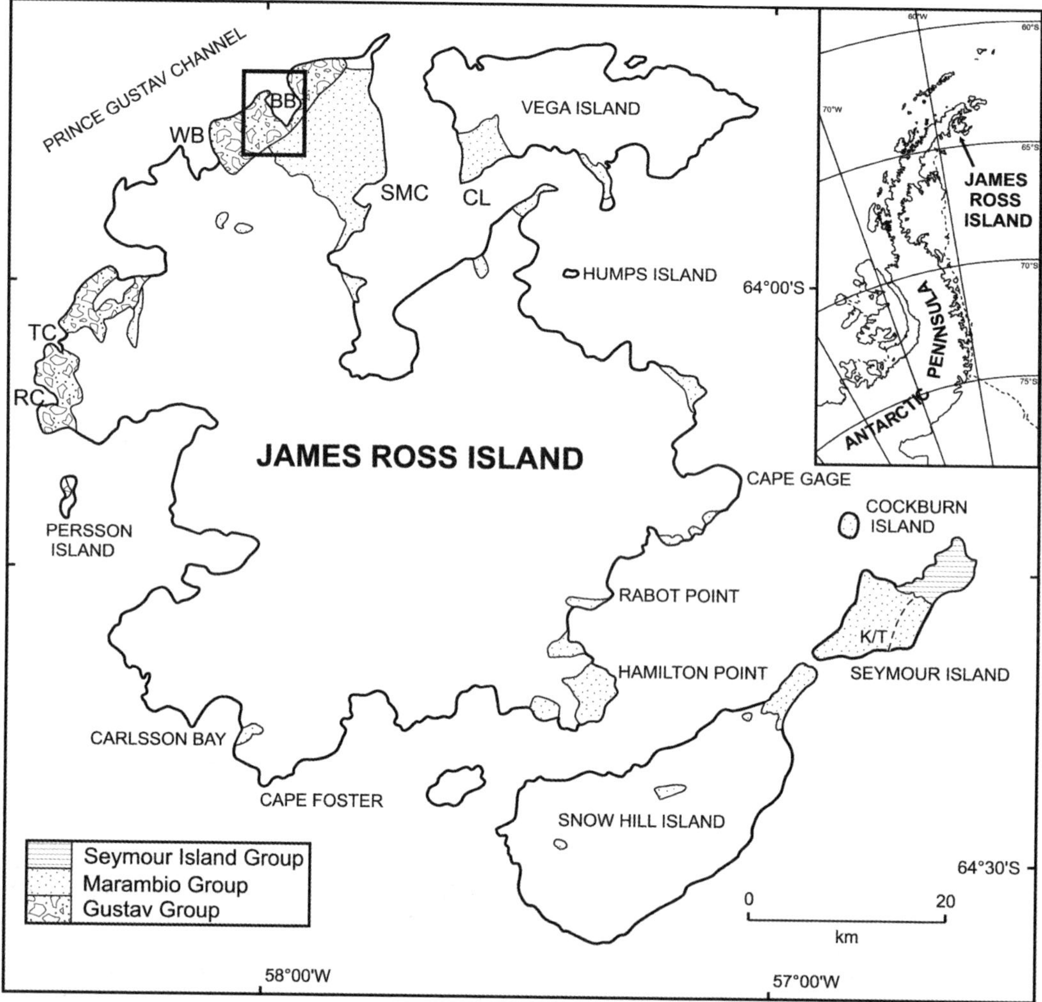

Fig. 1. Schematic geological map of the James Ross Island area. White areas are either the James Ross Island Volcanic Group or snow/ice cover. BB, Brandy Bay; CL, Cape Lamb; RC, Rum Cove; SMC, Santa Marta Cove; TC, Tumbledown Cliffs; WB, Whisky Bay. Area of outcrop of the Pedersen Formation on the adjacent Antarctic Peninsula shown in Hathway & Riding (2001, fig. 1). Position of the K/T (Cretaceous–Tertiary) boundary on Seymour Island is indicated.

and palaeobiological events in Antarctica (Huber 1998; Cantrill & Poole 2002).

Background and methods

The James Ross Basin, which may in turn be a component of the larger Larsen Basin, was one of a series of extensive back-arc basins that formed in the Patagonia–Antarctic Peninsula region during the mid-Mesozoic–early Cenozoic (Hathway 2000). The Cretaceous basin fill comprises a regressive mega-sequence of arc-derived clastic and volcaniclastic marine rocks that has been subdivided into the older Gustav Group (Aptian–Coniacian) and the younger Marambio Group (Coniacian–Danian) (Hathway 2000; Hathway & Riding 2001; Riding & Crame 2002 and references therein).

The Gustav Group is confined to the NW coast of James Ross Island and certain isolated outcrops on the adjacent margins of the NE Antarctic Peninsula (Fig. 1) (Riding & Crame 2002). Reaching a maximum thickness of 2.6 km, it is characterized by coarse-grained lithologies such as pebble–boulder conglomerates, breccias and coarse- to fine-grained pebbly

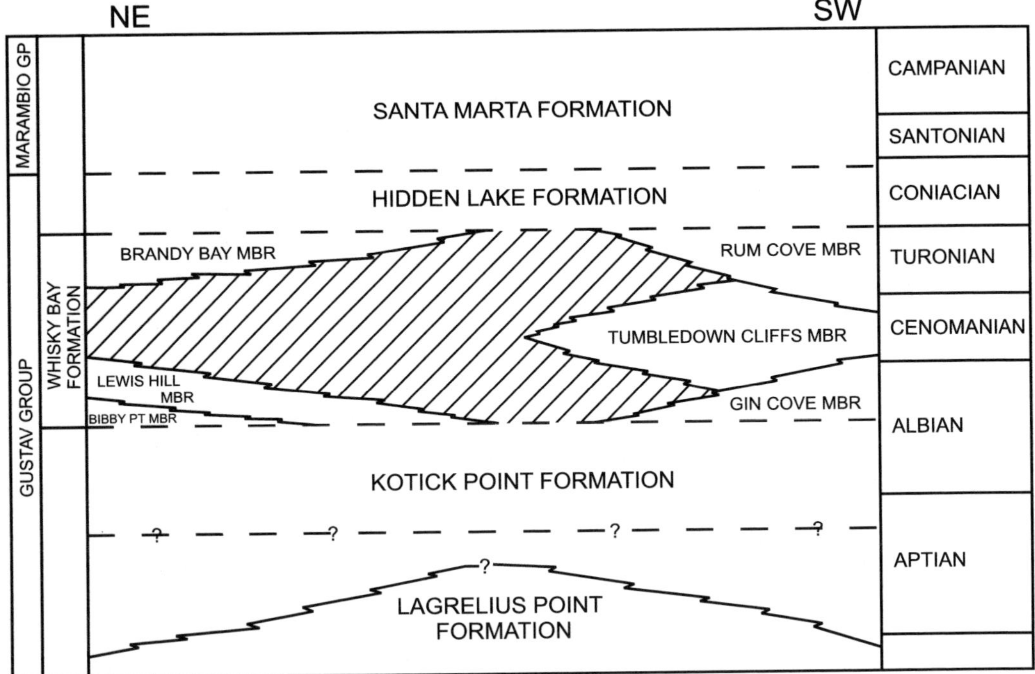

Fig. 2. Stratigraphic correlation diagram for the Lagrelius Point, Kotick Point, Whisky Bay, Hidden Lake and Santa Marta formations. Note the considerable lateral variation within the Whisky Bay Formation. Age assignments are based on a combination of micro- and macrofossil biostratigraphy, and strontium isotope stratigraphy.

sandstones, together with subordinate sandstones, siltstones and mudstones. It has been formally subdivided into five component formations (the Pedersen, Lagrelius Point, Kotick Point, Whisky Bay and Hidden Lake formations) and is generally interpreted to represent a variety of slope-apron and deep-water submarine-fan environments (Ineson 1989; Buatois & Medina 1993). The Gustav Group dips SE and passes conformably upwards into the finer grained Marambio Group (Fig. 1). The latter unit, which is up to 2.5 km thick and is exposed over the greater part of the James Ross Basin, comprises a variety of fine- to medium-grained sandstones, siltstones and silty mudstones, with minor coarser grained intervals, coquinas and other shell beds. The Marambio Group is in places intensely fossiliferous, with vertebrate, invertebrate, plant and microfossil assemblages that have been described in detail in recent years. Based on detailed field mapping and lithostratigraphy, the Marambio Group has been subdivided into three component formations (the Santa Marta, Snow Hill Island and López de Bertodano formations) (Pirrie *et al.* 1997). The Marambio Group was deposited in a variety of inner- to outer-shelf settings (Macellari 1988; Pirrie 1989; Pirrie *et al.* 1991*a*; Scasso *et al.* 1991).

The steepest structural dips in the James Ross Basin occur along the NW coastal margin, where they typically range between 45°SE and subvertical (Fig. 1). However, these dips decrease rapidly SE and within a horizontal distance of 5 km can be as low as 10°SE. The Gustav Group of NW James Ross Island is in effect exposed in a NE-trending monoclinal syncline that can be shown to be the product of syn- rather than post-depositional deformation (Whitham & Marshall 1988). In the Aptian–Coniacian this region was in close proximity to the fault-bounded basin margin where phases of arc uplift and related differential subsidence led to the accumulation of a deep-marine clastic wedge. In effect, this wedge was continually tilted to the SE and successive beds onlapped around a single, progressive unconformity (Whitham & Marshall 1988, fig. 6; Hathway 2000, fig. 9).

The reference section through both the Whisky Bay and Hidden Lake formations on the SW shore of Brandy Bay (Ineson *et al.* 1986;

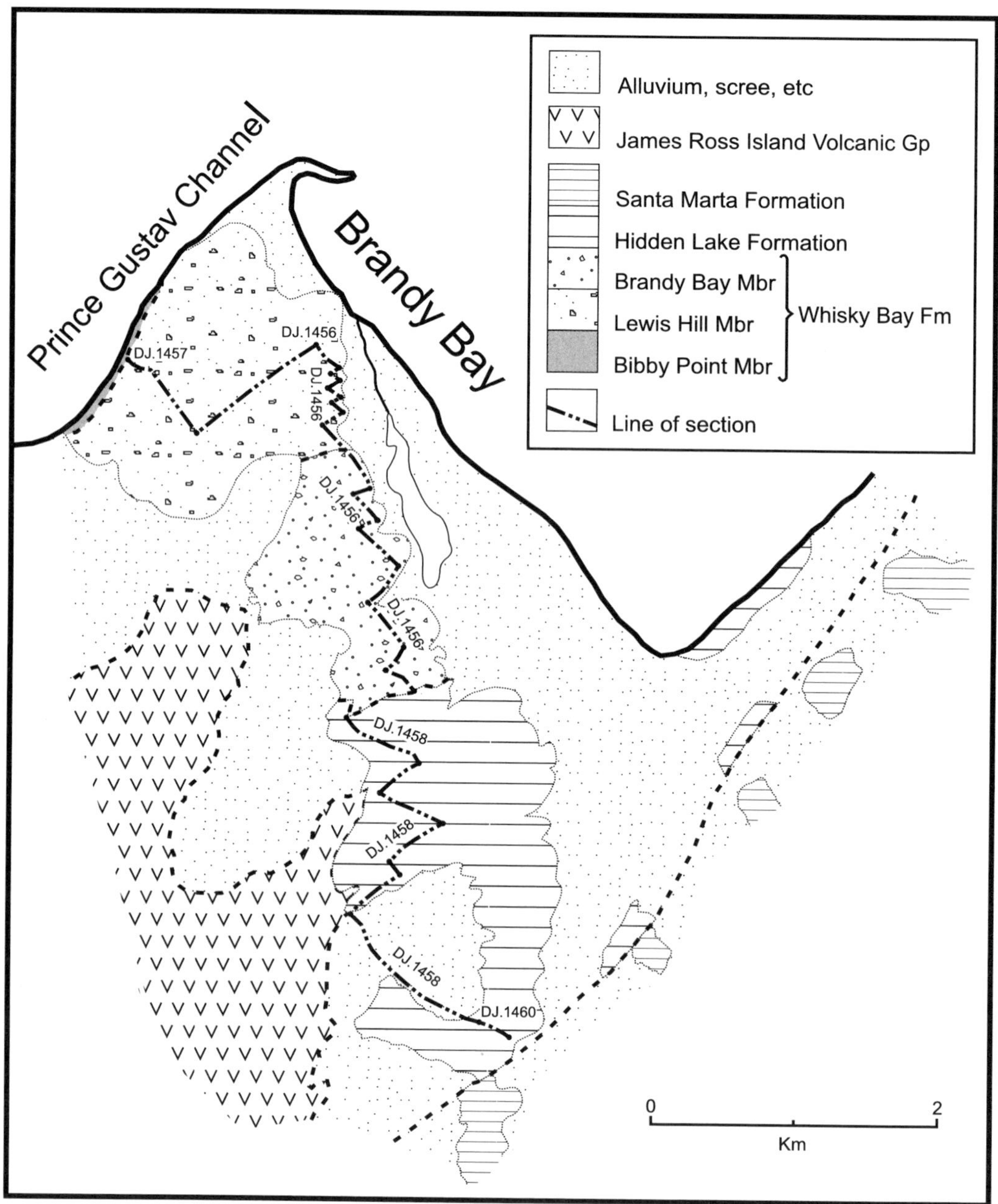

Fig. 3. Geological map of the Brandy Bay area showing the position of the key reference section measured through the Whisky Bay and Hidden Lake formations. Location map for Figure 3 is given in Figure 1.

McArthur *et al.* 2000) was re-examined by two of us (J. A. Crame and D. Pirrie) in early 2002. Wherever possible, the line of the old section (D.8228) was followed and its position plotted on a topographical map by GPS (using the new station numbers DJ.1456–DJ.1460; Fig. 3). Bed thicknesses were measured using a combination of Jacob staff/abney level and tape measure techniques, and detailed collections made for macro- and micropalaeontology. Samples for the latter were analysed by one of us (J. B. Riding) in the laboratories of the British Geological Survey, Keyworth, UK. Ammonite determinations reported here are essentially provisional ones and will be the subject of a forthcoming taxonomic review (led by W. J. Kennedy). It is also hoped that some macrofossil samples will prove amenable to dating by $^{87}Sr/^{86}Sr$ analysis.

Whisky Bay Formation

Bibby Point Member. The Bibby Point Member constitutes the basal 79 m of the measured section (Figs 3 & 4). It is composed of dark green, channelized, normally graded pebble conglomerates interbedded with pebbly sandstones, sandstones and scarce mudstones. Clasts in the conglomerates are dominated by angular blocks of interbedded radiolarian mudstones and volcanic ash, interpreted as being derived from the Nordenskjöld Formation (Late Jurassic–Early Cretaceous), along with well-rounded metasedimentary clasts derived from the Trinity Peninsula Group (? uppermost Carboniferous–Triassic) and dark green, chloritic mudstones. Three dimitobelid belemnites (DJ.1457.1–DJ.1457.3) can be added to the previously known mid–late Albian molluscan fauna (Ineson *et al.* 1986; Riding & Crame 2002).

Lewis Hill Member. The Bibby Point Member passes conformably upwards into the 552 m-thick Lewis Hill Member (Figs 3 & 4). The lower 140 m of this unit is dominated by cobble conglomerates with minor intercalated pebbly sandstones and mudstones. The conglomerates are clast supported, with abundant silicic volcanic (Antarctic Peninsula Volcanic Group) and metasedimentary clasts (Trinity Peninsula Group), together with some Nordenskjöld Formation blocks and intraformational siltstone clasts up to 2 m across. The section then shows a gradual fining- and thinning-upwards trend with thinner bedded small pebble conglomerates interbedded with siltstones. A distinctive sequence of fossiliferous, pale grey weathering medium- to coarse-grained sandstones occurs between approximately 320 and 370 m (Fig. 4), and this proved to be a useful marker horizon for lateral correlations. The upper 250 m of the Lewis Hill Member is only intermittently exposed. Cobble–boulder conglomerates initially predominate, but at higher levels there are thick bioturbated mudstone units with subordinate fine- to medium-grained sandstones and only rare pebble conglomerates.

The most prominent macrofossil occurring within the Lewis Hill Member is a small inoceramid bivalve that is locally abundant between 320 and 370 m (Fig. 4). This has been referred to *Actinoceramus concentricus* (Parkinson), *sensu lato*, and has strong middle–late Albian age affinities (Ineson *et al.* 1986; Crampton 1996*a*). Other macrofossils include gaudryceratid ammonites, the bivalve *Aucellina*, encrusting bryozoans and indeterminate gastropod moulds. Smooth terebratulid brachiopods are associated with the uppermost 135 m of the member (496–631 m, Fig. 4).

The Lewis Hill Member is also characterized by abundant and well-preserved palynofloras. Whereas miospores are dominated by *Cyathidites* spp. and bisaccate pollen grains, the dinoflagellate cysts are significantly more diverse and include types such as *Ascodinium* spp., chorate cysts, *Cribroperidinium edwardsii*, *Diconodinium multispinum*, *Odontochitina operculata* and various reworked taxa interpreted to be derived from the Late Jurassic–Early Cretaceous Nordenskjöld Formation. This flora is of unequivocal Australasian affinity and is consistent with a latest Albian age for almost the entire section (Morgan 1980; Riding & Crame 2002). Nevertheless, the three highest palynological samples (DJ.1504.15, DJ.1504.17 and DJ.1504.18; Fig. 4) yielded the first records of *Ascodinium serratum*, *sensu stricto*, in Antarctica and this distinctive species is believed to be confined to the early Cenomanian (Morgan 1980). Therefore the Albian–Cenomanian boundary can be placed for the first time in Antarctica between palynological samples DJ.1504.10 and DJ.1504.15 in our reference section (Fig. 4). The same level would also mark the boundary between the *Xenascus asperatus* and *Diconodinium multispinum* interval zones of Helby *et al.* (1987). Previous records of Turonian taxa such as *Isabelidinium glabrum* from the uppermost Lewis Hill Member (Riding & Crame 2002) are now thought to belong to Brandy Bay Member samples.

Brandy Bay Member. The boundary between the Lewis Hill and Brandy Bay members is marked by a sharp lithological transition from the mudstones, muddy sandstones and green-weathering volcaniclastic sandstones of the former to the rusty red–brown pebbly sandstones and conglomerates of the latter. It is also accompanied by a pronounced change in dip, from values as high as 35°–40°SE in the uppermost Lewis Hill Member to approximately 19°SE in the basal Brandy Bay Member.

The lower 75 m of the Brandy Bay Member is poorly exposed but dominated by slabby weathering coarse- to very-coarse-grained

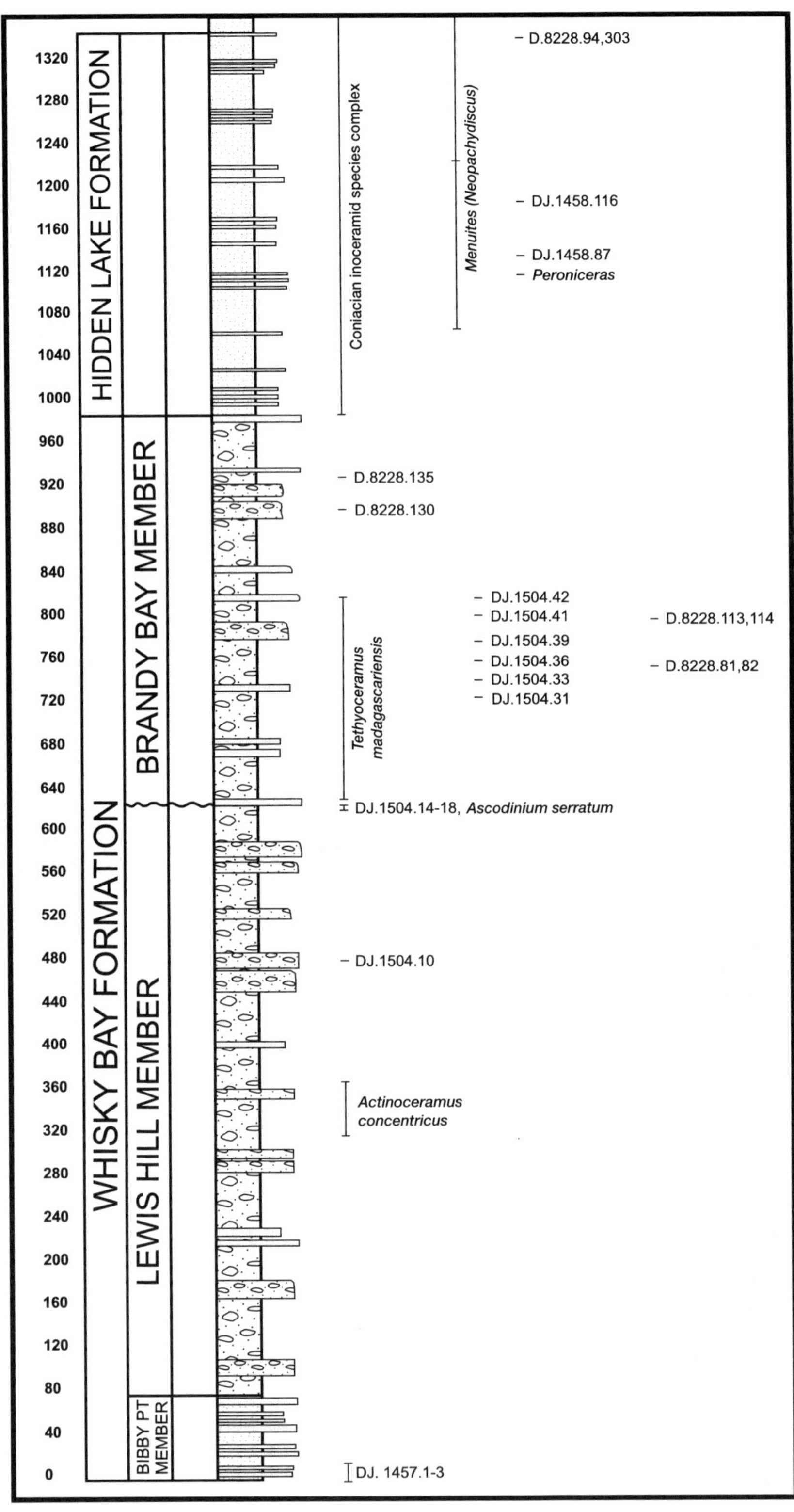

HIDDEN LAKE FORMATION
WHISKY BAY FORMATION
BRANDY BAY MEMBER
LEWIS HILL MEMBER
BIBBY PT MEMBER
1320
1280
1240
1200
1160
1120
1080
1040
1000
960
920
880
840
800
760
720
680
640
600
560
520
480
440
400
360
320
280
240
200
160
120
80
40
0
Coniacian inoceramid species complex
Menuites (Neopachydiscus)
– D.8228.94,303
– DJ.1458.116
– DJ.1458.87
– Peroniceras
– D.8228.135
– D.8228.130
Tethyoceramus madagascariensis
– DJ.1504.42
– DJ.1504.41
– DJ.1504.39
– DJ.1504.36
– DJ.1504.33
– DJ.1504.31
– D.8228.113,114
– D.8228.81,82
DJ.1504.14-18, Ascodinium serratum
– DJ.1504.10
Actinoceramus concentricus
DJ. 1457.1-3

pebbly sandstones, interbedded with granule–small pebble conglomerates. The latter are typically clast supported, with the clasts being well rounded and overwhelmingly of vein quartz and metasedimentary rocks derived from the Trinity Peninsula Group. Interbedded mudstones become progressively more abundant higher in the section until they predominate between approximately 808 and 987.5 m (Fig. 4). The silty mudstones are dark grey, planar laminated and bioturbated. They frequently contain carbonate concretions and in places these concretions are preferentially nucleated around networks of *Thalassinoides* burrows. Synsedimentary deformation is common in these upper mudstones, with rotational slump scars, slump sheets and isolated displaced blocks.

The base of the Brandy Bay Member is marked by the abrupt incoming of inoceramid bivalves that have traditionally been assigned to *Tethyoceramus madagascariensis* (Heinz); these can then be traced up to about the 820 m level in the section (Fig. 4). *Tethyoceramus madagascariensis* is an essentially Southern Hemisphere species with strong late Turonian or early Coniacian affinities (Ineson *et al.* 1986; Crampton 1996*b*). However, the type material, from Manasoa, SW Madagascar, has recently been revised and it is clear that *T. madagascariensis* is very closely related to three other late Turonian–early Coniacian species of *Tethyoceramus* (Walaszczyk *et al.* 2004). Indeed, the suggestion has even been made that the Antarctic specimens may be closer to *Inoceramus*? *nukeus* Wellman from the upper part of the Lower Coniacian in Madagascar and the undifferentiated Coniacian Teratan stage of New Zealand (Crampton 1996*b*). Until further taxonomic comparisons can be carried out, we prefer to leave the Antarctic material within *Tethyoceramus madagascariensis*, *sensu lato*.

Tethyoceramus madagascariensis is typically accompanied by a distinctive flat oyster (usually fragmented), numerous terebratulid brachiopods assigned provisionally to *Rectithyris whiskyi* Sandy (1991), and pleurotomariid and naticid gastropods. A large, thick-shelled astartid bivalve first occurs at 740 m and small, pieces of fossil wood are relatively common throughout. Although ammonites are rare, notable occurrences include a possible kossmaticeratid from the 900 m level (D.8228.130; Fig. 4) (? *Marshallites*; Thomson 1984) and a small heteromorph from 930 m (D.8228.135; ? diplomoceratid). A particularly distinctive faunal assemblage from a pebbly conglomerate at 755 m yielded small dimitobelid belemnites (*Dimitobelus cf. stimulus* (P. Doyle pers. comm. 2002)), a colonial coral (D.8228.83), encrusting bryozoans, astartids, oysters and a fragment from the hinge region of a thick-shelled bakevelliid bivalve (DJ.1456.192). Such an assemblage has a distinctly shallower water aspect to it than those encountered elsewhere within the Whisky Bay Formation.

Six samples from the Brandy Bay Member (DJ.1504.31–DJ.1504.42; Fig. 4) yielded variably productive palynofloras. These were noticeably less rich in organic material than those from the Lewis Hill Member, but wood fragments and various other plant tissues are common–abundant. New dinoflagellate cyst taxa from this interval include *Aptea* sp., *Oligosphaeridium* spp., and the key marker forms *Isabelidinium acuminatum* (late Cenomanian–early Turonian according to Morgan 1980 and Marshall 1984) and *I. glabrum* (early Turonian–mid-Coniacian). A consensus palynological age for the Brandy Bay Member would therefore appear to be early Turonian (Morgan 1980; Marshall 1984; Barreda *et al.* 1999), with a correlation to the *Palaeohystrichophora infusuroides* Interval Zone of Helby *et al.* (1987). Nevertheless, a suite of four oyster samples from approximately the 750–800 m-level in the section (D.8228.81, D.8228.82, D.8228.113 and D.8228.114; Fig. 4) gave consistent late Turonian $^{87}Sr/^{86}Sr$ ages (McArthur *et al.* 2000). This in turn indicates that at least the upper half of the informal *T. madagascariensis* zone must be of this age, as is the incoming of *Isabelidinium* spp. (Fig. 4). There is a possibility that there is a disparity in the age ranges of some Australian palynomorphs (Riding & Crame 2002); this is because of the inherent difficulties of making accurate correlations from Australasia to NW Europe, given relatively small numbers of cosmopolitan taxa. For example, the overlapping ranges of *Isabelidinium acuminatum* and *I. glabrum* may be younger than early Turonian as envisaged, for example, by Marshall (1984) and McMinn (1988). It is, of course, possible that the basal 125 m of the Brandy Bay Member is older than late Turonian and the topmost levels could be basal Coniacian. However, the Turonian–Coniacian boundary is probably best placed at the moment at the junction between the Brandy Bay Member and

Fig. 4. Summary sedimentary log through the Whisky Bay and Hidden Lake formations. Key biostratigraphic markers are indicated.

the overlying Hidden Lake Formation (see below).

Hidden Lake Formation

The lower 10 m of the Hidden Lake Formation in this region (Figs 3 & 4) comprises medium- to coarse-grained sandstones, largely derived from reworked pyroclastic tuffs, interbedded with wavy–planar laminated very-fine- to fine-grained sandstones and bioturbated mudstones. Overlying medium- to coarse-grained sandstones characteristically show lenticular mega-ripple bedforms thought to be tidal in origin (cf. Whitham *et al.* 2006). At this locality the basal Hidden Lake Formation is interpreted as representing the toesets of a substorm-wave base volcaniclastic fan delta succession passing laterally and vertically into a basin floor facies association (Whitham *et al.* 2006). The formation reflects a partial basin inversion event, separating the deeper water submarine-fan and slope-apron environments below from the overlying shallow-marine-shelf facies of the Santa Marta Formation. Sedimentation was intimately linked to coeval calc-alkaline pyroclastic eruptions on the adjacent volcanic arc. Between 998 and 1078 m (Fig. 4) the Hidden Lake Formation is characterized by interbedded parallel–wavy laminated bioturbated mudstones, siltstones, ripple cross-laminated and tabular cross-bedded sandstones, and normally graded sandstones. Several slump sheets, individually up to 3.5 m in thickness, are present between 1078 and 1108 m (Fig. 4), and at higher levels parallel to wavy–planar laminated medium-grained sandstones, siltstones and laminated mudstones become increasingly abundant. Broadly similar lithologies occur in the middle and upper levels of the Hidden Lake Formation, although in places exposure is rather poor. At 1330 m (Fig. 4) there is a distinctive change in lithologies to bioturbated silty sandstones yielding abundant carbonate concretions nucleated around macrofossils and *Thalassinoides*-type burrow networks. At approximately the 1345 m level these silty sandstones grade up into pale grey weathering siltstones and mudstones of the basal Santa Marta Formation (Figs 3 & 4).

The base of the Hidden Lake Formation is characterized by the sudden incoming of a distinctive group of inoceramid bivalves with strong Coniacian affinities. These typically show a pattern of *Anwachsringreifen* ornament, whereby narrow, regular and evenly spaced concentric rings are superimposed on low, open folds. Initial types typically have an erect valve profile and resemble European Lower Coniacian species such as *Inoceramus* (*Inoceramus*) *inaequivalvis* Schlüter and *I.* (*I.*) *koegleri* Andert (Crame 1983); they in turn grade up into more mytiloid forms provisionally assigned to *Inoceramus neocaledonicus* Jeannet. This species too has Coniacian age affinities in Europe, but may be Coniacian–Santonian in Madagascar (Sornay 1964; Herm *et al.* 1979; Walaszczyk 1992; Walaszczyk *et al.* 2004). As the *I. neocaledonicus* species group also shows some overlap with the highly variable *Inoceramus australis* Woods from the Piripauan stage (uppermost Coniacian–middle Santonian) of New Zealand (Crampton 1996*b*), it is clear that this whole species complex is in need of careful taxonomic revision (McArthur *et al.* 2000; Walaszczyk *et al.* 2004).

Terebratulid brachiopods, again assigned provisionally to *Rectithyris whiskyi*, oyster fragments and large astartid bivalves similar to those of the Brandy Bay Member still occur within the basal 100 m of the Hidden Lake Formation, but are noticeably less abundant. They are accompanied by bryozoans, pleurotomariid and patelliform gastropods, and dimitobelid belemnites (*Dimitobelus cf. ongleyi* (P. Doyle pers. comm. 2002)), together with a sparse vertebrate assemblage comprising reptile bones, along with shark teeth and vertebrae. Charcoalified wood fragments are particularly characteristic of the lower Hidden Lake Formation and are largely responsible for giving the formation its distinctive rusty-orange weathering hue; some of these fragments are in excess of 30 cm in length. A distinctive bed at the 1118 m level (Fig. 4) contains abundant compressed and charcoalified wood fragments, together with small angiosperm leaves retaining veination (cf. Hayes *et al.* 2006).

Ammonites are generally sparse in the lower–middle-levels of the Hidden Lake Formation, but there are occasional gaudryceratids, and at 1068 m (Fig. 4) there is the first occurrence of a small–medium pachydiscid referable to *Menuites* (*Neopachydiscus*) (W. J. Kennedy pers. comm. 2004). This taxon can then be traced up-section for approximately 100 m, but by 1218 m it has been replaced by a smaller, more tumid species of the same subgenus. This pachydiscid zonation may have at least local stratigraphic utility as *M.* (*N.*) sp. 1 in the Whisky Bay region (Fig. 1) is confined to a zone between 125 and 205 m above the base of the formation (Ineson *et al.* 1986). A series of strongly keeled fragments from approximately the 1120 m level (Fig. 4) has been referred to *Peroniceras* (Thomson 1984), and a small, straight-shafted heteromorph from 1138 m

(DJ.1458.87; Fig. 4) may be a diplomoceratid. A second, more complex heteromorph from approximately 1188 m (DJ.1458.116) resembles *Eubostrychoceras* from the basal Santa Marta Formation (Olivero 1988).

Although the exposure is typically poor, the uppermost 150 m of the Hidden Lake Formation is more fossiliferous. Ammonites include *Gaudryceras* and *Menuites* (*Neopachydiscus*), and there are numerous large forms of *Inoceramus neocaledonicus–Inoceramus australis*. A medium–large *Baculites* ammonite is present at 1330 m and clearly ranges up into the overlying Santa Marta Formation (Olivero 1988). $^{87}Sr/^{86}Sr$ isotope ages based on large inoceramids from the uppermost Hidden Lake Formation (D.8228.94 and D.8228.303; Fig. 4) and lowermost Santa Marta Formation (D.8228.326, D.8228.331 and D.8228.333) are indistinguishable (87.0–87.1 Ma; late Coniacian) (McArthur *et al.* 2000). Further strontium dates indicate that the Coniacian–Santonian boundary is best placed at the 150 m level in the Santa Marta Formation (McArthur *et al.* 2000, p. 635). In this study the Turonian–Coniacian boundary is placed at the base of the Hidden Lake Formation, commensurate with the incoming of the distinctive Coniacian inoceramid species complex (Fig. 4).

Eleven samples of the Hidden Lake Formation produced variably productive palynofloras in which spores and pollen are more abundant than marine microplankton. The dinoflagellate cysts are of Australasian affinity and include key markers such as ?*Actinotheca aphroditae*, *Conosphaeridium striatoconus*, *Spinidinium echinoideum* subsp. *rhombicum* and *Xenascus australensis*, which are all consistent with a Coniacian age, although it should be noted that these species are not restricted to the Coniacian stage (Marshall 1984; Helby *et al.* 1987; McMinn 1988). This is the first report of the index species *Spinidinium echinoideum* subsp. *rhombicum* from Antarctica; the range base of this form is intra Coniacian (Marshall 1984). The Hidden Lake Formation is within the largely Coniacian *Conosphaeridium striatoconus* Interval Zone of Wilson (1984) and Helby *et al.* (1987).

Discussion

Stratigraphical synthesis

The early Aptian age of the exposed base of the Lagrelius Point Formation corresponds with a widespread pulse of marine sedimentation at this time in the Antarctic Peninsula–Scotia Arc region (Howlett 1989; Riding *et al.* 1998). However, the precise lateral and vertical extent of this coarse clastic, deep-marine depositional unit is unknown, as is its exact relationship to the overlying Kotick Point Formation (Fig. 2). The latter comprises up to 1000 m of interbedded breccias, conglomerates, sandstones and mudstones that are characterized by small to extremely large allochthonous blocks of the Nordenskjöld Formation (Ineson *et al.* 1986). The formation is generally interpreted to represent a series of slope apron and submarine fan deposits that were sourced from the adjacent fault-bounded basin margin (Ineson 1989; Hathway 2000). The lowest macrofaunas from the Kotick Point Formation are not age-diagnostic (Ineson *et al.* 1986), but coeval dinoflagellate cyst floras are unequivocally of early Albian age and are attributable to the *Muderongia tetracantha* Interval Zone of Helby *et al.* (1987) (Riding & Crame 2002). It is concluded that the Aptian–Albian boundary is not exposed in NW James Ross Island and thus there is no way of knowing whether the sequence is stratigraphically complete (Fig. 2).

On palynological evidence, a suite of taxa representing the *Canninginopsis denticulata* Interval Zone of Helby *et al.* (1987) suggests that the early–mid Albian transition occurs in the uppermost levels of the Kotick Point Formation (Fig. 2) (Riding *et al.* 1992; Riding & Crame 2002). Whereas the overlying Bibby Point Member of the Whisky Bay Formation may be mid–late Albian in age, the bulk of available macro- and micropalaeontological evidence suggests that the greater part of the Lewis Hill Member, the entire Gin Cove Member and the Lower Tumbledown Cliffs Member are all late Albian in age. Dinoflagellate cyst associations from these various units are indicative of subzone a of the *Endoceratium ludbrookiae* Interval Zone of Helby *et al.* (1987) (Riding & Crame 2002). Such age determinations indicate that the Gin Cove Member, from the SW region of the Gustav Group coastal outcrop (Fig. 1), must be laterally equivalent to the Bibby Point Member and the bulk of the Lewis Hill Member from the NE region (Fig. 2). Nevertheless, it is important to note that there are some significant lithological differences between the two members, with the Gin Cove Member being somewhat thinner (315 m in total thickness) and predominantly composed of finer grained rock types such as bioturbated medium- to fine-grained silty sandstones, with only subordinate graded conglomerates and pebbly sandstones (Ineson *et al.* 1986). In addition, two prominent bivalves from the Gin Cove Member, the medium–large *Inoceramus carsoni* McCoy and the distinctive oxytomid genus, *Maccoyella*, are entirely absent from the Lewis Hill Member.

Although the Albian–Cenomanian boundary is necessarily placed on palynological grounds in the uppermost levels of the Lewis Hill Member, its precise position within the Tumbledown Cliffs Member is uncertain (Fig. 2). The latter unit comprises an approximately 300 m-thick succession of parallel-bedded and channelized, graded sandstones, pebbly sandstones and conglomerates that are locally affected by slumping (Ineson *et al.* 1986). The lower levels contain the characteristic *Actinoceramus concentricus* fauna, but in this particular instance the presence of a turrilitid ammonite referable to *Mariella* may also indicate that the beds range up into the early Cenomanian (Thomson 1984). A second distinctive molluscan fauna from the upper levels of the Tumbledown Cliffs Member contains the ammonites *Newboldiceras* sp., *Sciponoceras* sp. *Desmoceras* aff. *latidorsatum* Michelin, *Gaudryceras* cf. *stefanini* Venzo and *Pseudouhligella* sp, together with *Inoceramus pictus* Sowerby (Thomson 1984; Ineson *et al.* 1986). This fauna has strong mid–late Cenomanian age affinities and indicates that this stage is much more fully represented in this region than it is to the NE. There is no direct equivalent of the upper Tumbledown Cliffs Member in the Brandy Bay–Whisky Bay region (Figs 1 & 2).

With the recalibration of the reference section presented in this study, it is clear that the base of the Brandy Bay Member is marked by the sudden incoming of both *Tethyoceramus madagascariensis* and dinoflagellate cyst taxa such as *Isabelidinium* spp. Strontium isotope dating strongly suggests that these events are late Turonian in age, and, as the Lewis Hill Member–Brandy Bay Member junction is marked by a sharp change in dip, there is good evidence for an early Cenomanian–late Turonian unconformity in the Brandy Bay region (Fig. 2). However, it is not possible to trace this hiatus to the SE and the full stratigraphical relations between the Tumbledown Cliffs and Rum Cove members are unclear. The base of the largely mudstone-dominated Rum Cove Member may be, in part at least, laterally equivalent to the top of the Tumbledown Cliffs Member but in its upper levels it contains a *T. madagascariensis* fauna (Fig. 2) (Ineson *et al.* 1986). A detailed palynostratigraphy for the Rum Cove Member has yet to be established (Riding & Crame 2002).

The nature of the base of the Hidden Lake Formation is a matter of some conjecture. If the Turonian–Coniacian boundary is placed at the Whisky Bay Formation–Hidden Lake Formation transition, then it is clear that there is a greater thickness of Turonian strata in the Brandy Bay region than at Rum Cove (Figs 1 & 2). Nevertheless, in both these localities the contact appears to be conformable and it is only in the Whisky Bay area that an angular unconformity at this boundary can be demonstrated (Ineson *et al.* 1986). The top of the Hidden Lake Formation, and thus of the Gustav Group, is well dated by strontium isotopes as late Coniacian.

Wider implications

As the boundary between the Lagrelius Point and Kotick Point formations is not exposed, there is no way of knowing whether the extensive Aptian–Albian succession is stratigraphically complete (Fig. 2). Thick mid–late Albian successions are present in both the NE and the SW areas of the Gustav Group outcrop (Figs 1 & 2), but in the former of these the highest levels of the Lewis Hill Member only just extend into the earliest Cenomanian. With the recalibration of samples achieved in this study, the junction between the Lewis Hill and Brandy Bay members assumes much greater stratigraphical significance, and can be regarded as an unconformity of probable early Cenomanian–late Turonian extent. However, it is clear that this discontinuity cannot be traced into the Tumbledown Cliffs–Rum Cove area where this time interval is represented by the upper part of the Tumbledown Cliffs and the lower part of the Rum Cove members (Fig. 2). There is some evidence to suggest that the base of the Hidden Lake Formation is at least locally transgressive (Ineson *et al.* 1986) and the Rum Cove Member would appear to be significantly thinner than its partial lateral equivalent to the NE, the Brandy Bay Member (Fig. 2). Nevertheless, it is not possible, at present, to demonstrate a regional unconformity at the base of the Hidden Lake Formation. The Hidden Lake Formation–Santa Marta Formation boundary is traceable laterally over a considerable distance and is entirely conformable (Fig. 1).

The discontinuous unconformities between the Lewis Hill and Brandy Bay members, and Whisky Bay and Hidden Lake formations, can be directly related to the proposed style of deep-water clastic sedimentation close to the basin margin. Although uplift of the Antarctic Peninsula led to the progressive tilt of the clastic wedge to the SE (Whitham & Marshall 1988), it is likely that this was at an irregular rather than constant rate. Successive beds would indeed have onlapped onto a single progressive unconformity (Whitham & Marshall 1988, fig. 6; Hathway 2000, fig. 9), but the precise location of this feature would have shifted in both time and space. Even over a horizontal distance of as little as 25 km, there could have been a change from areas of relatively deep-water sedimentation to areas of non-deposition and erosion. The stratigraphic correlations established in this study (Fig. 2) indicate that an active fault zone located along the Prince Gustav Channel (Fig. 1) exerted strong local control over adjacent sediment accumulation from at least the early Albian through to early Coniacian times.

Global sea level fluctuated during the early Albian–early Coniacian with a marked mid Turonian highstand followed by a mid–late Turonian regression (Haq *et al.* 1987; Hathway 2000; Hart *et al.* 2001). Deposition of the shallower water Hidden Lake Formation can be linked to a Coniacian phase of partial basin inversion coupled to reduced subsidence rates along the basin margin (Pirrie 1991). The bulk of the Marambio Group accumulated at shelf depths and only in its uppermost (Maastrichtian) levels is there evidence of base-level changes that may represent global sea-level events (Pirrie *et al.* 1991*b*; Crame *et al.* 2004).

The results presented here also indicate the potential importance of the Gustav Group for establishing the full biostratigraphic ranges of certain key taxonomic groups within the James Ross Basin. For example, the earliest representative of the kossmaticeratid ammonites, a group that characterizes the overlying Marambio Group, may be a specimen from the upper Brandy Bay Member referred to ?*Marshallites* (Figs 2 & 4). Similarly, a pronounced radiation of heteromorph ammonite taxa in the lower Santa Marta Formation (Olivero 1988) may be traced back to a series of specimens in both the Hidden Lake Formation and upper Brandy Bay Member of the Whisky Bay Formation (Fig. 4). The benthic marine faunas are still under taxonomic investigation, but it is interesting to note the concentration of a number of typically shallow-water (i.e. shelf depth) groups (i.e. oysters, astartids, bakevelliid, etc) that occurs as early as the mid-levels of the Brandy Bay Member – a time of widely recognized global sea-level fall (Fig. 4). There are indications that the radiation of certain plant taxa, including some angiosperms, may be traced back to the same level too (Keating *et al.* 1992; Hayes *et al.* 2006). Thus, the pronounced Late Cretaceous expansion of both shallow-marine and terrestrial biotas in Antarctica may possibly be traced back to at least the Turonian stage. Fuller definition of the Cenomanian–Coniacian biostratigraphy will also aid the interpretation of Late Cretaceous extinction patterns in Antarctica (Crame *et al.* 1996).

This project was funded by NERC Antarctic Funding Initiative grant GR3/AFI2/38 to J. M. McArthur, J. A. Crame, M. F. Thirlwall and W. J. Kennedy. D. Pirrie acknowledges study leave from the University of Exeter, Camborne School of Mines. Logistic support from HMS *Endurance* and the BAS Field Assistants C. Day, A. Taylor and T. O'Donovan is gratefully acknowledged. W. J. Kennedy and P. Doyle are thanked for their help with the ammonite and belemnite determinations, respectively. J. B. Riding publishes with the permission of the Executive Director, British Geological Survey (NERC). Referees' reports by M. Hart, R. Davey and P. Doyle are gratefully acknowledged.

References

BARREDA, V., PALAMARCZUK, S. & MEDINA, F. 1999. Palinologia de la Formación Hidden Lake (Coniaciano–Santoniano), Isla James Ross, Antártida. *Revista Española de Micropaleontología*, **31**, 53–72.

BUATOIS, L.A. & MEDINA, F.J. 1993. Stratigraphy and depositional setting of the Lagrelius Point Formation from the Lower Cretaceous of James Ross Island, Antarctica. *Antarctic Science*, **5**, 379–388.

CANTRILL, D.J. & POOLE, I. 2002. Cretaceous patterns of floristic change in the Antarctic Peninsula. *In*: CRAME, J.A. & OWEN, A.W. (eds) *Palaeobiogeography and Biodiversity Change: the Ordovician and Mesozoic–Cenozoic Radiations*. Geological Society, London, Special Publications, **194**, 141–152.

CRAME, J.A. 1983. Cretaceous inoceramid bivalves from Antarctica. *In*: OLIVER, R.L., JAMES, P.R. & JAGO, J.B. (eds) *Antarctic Earth Science*. Australian Academy of Science, Canberra & Cambridge University Press, Cambridge, 298–302.

CRAME, J.A., FRANCIS, J.E., CANTRILL, D.J. & PIRRIE, D. 2004. Maastrichtian stratigraphy of Antarctica. *Cretaceous Research*, **25**, 411–423.

CRAME, J.A., LOMAS, S.A., PIRRIE, D. & LUTHER, A. 1996. Late Cretaceous extinction patterns in Antarctica. *Journal of the Geological Society, London*, **153**, 503–506.

CRAME, J.A., MCARTHUR, J.M., PIRRIE, D. & RIDING, J.B. 1999. Strontium isotope correlation of the basal Maastrichtian Stage in Antarctica to the European and US biostratigraphic schemes. *Journal of the Geological Society, London*, **156**, 957–964.

CRAMPTON, J.S. 1996*a*. *Biometric Analysis, Systematics and Evolution of Albian* Actinoceramus *(Cretaceous Bivalvia, Inoceramidae)*. Institute of Geological & Nuclear Sciences Monograph, **15**.

CRAMPTON, J.S. 1996*b*. *Inoceramid Bivalves From the Late Cretaceous of New Zealand*. Institute of Geological & Nuclear Sciences Monograph, **14**.

FELDMANN, R.M. & WOODBURNE, M.O. (eds). 1988. *Geology and Paleontology of Seymour Island, Antarctic Peninsula*. Geological Society of America, Memoirs, **169**.

HAQ, B.U., HARDENBOL, J. & VAIL, P.R. 1987. Chronology of fluctuating sea levels since the Triassic. *Science*, **235**, 1156–1167.

HART, M.B., JOSHI, A. & WATKINSON, M.P. 2001. Mid–Late Cretaceous stratigraphy of the Cauvery Basin and the development of the Eastern Indian Ocean. *Journal of the Geological Society of India*, **58**, 217–229.

HATHWAY, B. 2000. Continental rift to back-arc basin: Jurassic–Cretaceous stratigraphical and structural evolution of the Larsen Basin, Antarctic Peninsula. *Journal of the Geological Society, London*, **157**, 417–432.

HATHWAY, B. & RIDING, J.B. 2001. Stratigraphy and age of the Lower Cretaceous Pedersen Formation, northern Antarctic Peninsula. *Antarctic Science*, **13**, 67–74.

HAYES, P., FRANCIS, J.E., CANTRILL, D.J. & CRAME, J.A. 2006. Palaeoclimate analysis of Late Cretaceous angiosperm leaf floras, James Ross Island, Antarctica. *In*: FRANCIS, J.E., PIRRIE, D. & CRAME, J.A. (eds) *Cretaceous–Tertiary High-latitude Palaeoenvironments, James Ross Basin, Antarctica.* Geological Society, London, Special Publications, **258**, 49–62.

HELBY, R., MORGAN, R. & PARTRIDGE, A.D. 1987. *A Palynological Zonation of the Australian Mesozoic*. Memoirs of the Association of Australasian Palaeontologists, **4**, 1–94.

HERM, D., KAUFFMAN, E.G. & WEIDMAN, J. 1979. The age and depositional environment of the 'Gosau' Group (Coniacian–Santonian), Brandenburg/Tirol, Austria. *Mitteilungen der Bayerischen Staatssammlung für Paläontologie und Historische Geologie, München*, **19**, 27–92.

HOWLETT, P.J. 1989. *Late Jurassic–Early Cretaceous Cephalopods of Eastern Alexander Island*. Special Papers in Palaeontology, **41**.

HUBER, B.T. 1998. Tropical paradise at the Cretaceous poles? *Science*, **282**, 2199–2200.

INESON, J.R. 1989. Coarse-grained submarine fan and slope apron deposits in a Cretaceous back-arc basin, Antarctica. *Sedimentology*, **36**, 793–819.

INESON, J.R., CRAME, J.A. & THOMSON, M.R.A. 1986. Lithostratigraphy of the Cretaceous strata of west James Ross Island, Antarctica. *Cretaceous Research*, **7**, 141–159.

KEATING, J.M., SPENCER-JONES, M. & NEWHAM, S. 1992. The stratigraphical palynology of the Kotick Point and Whisky Bay formations, Gustav Group (Cretaceous), James Ross Island. *Antarctic Science*, **4**, 279–292.

MACELLARI, C.E. 1988. Stratigraphy, sedimentology and paleoecology of Upper Cretaceous/Paleocene shelf-deltaic sediments of Seymour Island (Antarctic Peninsula). *In:* FELDMANN, R.M. & WOODBURNE, M.O. (eds) *Geology and Paleontology of Seymour Island, Antarctic Peninsula.* Geological Society of America, Memoirs, **169**, 25–53.

MARSHALL, N.G. 1984. *Late Cretaceous dinoflagellates from the Perth Basin, Western Australia*. Unpublished PhD thesis, University of Western Australia, Perth.

MCARTHUR, J.M., CRAME, J.A. & THIRLWALL, M.F. 2000. Definition of Late Cretaceous Stage boundaries in Antarctica using strontium isotope stratigraphy. *Journal of Geology*, **108**, 623–640.

MCARTHUR, J.M., THIRLWALL, M.F., ENGKILDE, M., ZINSMEISTER, W.J. & HOWARTH, R.J. 1998. Strontium isotope profiles across K/T boundary sequences in Denmark and Antarctica. *Earth and Planetary Science Letters*, **160**, 179–192.

MCMINN, A. 1988. Outline of a Late Cretaceous dinoflagellate zonation of northern Australia. *Alcheringa*, **12**, 137–156.

MORGAN, R. 1980. *Palynostratigraphy of the Australian Early and Middle Cretaceous*. Memoirs of the Geological Survey of New South Wales, Palaeontology, **18**.

OLIVERO, E.B. 1988. Early Campanian heteromorph ammonites from James Ross Island, Antarctica. *National Geographic Research*, **4**, 259–271.

OLIVERO, E.B., SCASSO, R.A. & RINALDI, C.A. 1986. Revision of the Marambio Group, James Ross Island, Antarctica. *Contribución Instituto Antártico Argentino*, **331,** 1–28.

PIRRIE, D. 1989. Shallow-marine sedimentation within an active margin basin, James Ross Island, Antarctica. *Sedimentary Geology*, **63**, 61–82.

PIRRIE, D. 1991. Controls on the petrographic evolution of an active margin sedimentary sequence: the Larsen Basin, Antarctica. *In*: MORTON, A.C., TODD, S.P. & HAUGHTON, P.D.W. (eds) *Developments in Sedimentary Provenance Studies*. Geological Society, London, Special Publications, **57**, 231–249.

PIRRIE, D., CRAME, J.A., LOMAS, S.A. & RIDING, J.B. 1997. Late Cretaceous stratigraphy of the Admiralty Sound region, James Ross Basin, Antarctica. *Cretaceous Research*, **18**, 109–137.

PIRRIE, D., CRAME, J.A. & RIDING, J.B. 1991*a*. Late Cretaceous stratigraphy and sedimentology of Cape Lamb, Vega Island, Antarctica. *Cretaceous Research*, **12**, 227–258.

PIRRIE, D., WHITHAM, A.G. & INESON, J.R. 1991*b*. The role of tectonics and eustasy in the evolution of a marginal basin: Cretaceous–Tertiary Larsen Basin, Antarctica. *In*: MACDONALD, D.I.M. (ed.) *Sedimentation, Tectonics and Eustasy*. International Association of Sedimentologists, Special Publications, **12**, 293–305.

RIDING, J.B. & CRAME, J.A. 2002. Aptian to Coniacian (Early–Late Cretaceous) palynostratigraphy of the Gustav Group, James Ross Basin, Antarctica. *Cretaceous Research*, **23**, 739–760.

RIDING, J.B., CRAME, J.A., DETTMANN, M.E. & CANTRILL, D.J. 1998. The age of the base of the Gustav Group in the James Ross Basin, Antarctica. *Cretaceous Research*, **19**, 87–105.

RIDING, J.B., KEATING, J.M., SNAPE, M.G., NEWHAM, S. & PIRRIE, D. 1992. Preliminary Jurassic and Cretaceous dinoflagellate cyst stratigraphy of the James Ross Island area, Antarctic Peninsula. *Newsletters on Stratigraphy*, **26**, 19–39.

RINALDI, C.A. (ed.) 1992. *Geología de la Isla James Ross*. Instituto Antártico Argentino, Buenos Aires.

SANDY, M.R. 1991. Cretaceous brachiopods from James Ross Island, Antarctic Peninsula and their paleobiogeographic affinities. *Journal of Paleontology*, **65**, 396–411.

SCASSO, R.A., OLIVERO, E.B. & BUATOIS, L. 1991. Lithofacies, biofacies and ichnoassemblage evolution of a shallow submarine volcaniclastic fan-shelf depositional system (Upper Cretaceous–James Ross Island, Antarctica). *South American Journal of Earth Sciences*, **4**, 239–260.

SORNAY, J. 1964. Sur quelques nouvelles espèces d'Inocérames du Sénonien de Madagascar. *Annales de Paléontologie. Invertebrés*, **50**, 167–197.

THOMSON, M.R.A. 1984. Cretaceous ammonite biostratigraphy of western James Ross Island, Antarctica. *In:* PERILLIAT, M. DE C. (ed.) *Memoria III Congreso Latinoamericano de Paleontología, Mexico*, 1984. Universidad Nacional Autonoma de Mexico, Instituto de Geología, Mexico City, 308–314.

WALASZCZYK, I. 1992. Turonian through Santonian deposits of the central Polish Uplands, their facies development, inoceramid paleontology and stratigraphy. *Acta Geologica Polonica*, **42**, 1–122.

WALASZCZYK, I., MARCINOWSKI, R., PRASZKIER, T., DEMBICZ, K. & BIEŃKOWSKA, M. 2004. Biogeographical and stratigraphical significance of the latest Turonian and Early Coniacian inoceramid/ammonite succession of the Mansoa section on the Onilahy River, south-west Madagascar. *Cretaceous Research*, **25**, 543–576.

WHITHAM, A.G. & MARSHALL, J.E.A. 1988. Syndepositional deformation in a Cretaceous succession, James Ross Island, Antarctica. Evidence from vitrinite reflectivity. *Geological Magazine*, **125**, 583–591.

WHITHAM, A.G., INESON, J.R. & PIRRIE, D. 2006. Marine volcaniclastics of the Hidden Lake Formation (Coniacian) of James Ross Island, Antarctica: an enigmatic element in the history of a back-arc basin. *In*: FRANCIS, J.E., PIRRIE, D. & CRAME, J.A. (eds) *Cretaceous–Tertiary High-latitude Palaeoenvironments, James Ross Basin, Antarctica.* Geological Society, London, Special Publications, **258**, 21–47.

WILSON, G.J. 1984. New Zealand Late Jurassic to Eocene dinoflagellate biostratigraphy: a summary. *Newsletters on Stratigraphy*, **13**, 104–117.

Marine volcaniclastics of the Hidden Lake Formation (Coniacian) of James Ross Island, Antarctica: an enigmatic element in the history of a back-arc basin

ANDREW G. WHITHAM[1,2], JON R. INESON[1,3] & DUNCAN PIRRIE[1,4]

[1]*British Antarctic Survey, High Cross, Madingley Road, Cambridge CB3 0ET, UK*

[2]*CASP, Department of Earth Sciences, University of Cambridge, 181a Huntingdon Road, Cambridge CB3 ODH, UK (e-mail: andy.whitham@casp.cam.ac.uk)*

[3]*Geological Survey of Denmark and Greenland (GEUS), Øster Voldgade 10, DK-1350 Copenhagen, Denmark (e-mail: ji@geus.dk)*

[4]*Camborne School of Mines, School of Geography, Archaeology and Earth Resources, University of Exeter, Cornwall Campus, Penryn, Cornwall TR10 9EZ, UK (e-mail: D.Pirrie@exeter.ac.uk)*

Abstract: The Coniacian Hidden Lake Formation of James Ross Island, Antarctica is a 300–400 m-thick succession of marine volcaniclastic conglomerates, sandstones and mudstones. It occurs at a point of transition in the evolution of the James Ross Basin, as it is underlain by deep-marine strata and overlain by shallow-marine strata. The succession reflects the two main factors controlling the deposition of the formation: (1) the influx of large quantities of volcaniclastic sediment; and (2) a pronounced inversion event in the early Coniacian heralding the cessation of transpressive tectonic activity in the James Ross Basin. The succession is dominated by a range of sediment density-flow deposits, which, combined with the limited faunas and the lack of wave-induced structures, suggest deposition in a relatively deep-marine environment below storm-wave base. Three main facies associations are recorded representing base-of-slope, fan-delta and basin-floor depositional environments. The volcaniclastic fan-delta association is dominated by fresh pyroclastic detritus and was deposited in response to volcanic eruptions on the adjacent arc. Thick beds of parallel-stratified sandstone record deposition from sustained, concentrated sediment density flows. The conditions immediately following pyroclastic eruptions lend themselves to the deposition of such deposits, as vegetation cover is destroyed and large amounts of poorly consolidated sediment are available for reworking. An enigmatic feature of the succession is the presence of units of cross-bedded sandstones thought to be of tidal origin that are locally abundant and are intimately interbedded with sediment density-flow deposits. The occurrence of tidal sediments in a substorm-wave base setting is explained by appealing to partial basin inversion during the final phases of strike-slip tectonic activity in the basin creating an irregular basin floor that focused and amplified tidal currents. The infilling of this basin topography by sediment and waning intrabasinal tectonism during the Coniacian resulted in the progressive elimination of this basin-floor topography and the onset of shallow-marine shelf sedimentation.

The marine volcaniclastic sediments of the Hidden Lake Formation of western James Ross Island, Antarctica (Fig. 1) represent a period of transition within a back-arc basin. This formation, the uppermost of the Gustav Group, is underlain by syndepositionally deformed deep-marine coarse-grained sediments of the lower Gustav Group (Ineson 1985, 1989; Whitham & Marshall 1988) and is overlain by largely undeformed finer grained shallow-marine deposits of the Marambio Group (Pirrie 1989). Although preliminary interpretations of the depositional environment of the Hidden Lake Formation have been presented (Macdonald *et al.* 1988; Ineson 1989; Pirrie *et al.* 1991), the sedimentology of the formation has not been fully described. This paper documents the sedimentology of the Hidden Lake Formation, not only to provide a framework for detailed chronostratigraphic studies, but also importantly for palaeoclimatic studies of fossil leaf floras that are developed at several levels within the formation and record significant radiation of the angiosperms (Cantrill & Poole 2002; Hayes *et al.* 2006). In addition, we highlight some interesting and unusual facies and facies associations. The

From: FRANCIS, J. E., PIRRIE, D. & CRAME, J. A. (eds) 2006. *Cretaceous–Tertiary High-Latitude Palaeoenvironments, James Ross Basin, Antarctica.* Geological Society, London, Special Publications, **258**, 21–47. 0305–8719/06/$15

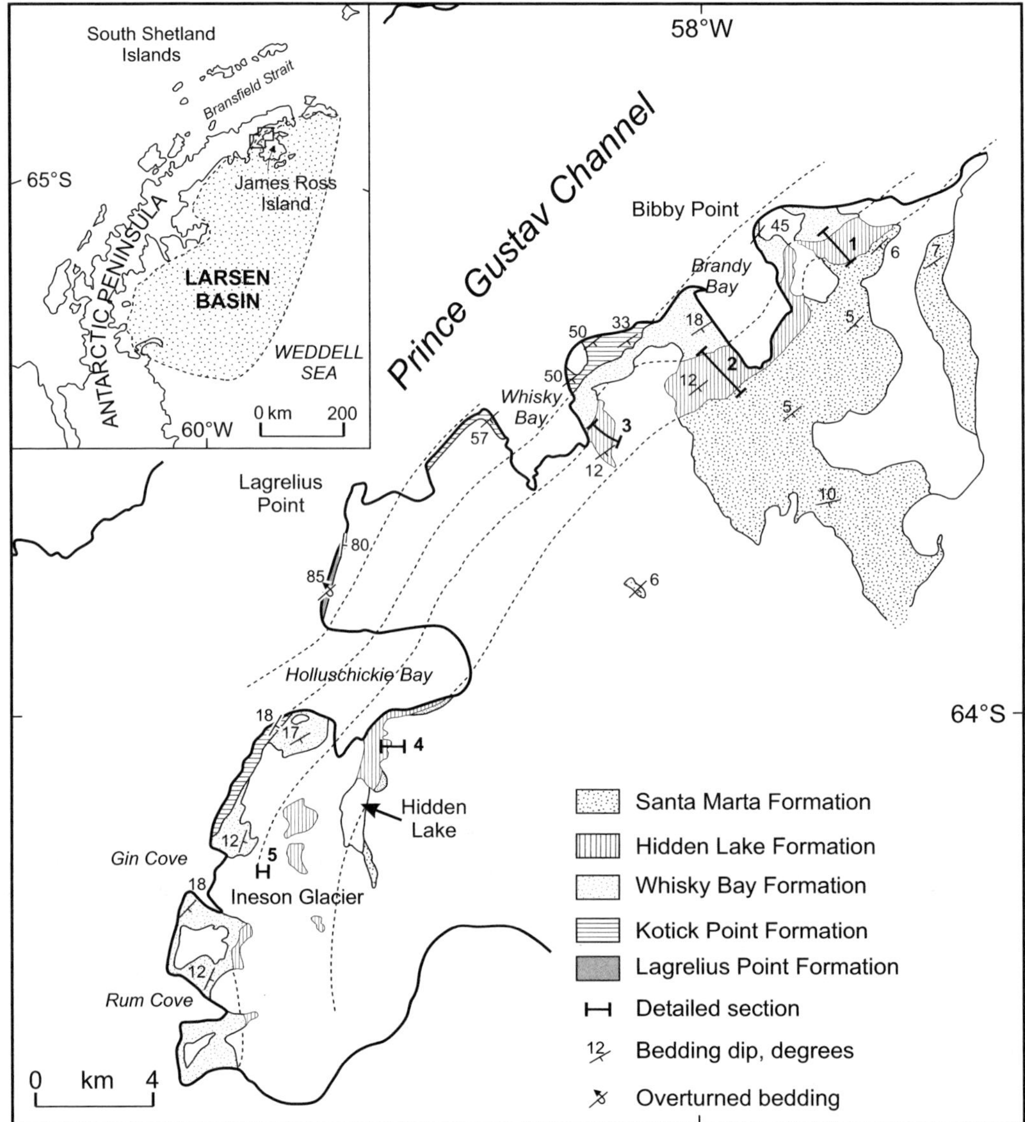

Fig. 1. Geological sketch map showing the exposure pattern of Cretaceous strata on NW James Ross Island; note the marked increase of dips towards the NW coast of the island. Localities 1–5 indicate the position of the main sections mentioned in the text. The dashed lines trace formation boundaries across sea and ice cover to illustrate the regional structure.

succession is dominated by the deposits of sediment density flows and there are no indicators of wave activity in the succession; yet, cross-bedded sandstones thought to be of tidal origin are found at many localities. Furthermore, at the northern end of the outcrop, thick units of parallel-stratified volcaniclastic sandstones are interpreted to represent deposition from quasi-steady concentrated sediment density flows. Such deposits are poorly documented in the sedimentological literature. The implications of the sedimentology of the formation, for understanding the evolution of the Larsen Basin in the James Ross Island region, are also discussed.

Geological setting

The Larsen Basin is a major sedimentary basin on the east side of the Antarctic Peninsula

(Macdonald *et al.* 1988; Hathway 2000). It developed in a back-arc setting relative to a volcanic arc that was formed by subduction of proto-Pacific oceanic crust beneath Gondwana. The eroded roots of this arc now forms the Antarctic Peninsula. The tectonic history and affinities of the Larsen Basin are still poorly understood (see discussions in Macdonald *et al.* 1988; Macdonald & Butterworth 1990; Storey 1991; Hathway 2000). This is, in part, due to inaccessibility and extensive ice cover, but is also due to the complex tectonic situation of the basin in a region influenced both by subduction-related processes and by the opening of the Weddell Sea (Fig. 1). During the break-up of Gondwana, the Weddell Sea region underwent stretching and subsidence in Middle and Late Jurassic times, culminating in sea-floor spreading along a ridge with an east–west orientation. This ridge rotated to a NE–SW orientation before being subducted beneath the Scotia Arc. Opening of the Weddell Sea may thus have been responsible for strike-slip movement or oblique extension along the eastern margin of the Antarctic Peninsula volcanic arc and influenced the evolution of back-arc sedimentary basins from the Late Jurassic to the Late Cretaceous (Storey & Nell 1988; Storey *et al.* 1996). Hence, although the Larsen Basin developed in a back-arc setting and its sediment fill was derived wholly from the volcanic arc, the relative influences of back-arc extension and oblique extension/lateral slip during opening of the Weddell Sea are poorly known.

The Larsen Basin was initiated in Jurassic times as a result of continental rifting during the early stages of Gondwana break-up, and the basin-fill is divided into three or possibly four megasequences (Hathway 2000). The fill is best exposed on and around James Ross Island (Fig. 1), where a nearly complete Aptian–Eocene section crops out (the uppermost megasequence of Hathway 2000). This part of the Larsen Basin has been called the James Ross Basin by del Valle *et al.* (1992). In NW James Ross Island, the Cretaceous succession is represented by the Gustav Group and the basal strata of the Marambio Group (Fig. 2). Throughout much of the area, gentle basinward (SE) dips are recorded. However, in a 4 km-wide zone along the NE–SW-trending coastline, parallel and proximal to the basin margin, the strata describe a NE–SW-trending monoclinal syncline (Fig. 1). This structure is of Cretaceous age and represents the cumulative effect of a number of episodes of flexure or tilting of basin-margin strata (Whitham & Marshall 1988). The last demonstrable episode of syndepositional flexure immediately preceded the deposition of the Hidden Lake Formation (see discussion below).

The marine sediments of the Gustav Group are exposed within this marginal flexured zone along the NW coast of James Ross Island (Fig. 1). The lower three formations of the group (Fig. 2) were deposited in a deep marine setting, and represent proximal submarine-fan and slope-apron depositional systems (Ineson 1989). The volcaniclastics of the Hidden Lake Formation, the subject of this paper, form the uppermost formation of the Gustav Group and are succeeded by fine-grained marine sediments of the Santa Marta Formation that accumulated in a low-energy marine-shelf setting (Pirrie 1989; Pirrie *et al.* 1991).

Hidden Lake Formation

Stratigraphy

The Hidden Lake Formation is a suite of distinctive brown-weathering, volcaniclastic sediments that is readily traced along the length of NW James Ross Island (Fig. 1) (Bibby 1966; Ineson *et al.* 1986). The formation is generally quite poorly exposed, but a number of good stream-cut and coastal sections are found along the length of the outcrop. It is from these sections that most of the observations in the following paper have been made. The formation is at least 400 m thick in the SW, thinning north of Whisky Bay to around 300 m at the northernmost limit of its outcrop (Fig. 1). In most areas, the base of the formation is defined by an abrupt facies shift and is apparently conformable at most localities. The boundary is placed where thinly interbedded marine mudstones and sandstone turbidites with rare channelized conglomerates are succeeded by volcaniclastic conglomerates and cross-bedded sandstones. In Whisky Bay (Fig. 1) the boundary is an angular unconformity with a discordance of 6°; the basal beds of the Hidden Lake Formation onlap westwards onto the unconformity surface. The formation shows an overall fining-upwards trend throughout its outcrop, being dominated by sandstones and conglomerates in its lower half, grading up into a succession of interbedded sandstones and mudstones. These in turn pass up into the heavily bioturbated, fossiliferous mudstones and subordinate sandstones of the Santa Marta Formation (Fig. 2).

The invertebrate macrofauna of the Hidden Lake Formation is generally sparse, although locally it is rich and diverse. The faunal list includes gastropods, ammonites, various bivalves (oysters, inoceramids, mytilids), brachiopods,

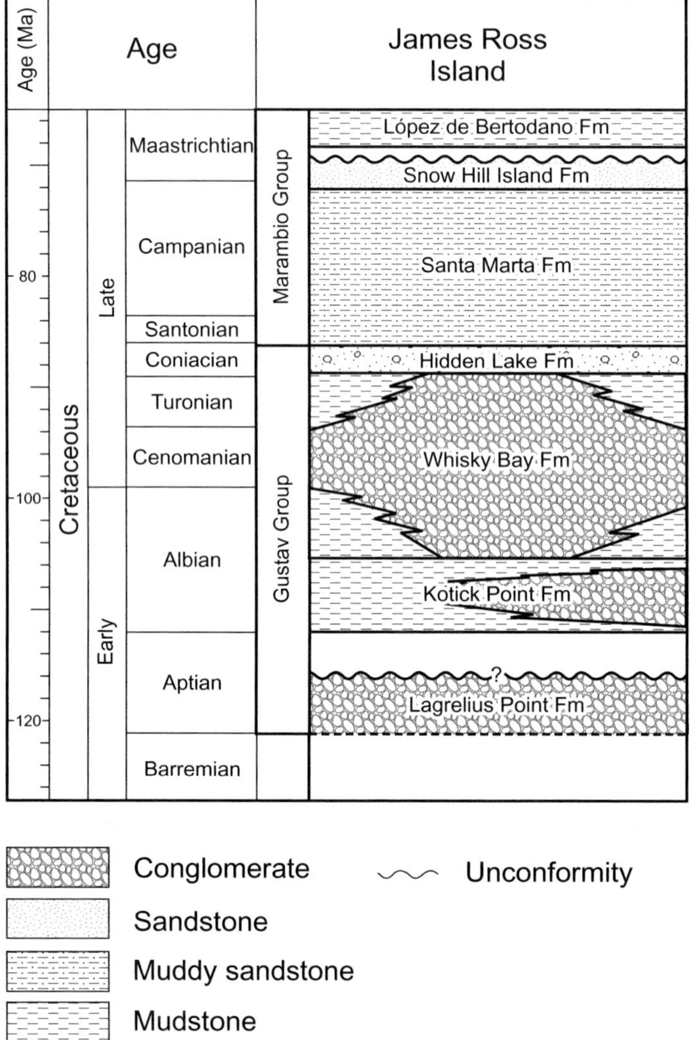

Fig. 2. Lithostratigraphy of the Cretaceous sedimentary succession on NW James Ross Island, based on Ineson *et al.* (1986) and Pirrie *et al.* (1997), modified according to McArthur *et al.* (2000) and Riding & Crame (2002).

bryozoans and corals along with rare shark vertebrae. Of these, only the inoceramid bivalves, the ammonites and the terebratulid brachiopods are thought to be *in situ*. The typical shallow-water marine elements (oysters, corals and bryozoans) are fragmented, occur in mass-flow deposits and are clearly derived. A relatively diverse suite of trace fossils is also present in the formation and includes the ichnogenera *Asterosoma*, *Didymaulichnus*, *Ophiomorpha*, *Palaeophycus*, *Planolites*, *Skolithos*, *Teredolites*, *Thalassinoides* and *Tissoa* (Buatois & López-Angriman 1992*a*), along with a decapod trackway, *Foersterichnus rossensis* (Pirrie *et al.* 2004). In addition to the invertebrate fauna, the Hidden Lake Formation contains very abundant fossil wood, which is commonly charcoalified. At several levels within the formation, there are also abundant and moderately diverse leaf floras, allowing a palaeobotanical assessment of palaeoclimate (Hayes *et al.* 2006).

On the basis of the macrofauna, the formation was initially assigned a Coniacian–Santonian age (Ineson *et al.* 1986). Subsequent work on both the macrofauna and the palynology of the Hidden Lake Formation yielded ages from

Cenomanian (Olivero & Palamarczuk 1987) to Turonian–Santonian (Baldoni & Medina 1989). In a preliminary palynological review of the James Ross Island succession, Riding *et al.* (1992) proposed a Coniacian–earliest Santonian age for the Hidden Lake Formation. This was compatible with the earlier assignment based on macrofaunal determinations (Ineson *et al.* 1986) and was confirmed in a subsequent restudy by Riding (1999). However, recent Sr isotopic age determinations indicate a very tight age range for the formation, with a date of 88 Ma for the base of the formation and 87.4 Ma for the top (McArthur *et al.* 2000; J.M. McArthur pers. comm. 2003). These data suggest that the Hidden Lake Formation was deposited entirely during the Coniacian (see review of the biostratigraphic data by Riding & Crame 2002).

Provenance

A common feature of the succession is the presence of flattened pumice fragments and accretionary lapilli (Fig. 3A), while reworked bread-crust bombs occur in places (Fig. 3B). The significance of these clasts is that they imply contemporaneous explosive volcanic activity. Pyroclastic deposits must have been rapidly eroded and redeposited in the basin, with limited residence time prior to reworking, as such clast types are particularly prone to alteration. A contribution of sediment from other rock types is indicated by the presence of volcanic, metasedimentary and sedimentary clasts, a suite that is typical of the underlying formations of the Gustav Group (Ineson 1989; Pirrie 1991; Browne & Pirrie 1995). The occurrence of acid plutonic clasts in discrete conglomerate beds at locality 5 is noteworthy; such clasts are extremely rare in the underlying formations of the Gustav Group. This suggests that significant unroofing of plutonic intrusions first occurred in Coniacian times.

Petrographic studies of medium-grained sandstones (Fig. 4), following the methods of Ingersoll & Suczek (1979) and Dickinson (1985), confirm derivation from an active volcanic arc through the overwhelming dominance of volcanic lithic arenites, *sensu* Pettijohn *et al.* (1972). There is no significant compositional variation, neither geographically, stratigraphically nor between facies (see also Pirrie 1991). Zoned plagioclase, typically of andesine composition, forms 18–63% of grains. The lithic component (37–79%) is dominated by fragments of aphyric and plagioclase phyric lava and altered volcanic glass, now replaced by zeolites and chlorite. Quartz is present but of minor importance (<11%). Clinopyroxene, hornblende and opaque minerals (ilmenite, titanoaugite) are generally present in minor amounts, although locally abundant. The clinopyroxene is typically fresh and unaltered, and has an augite composition with a high content of Na_2O and MnO and a low CrO content (Pirrie 1991).

The large proportion of plagioclase crystals and volcanic lithic clasts, and the relative scarcity of quartz grains, suggest that the Hidden Lake Formation was derived from an immature arc source (Fig. 4). The presence of zoned andesine plagioclase, the lack of sodic and potassic feldspars, and the augite composition suggest an andesitic, stratovolcanic source (Enkeboll *et al.* 1982; Dickinson 1985). Thus, the composition of the sediments forming the

Fig. 3. (**A**) Photograph of a bedding plane showing impressions of flattened clasts of pumice, locality 1; scale bar is 1 cm. (**B**) Bread-crust bomb showing characteristic cracked surface infilled by sandy matrix, locality 1; scale bar is 1 cm.

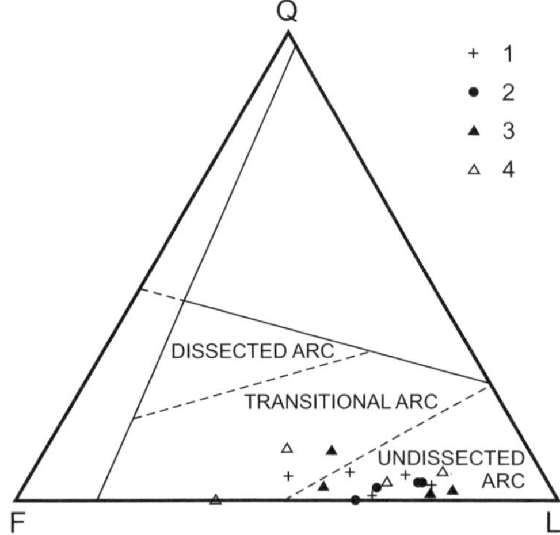

Fig. 4. Quartz–feldspar–lithic (QFL) plot summarizing the petrology of the Hidden Lake Formation sandstones; fields after Dickinson (1985), numbers (1–4) refer to localities (see Fig. 1).

Hidden Lake Formation testifies to its derivation from the volcanic arc to the NW.

Facies analysis

Eight lithofacies are recognized in the Hidden Lake Formation. The facies are dominated by sediment density-flow deposits (six facies), but there is an important component of deposits formed by episodic bottom currents (one facies) and deposits formed by suspension settling (one facies). In the following section the sediment density-flow deposits are described and interpreted first, from coarse to fine. Facies groups or associations are described in the following section.

Facies 1: Conglomerates

Facies 1 consists predominantly of coarse pebble–cobble conglomerates, but also includes boulder conglomerates, locally containing outsized blocks up to several metres in diameter that may protrude above the upper bed boundary. Both clast- and matrix-supported conglomerates are represented. The matrix typically comprises medium- to coarse-grained, poorly sorted, lithic sandstone; clay is rare to absent in this facies. Most beds are 1–3 m thick (maximum of 4.5 m), lenticular in cross-section, with erosional bases and wedge out over a distance of 10–20 m normal to palaeocurrent (current-parallel sections are not exposed). Some beds are wholly ungraded, but typically the conglomerates show normal, coarse-tail grading. In some cases this grading is developed from base to top, but is more typically restricted to the upper third or half of the bed. Inversely graded basal divisions are developed in only a few beds. Tabular or elongate clasts are either randomly oriented or, less commonly, are aligned parallel to bedding.

The characteristics of these deposits indicate deposition from a range of sediment density flows. Scoured, erosional basal contacts indicate flow turbulence, which, combined with the poorly stratified, thick-bedded nature of these deposits, suggests rapid sedimentation from concentrated density flows in the sense of Mulder & Alexander (2001), also classified as high-density turbidity currents by Walker (1975) and Lowe (1982). Ungraded or poorly graded beds probably represent conditions transitional between hyperconcentrated flows and concentrated sediment density flows (Mulder & Alexander 2001).

Facies 2: Pebbly sandstones

Facies 2 consists of medium- to very-coarse-grained pebbly sandstone in beds 0.15–4.5 m thick (Fig. 5). Basal surfaces are typically highly erosional and concave-up. Beds are commonly lenticular at an outcrop scale. Most beds show normal, coarse-tail grading in their upper parts, but some are normally graded throughout (Fig. 5A) whilst others are ungraded. Inversely graded basal divisions are observed only rarely. Although typically matrix-rich, pebbles may be concentrated near bed bases and locally form discontinuous, clast-supported, imbricated, conglomerate intervals up to 30 cm thick (Fig. 5B). Imbricated pebbles are found in these intervals with *a*-axes parallel to the direction of flow, which was towards the SE, away from the arc (Fig. 6). Bed-parallel clast fabrics are locally developed. Parallel-stratified intervals (maximum of 0.8 m thick) are sometimes developed in the upper sand-rich portions (Fig. 5A), suggesting that this facies may be transitional to Facies 3.

Deposition of this facies was from concentrated density flows (*sensu* Mulder & Alexander 2001) or high concentration turbidity currents (*sensu* Walker 1975; Surlyk 1978; Lowe 1982). Flow turbulence is indicated by scouring at bed bases, and high particle concentrations are indicated by normal, coarse-tail grading and *a*-axis clast imbrication (indicating an absence of tractional rolling). The presence of grading implies deposition by a decelerating flow. Ungraded and poorly graded beds probably represent

Fig. 5. Facies 2: pebbly sandstones, locality 1. (**A**) Weakly coarse-tail graded bed with minor basal scours and parallel-stratified top; staff is 1.5 m long with 10 cm gradations (also in Figs 8, 9, 12 & 16). (**B**) Pebbly sandstone showing imbricated clast-rich basal interval, weak coarse-tail graded central portion and parallel-stratified top (the hammer is 35 cm long).

conditions transitional between hyperconcentrated flows and concentrated sediment density flows (Mulder & Alexander 2001).

Facies 3: Parallel-stratified sandstones

This facies is composed of very-coarse- to fine-grained poorly sorted sandstone, locally with dispersed pebbles and granules. It forms beds 0.9–5.2 m thick (Fig. 7), and is characterized by diffuse–well-developed parallel stratification (individual strata 1–10 cm thick) and transitional–parallel lamination (lamina thickness 0.5–1 cm thick). Bed boundaries are typically sharp and planar. In rare cases, a thin interval of cross-laminated sandstone caps the bed. Individual strata or laminae are typically laterally persistent at outcrop scale (i.e. several metres; Fig. 7A, B), but wedge-shaped terminations are also observed. The strata and laminae are defined by grain-size variation from poorly sorted coarse- or very-coarse-grained sandstone to fine- to medium-grained sandstone (Fig. 7C). This variation is typically subtle, despite the well-stratified appearance at outcrop. Some laminae show grading. Outsized, isolated clasts up to 10 cm across occur in places, sometimes flanked by composite small-scale scour-and-fill structures that truncate the parallel stratification and are infilled with a concentration of coarse detritus (Fig. 7D). Low-angle scour surfaces with a relief of a few tens of centimetres locally truncate the pervasive parallel stratification. In most cases, beds show no overall vertical trends in grain size, although some beds possess a thin (10–20 cm) pebbly basal layer that rapidly grades up into stratified sandstone.

This unusual facies was deposited from concentrated density flows (*sensu* Mulder & Alexander 2001). A sediment density-flow origin is suggested by the presence of rare, thin basal pebbly layers indicating that these beds may be transitional to Facies 2. Further support for a sediment density-flow origin is provided by thin, finer grained, ripple cross-laminated units at the tops of some beds suggesting that the final phases of deposition of this facies were from a waning turbulent cloud of suspended sediment. Flow turbulence is indicated by the erosive nature of bed bases, rare scours within beds and scours on the stoss side of rare outsize clasts found in the body of the flow deposits.

The diffuse lamination or stratification diagnostic of this facies is considered to have been formed by traction-related processes at the base of a sandy, concentrated density flow that deposited sediment primarily from suspension in an episodic yet regular fashion, probably

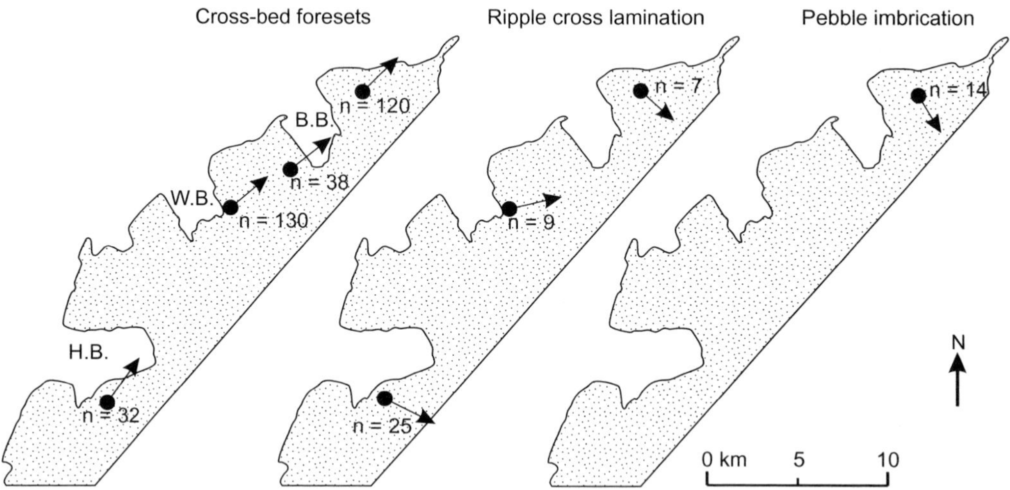

Fig. 6. Summary of palaeocurrent data from the Hidden Lake Formation; arrows are vector means of readings (n); B.B., Brandy Bay; W.B., Whisky Bay; H.B., Holluschickie Bay. The imbrication readings represent mean values for a bed based on the orientation of at least 10 pebbles.

controlled by turbulent burst–sweep cycles (Hiscott 1994). Hiscott's (1994) model explains stratification at the base of a highly concentrated current as being the result of a pulsating current that alternates between short episodes of local erosion and otherwise heavy suspension sedimentation. This is quite different to the steady-state mechanism for the formation of traction carpets as proposed by Hiscott & Middleton (1979, 1980) and Lowe (1982), which accounts for the stratification in terms of the relationship between basal shear stress against a rain of suspended sediment, building up a succession of traction deposits of differing thicknesses, depending on prevailing conditions. The thickness of the lamination and the lack of primary current lineation indicate that the stratification was not caused by fluid flow in the upper-stage plane-bed regime (Leeder 1999). Outsized clasts were transported at times when the sediment-laden flow possessed matrix strength.

Although Hiscott (1994) indicated that such 'spaced stratification' is indicative of strongly fluctuating hydrodynamic conditions and thus may not be produced beneath steady flows, it is notable that the average grain size of these beds in the Hidden Lake Formation is remarkably consistent through several metres of stratified sediment. Overall grading, as expected under unsteady flow conditions (Kneller 1995), is not observed. The suggestion is, therefore, that despite the internally pulsating nature of the turbulent flow, a steady flow state was maintained overall. This clearly requires a persistent sediment supply at source and invites comparison with horizontally stratified sand beds described from a glacio-lacustrine setting by Eyles *et al.* (1987) and the sustained flow turbidites described from the Eocene of Spitsbergen (Plink-Björklund & Steel 2004). The latter workers reported thick-bedded ungraded stratified sandy turbidites deposited by gradual aggradation during protracted flow events and demonstrated that such sustained turbidity currents were generated by hyperpycnal flows derived directly from sandy sources by fluvial input. Plink-Björklund & Steel (2004) suggested that such processes may be favoured in volcaniclastic settings due to the high rates of supply of coarse pyroclastic detritus. They also demonstrated that the development of this facies is influenced by its location on the delta slope. They noted that fully planar laminated ungraded beds are found on the upper slope. As beds are traced down-slope, structureless intervals appear and progressively form a greater proportion of the beds until there is virtually no lamination present in beds on the lower slope.

Facies 4: Medium- to thick-bedded sandstones

Facies 4 is represented by fine- to coarse-grained sandstones, in beds 5–60 cm thick (Fig. 8). Basal contacts are sharp, locally erosional and loaded. Most beds are normally graded and consist of a structureless basal interval overlain by a parallel-laminated interval. In a few cases, the

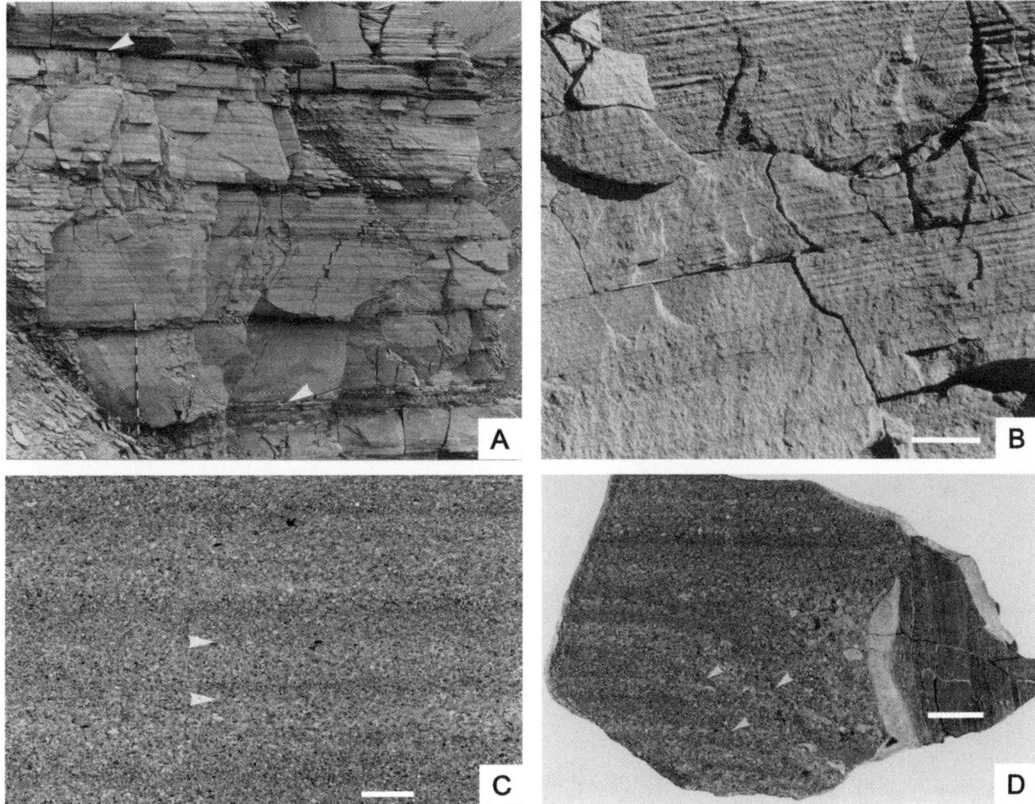

Fig. 7. Facies 3: parallel-stratified sandstone, locality 1 (all photographs). (**A**) A 4.5 m-thick interval of parallel-stratified sandstone (top and bottom arrowed); the staff is 1.5 m long with 10 cm gradations. Dark recessive intervals within the unit contain an increased proportion of volcanic lithics; note the lack of erosional surfaces. (**B**) Detail of well-developed parallel-stratification (the scale bar is 15 cm). (**C**) Polished slab; note the diffuse, gradational nature of the stratification. The arrowed unit shows a lower finer grained, weakly inverse graded zone (<1 cm thick) succeeded by an ungraded upper unit (the scale bar is 1 cm). (**D**) Polished slab showing scouring around an outsized clast. Three successive scour surfaces (arrowed) are infilled with a concentration of coarser debris (the scale bar is 1 cm).

parallel-laminated interval is succeeded by, or includes, an interval of ripple cross-laminated sandstone. The orientation of ripple foresets indicates sediment transport towards the SE, away from the arc (Fig. 6). This facies also includes rare, structureless, ungraded beds.

These beds were deposited by sediment density flows. Flow turbulence is indicated by scouring at the base of beds and grading of grains. The graded beds show Bouma T_{a-c} sequences indicating that they were deposited by concentrated density flows (T_a and T_b divisions), transitional to turbidity flows (T_c divisions) (*sensu* Mulder & Alexander 2001), or high to low concentration turbidity currents (*sensu* Lowe 1982; Pickering *et al.* 1986). Structureless ungraded beds probably represent the deposits of hyperconcentrated density flows (Mulder & Alexander 2001).

Facies 5: Thinly interbedded sandstones and mudstones

The proportion of mudstone to sandstone in this facies varies greatly, from 20:1 to 1:20. Sandstone beds are 1–7 cm thick and are composed of very-fine- to medium-grained sandstone (Fig. 9). Bed bases are sharp and may be load-casted. Many beds show well-developed normal grading, and parallel and ripple cross-lamination; isolated ripple trains occur locally. Ripple cross-laminae orientations show a wide degree of variation in this facies (Fig. 6). At locality 1 where the vector mean of observations indicates

Fig. 8. Facies 4: medium- to thick-bedded sandstones showing flat bases, thick, structureless lower portions and parallel-laminated tops, locality 1; the staff is 1.5 m long with 10 cm gradations.

transport towards 130°, readings vary between 085° and 180°. At locality 3 the vector mean of observations indicates transport towards 080°, with readings varying between 056° and 170°. At locality 4 the vector mean of observations indicates transport towards 117° with readings varying between 012° and 232°. Mudstone beds are up to 20 cm thick; locally they contain rich, monospecific inoceramid bivalve faunas. The facies shows various degrees of bioturbation. In places, the biogenic intermixing of sand and mud is such that only a vague parallel bedding is preserved. Recognizable ichnogenera include *Helminthoida*, *Planolites* and *Palaeophycus*.

This is a non-diagnostic facies that in isolation may be interpreted to record a range of different processes in a variety of depositional settings. Key features are the presence of marine bivalves and abundant bioturbation, and the absence of wave-induced sedimentary structures such as wave ripples and hummocky cross-stratification (HCS). One possible interpretation is that the sand beds are T_{bc} turbidites deposited by waning, turbulent sediment density flows, turbidity flows (*sensu* Mulder & Alexander 2001) or low concentration turbidity currents (*sensu* Pickering *et al.* 1986). These may have originated as sediment slides or may be the result of flood events or storms. The interbedded mudstones represent suspension deposits or the products of deposition from low concentration flows (Pickering *et al.* 1986). Alternatively, the sand beds may have been formed by sediment reworking by bottom currents in a subwave base environment (e.g. Shanmugam *et al.* 1994). Both origins are plausible given the presence of cross-bedded sandstones formed by inferred tidal currents (see

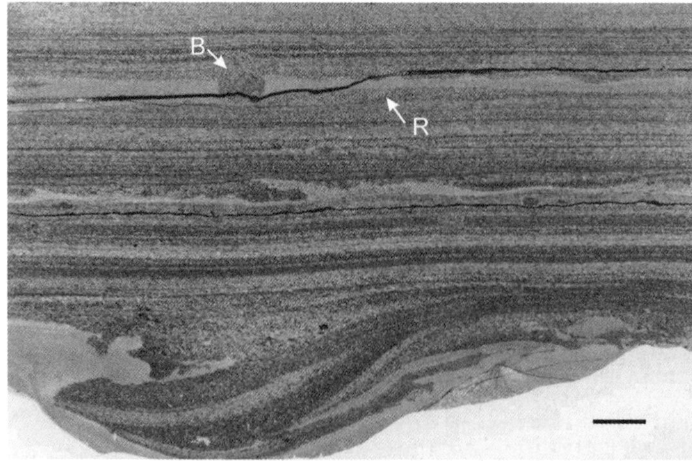

Fig. 9. Facies 5: polished slab of thinly interbedded sandstones and mudstones showing complex load casting and a predominance of parallel-lamination (locality 1). Three sandstone beds with intervening mudstones are shown. Note the local ripple cross-lamination (R) and bioturbation (B); the scale bar is 1 cm.

Facies 7) and sediment density-flow deposits in the succession (see facies 1–4).

Facies 6: Chaotic sediment sheets

These beds are up to 4.5 m thick and comprise discrete slabs or blocks of interbedded sandstone and mudstone (Facies 5) 'floating' in a matrix of structureless sandy mudstone. This muddy matrix typically dominates, forming up to 90% of individual beds. Bed bases are sharp, flat or weakly erosive, and individual beds are typically laterally persistent at an outcrop scale (up to 500 m). The bedded blocks may show internal deformation (Fig. 10) and juxtaposed slabs may show imbricate stacking. In some cases, disruption of the blocks has resulted in streaked-out lenses of sediment enveloped in a sandy mudstone (phacoids *sensu* Voigt 1962). Rooted slump folds were only rarely observed. Although scarce, kinematic indicators in the beds, such as fold-axis orientation and vergence and imbricate stacking of blocks, suggest derivation from the SE.

The complete spectrum from deformed but coherent stratal packets to isolated sediment rafts in a structureless mud matrix indicates that these beds were deposited by sediment density-flow processes ranging from sliding and slumping to cohesive debris flows (Dott 1963; Lowe 1979; Mulder & Alexander 2001). Imbricate stacking of stratal blocks suggests deposition at the toe of a slide sheet, possibly near the base of slope.

Facies 7: Cross-bedded sandstones

These tabular and trough cross-bedded, medium- to coarse-grained sandstones (Fig. 11) typically form sets about 0.3 m thick (maximum of 1.5 m thick). Set bases are commonly erosional with a basal lag of pebbles and siltstone clasts. Reactivation surfaces are common within sets. Foresets vary from low to high angle (up to 32°), and may be oversteepened, slumped and overturned. Most foreset laminae are 0.5–1 cm thick (maximum of 5 cm thick), and are defined by alternations of coarser and finer sandstone; some laminae are normally graded. Many foresets are draped by thin mudstone laminae that thicken towards the toes of foresets. These mudstone drapes are locally bioturbated and in some cases occur at regular intervals in cross-bed sets. Mudstone also occurs as discrete interbeds between cosets; such beds are typically lenticular, up to 20 cm thick,

Fig. 10. Facies 6: contorted, slump-folded sediment within a slide sheet, locality 3. The hammer is 35 cm long.

Fig. 11. Facies 7: large-scale tabular cross-bedding in volcaniclastic sandstones, locality 3. The staff is 1.5 m long with 10 cm gradations.

bioturbated and lack primary structures. Some mudstone beds contain an abundant, but low-diversity bivalve fauna. Units composed entirely of cross-bedded sandstone are up to 7 m thick and are typically laterally persistent; north of Whisky Bay, two such units can be traced approximately 500 m laterally without significant change in thickness. Foreset orientation indicates mainly NE-directed palaeocurrents (Fig. 6) parallel to the trend of the Antarctic Peninsula at all localities; sets with SW-orientated foresets, indicating flow reversal, are rare.

These sand beds were formed by the NE migration of medium- to large-scale bedforms (dunes *sensu* Ashley 1990). The episodic movement of these dunes is indicated by the presence of mudstone drapes on foresets. Reactivation surfaces indicate periodic modification of dune bedforms possibly by a subordinate opposing current (e.g. de Mowbray & Visser 1984). The scarcity of SW-directed foresets, however, indicates that the NE-directed currents were overwhelmingly dominant. Dunes are described from a number of different marine environments and are formed by the movement of bed load either by currents, such as storms, tides and geostrophic currents (Leeder 1999), or by sediment density flows (e.g. Broucke *et al.* 2004). A marine bottom current origin is favoured over a sediment density-flow origin first because of the thickness and lateral continuity of cosets, and secondly due to the fact that along the length of the outcrop the palaeocurrent trends are parallel to the Antarctic Peninsula. In contrast, palaeocurrents from the sediment density-flow deposits indicate sediment transport away from the Peninsula. Where cross-bedding is formed by sediment density flows, cross-bed orientation mirrors the orientation of other sediment density-flow sediment transport indicators (Broucke *et al.* 2004).

Tidal currents are favoured over other marine currents because of the presence of mud drapes, which in some cases are regularly spaced along with the presence of reactivation surfaces (Anderton 1976; Visser 1980; Allen & Homewood 1984). Modern contourite sediments deposited from geostrophic currents are typically fine grained, in the silt range, and transport of fine sand-sized material requires extreme flow conditions (McCave *et al.* 1995). Such conditions are rare but can be created by marked tectonic constrictions, such as in the Faroe–Shetland Channel where geostrophic bottom currents transport sediment up to gravel size and produce a range of bedforms (Knutz & Cartwright 2004; Masson *et al.* 2004). Such currents are also episodic in nature (Ramsay *et al.* 1996), although whether they are regularly so, like tidal currents, is unknown.

A storm-related origin for the cross-bedding in the Hidden Lake Formation can be ruled out because of the lack of wave ripples and HCS in the succession. It should be noted that the influence of tidal currents is not restricted to the shallow-marine environment (see discussion below). Given the fact that evidence for the interpretation of this facies is also derived from its associated facies, the issue of the origin of the cross-bedding is discussed further below. The occurrence of oversteepened, slumped and recumbent foresets has been attributed variously to current shear of water-saturated sands (Allen & Banks 1972), earthquake shock and rapid sedimentation (Roe 1987).

Facies 8: Graded structureless sandstone

Facies 8 is represented by rare thin beds (maximum of 8 cm thick) of very well sorted granule to very fine sand grade sediment that is dominated by volcanic lithic grains. These beds typically show well-developed normal distribution grading, but are otherwise structureless and have sharp, flat, non-erosional bases; a basal inverse graded zone is present in some beds (Fig. 12). Some beds contain accretionary lapilli. These thin beds can be traced laterally for considerable distances, sometimes in excess of 1 km.

The exceptionally good grading and sorting in these beds, together with the absence of basal erosion or internal sedimentary structures, argues against a sediment density-flow origin (cf. Facies 5). In view of the clast compositions, these beds are interpreted as subaqueous volcanic air-fall deposits formed by settling from suspension through the water column (Ledbetter & Sparks 1979; Whitham 1993). The basal inverse grading noted in some beds probably reflects the variation in density between vesicular ash and crystal fragments. Beds containing accretionary lapilli are also found in overlying and underlying formations (Ineson 1989; Pirrie 1989).

Fig. 12. Facies 8: polished slab of air-fall tuff showing well-developed normal distribution grading, locality 2. Horizontal lines (top) are saw-cut marks; the scale bar is 1 cm.

Facies associations and depositional environments

The lithofacies of the Hidden Lake Formation occur in three associations, that are interpreted to represent *base-of-slope*, *volcaniclastic fan-delta* and *basin-floor* depositional systems, respectively. The lateral and stratigraphic relationships between these systems parallel to the basin margin are shown in Figure 13 and are discussed further below.

Base-of-slope association

This association forms the lower half (*c.* 200 m) of the Hidden Lake Formation exposed between Holluschickie Bay and Rum Cove (Fig. 1). Continuous sections are rare in this area, however, and detailed study was only possible at locality 5, where the association is composed of amalgamated coarse conglomerates and pebbly sandstones (Facies 1 and Facies 2; Fig. 14). Bed bases are typically deeply scoured, and irregular and individual beds are laterally discontinuous. The most complete vertical section (*c.* 70 m thick) shows a marked subdivision into a lower interval of stacked fining-upwards cycles and an upper interval (*c.* 30 m) of disorganized, poorly graded, clast-rich conglomerate beds showing no discernible vertical grain-size trends. The fining-upwards cycles are 8–20 m thick and are composite units consisting of three to more than 10 beds (3.3–11.3 m; Fig. 14). Their lower portions consist of erosively based lenticular graded conglomerate beds (1–4 m thick). These are succeeded by sandy, matrix-rich conglomerates and pebbly sandstone beds showing both scoured and flat, non-erosional bases. Graded sandstones and bioturbated mudstones (Facies 4 and Facies 5) are preserved locally at the tops of cycles.

This thick stack of coarse-grained marine sediments records sedimentation from energetic, concentrated and hyperconcentrated density flows. The absence of wave-induced structures in the finer intervals suggests that deposition occurred beneath the range of storm-generated currents. Poor exposure at this stratigraphic level precludes detailed discussion of the lateral variation within this association, although detailed

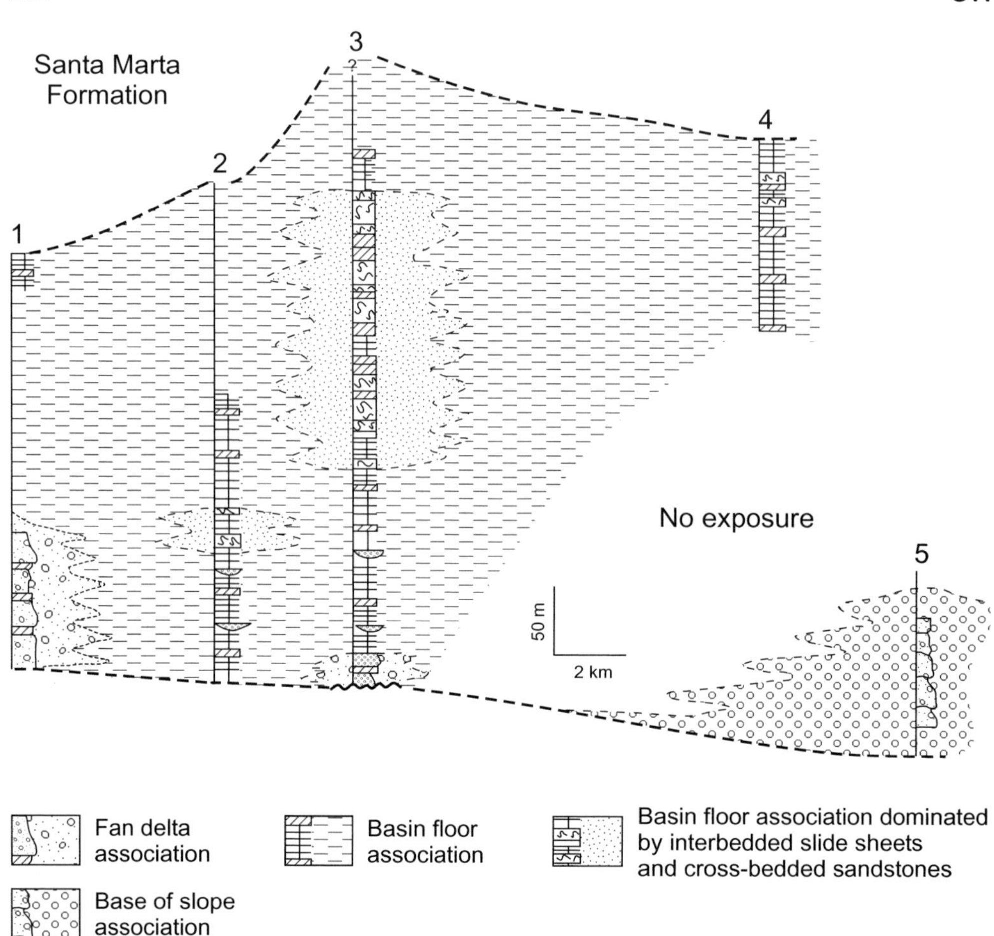

Fig. 13. Generalized log panel showing schematically the lateral distribution of facies associations in the Hidden Lake Formation along the NW coast of James Ross Island.

mapping indicates that the association is conglomerate-dominated throughout its outcrop. The degree of exposure, however, precludes confident interpretation of the precise depositional environment. The lenticular, scoured bed contacts and, in particular, the well-developed fining-upwards cycles suggest deposition in channels on a submarine fan. Equally, however, such features can be reproduced on coarse-grained slope aprons; the fining-upwards cycles may record the fill of chutes or large scours, or they may record deposition from waning, surging flows (e.g. Surlyk 1984). In the absence of data concerning the lateral construction and palaeocurrent pattern of this association, therefore, the setting cannot be narrowed down beyond a proximal, relatively deep marine environment characterized by high-energy sediment density flows – a generalized base-of-slope setting is inferred.

Volcaniclastic fan-delta association

This association is found at the base of the Hidden Lake Formation at localities 1 and 3 (Fig. 1), where it is about 100 and 30 m thick, respectively. It has an estimated maximum lateral extent of about 4 km and wedges out into mud-rich sediments assigned to the basin-floor association (Fig. 13). A lenticular form parallel to the basin margin is suggested by the available data, although incomplete exposure does not allow an accurate determination of the large-scale geometry.

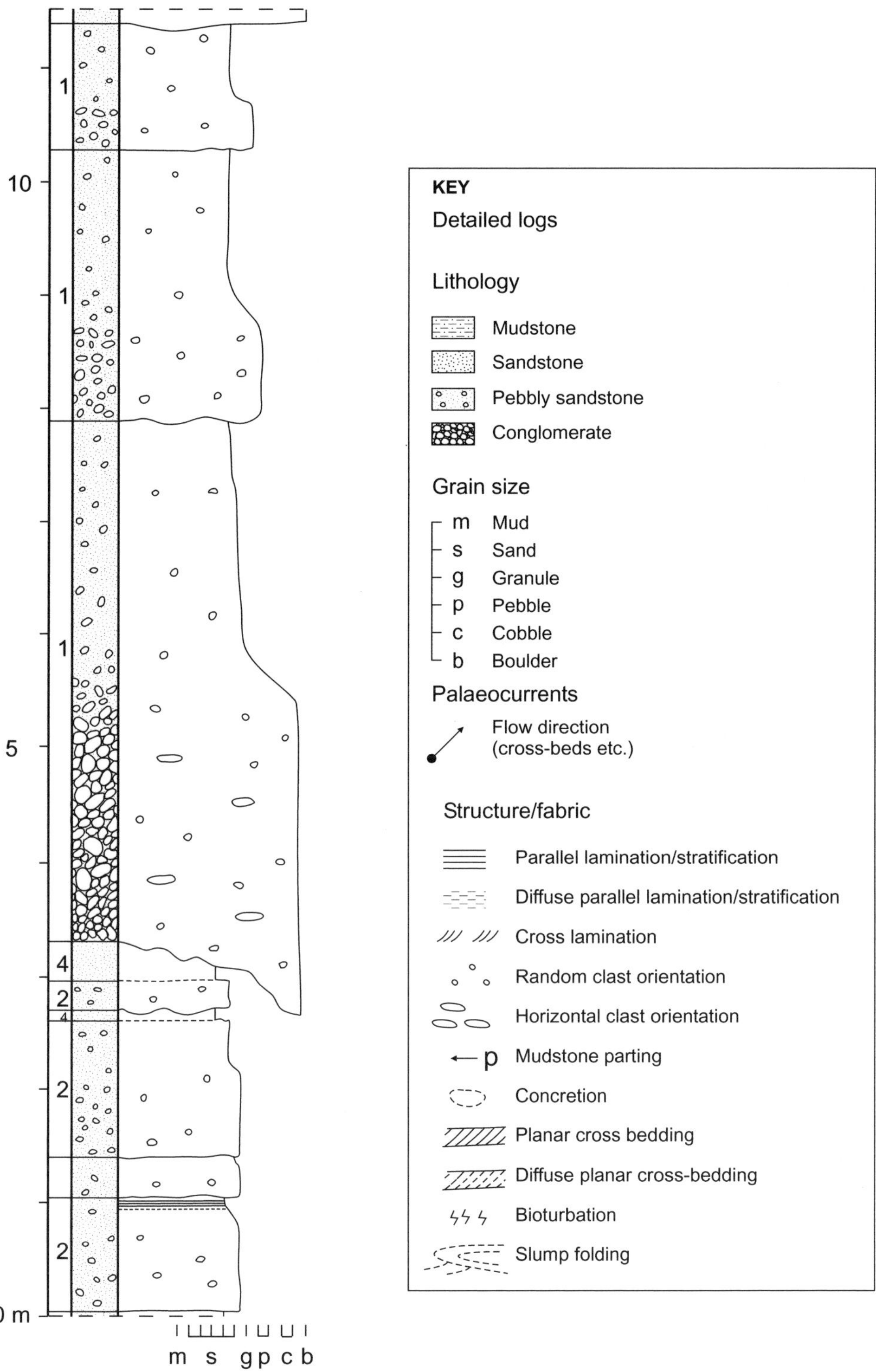

Fig. 14. Sedimentological log through part of the base-of-slope association from Ineson Glacier (locality 5). Facies numbers are indicated in the left-hand column.

The association is composed of an unusual combination of facies. It is dominated by graded pebbly sandstones of concentrated density-flow origin (Facies 2), cross-bedded sandstones (Facies 7) and parallel-stratified sandstones (Facies 3) (Figs 15 & 16). Intervals of medium- to thick-bedded sandstones (Facies 4) are also present. Thinly interbedded sandstones and mudstones (Facies 5), clast-rich conglomerates (Facies 1) and air-fall tuffs (Facies 8) occur only rarely. Reworked volcaniclastic material occurs throughout the Hidden Lake Formation, but is particularly abundant and fresh in appearance in this association. The sandstones contain an abundance of zoned plagioclase and fresh clinopyroxene, and laminae rich in magnetite occur locally. Whilst plant debris and wood fragments are common (many of which are charcoalified), macrofossils are scarce overall, being restricted to discrete mudstone intervals. The ichnofauna is of low diversity (Buatois & López-Angriman 1992*b*; Pirrie *et al.* 2004).

At locality 1 (Fig. 1), where the association is best displayed, the base of the association (and of the Hidden Lake Formation) is marked by the abrupt appearance of medium- to thick-bedded pebbly sandstone turbidites (Facies 4) containing abundant wood fragments. This facies dominates the lower 15 m of the succession and shows a weak thickening- and coarsening-upwards trend. The remaining approximately 80 m of the association is a complex alternation of lenticular–sheet-like pebbly sandstones or sandy conglomerates (Facies 2), parallel-stratified sandstones (Facies 3) and cross-bedded sandstones (Facies 7). The parallel-stratified sandstones typically form sheet-like bodies at the scale of outcrop (around 100 m; Fig. 17), but locally drape low-angle, concave-up erosion surfaces and wedge out laterally. Cross-bedded sandstones are intimately interbedded with the sediment density-flow deposits (Facies 2). They occur in units ranging from single sets to complex cosets up to 3 m thick and some directly overlie graded pebbly sandstones (Figs 16 & 17); in such cases they clearly record reworking of the sediment density-flow deposits by tractional bottom currents.

Although small-scale fining-upwards cycles can be recognized locally (e.g. 3.2–6.7 m; Fig. 15), the association as a whole generally shows no major systematic vertical grain-size trends, with the exception of the basal unit at locality 1 described above. However, the association is commonly divided into discrete 20–30 m-thick packets by thin (0.5–3 m), mudstone-rich intervals composed of bioturbated fossiliferous mudstones interbedded with

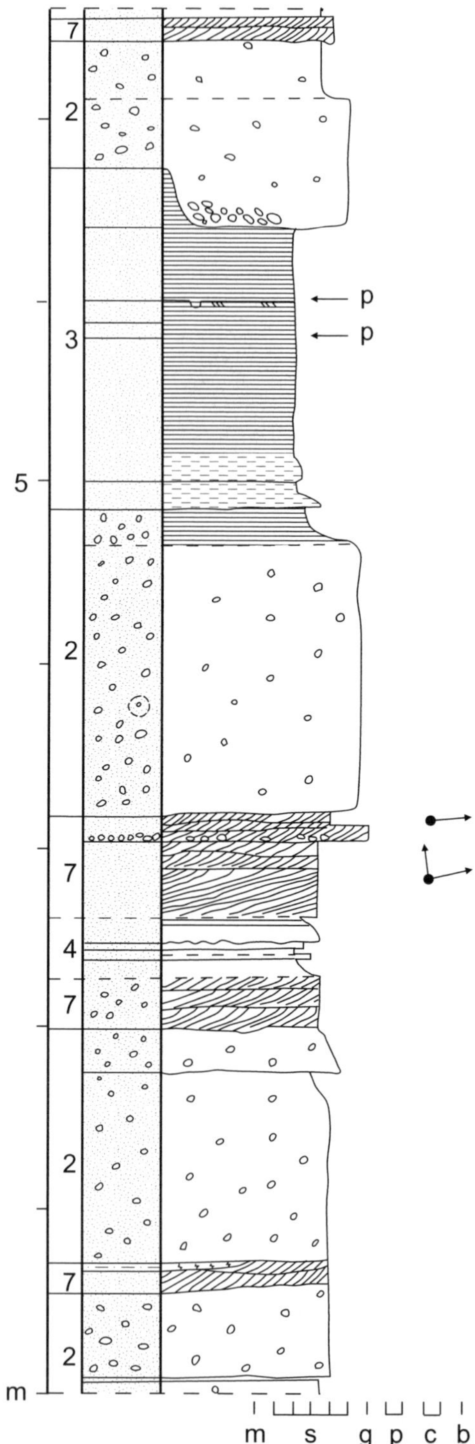

Fig. 15. Sedimentological log through part of the fan-delta association near Bibby Point, locality 1. For the key, see Figure 14.

Fig. 16. Fan-delta association (locality 1). Interbedded succession of volcaniclastic mass-flow deposits (m) and cross-bedded sandstones (c). Note the pebbly clast-rich base to the mass-flow deposit; pale concretionary haloes are developed around mudstone clasts in the mass-flow deposit; the staff is 1.5 m long with 10 cm gradations.

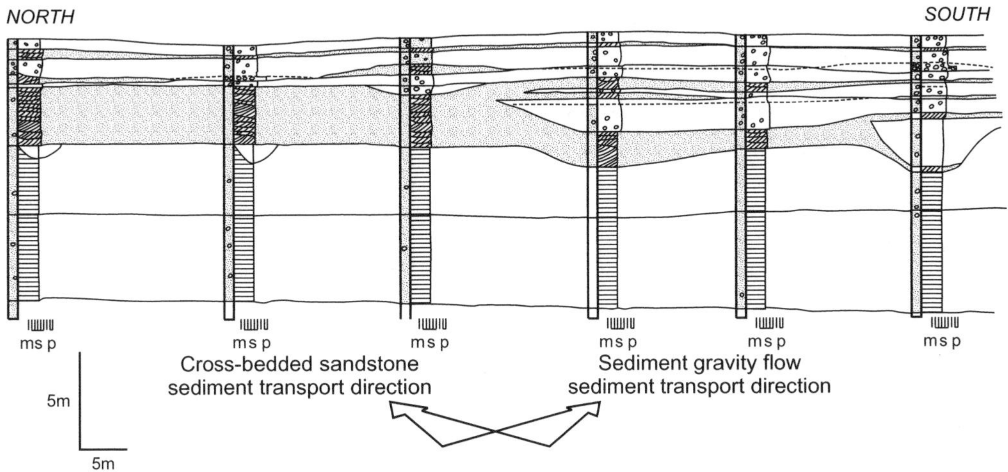

Fig. 17. Sedimentological log panel showing the lateral variation of facies in the fan-delta association (locality 1). Note the lateral persistence of the parallel-stratified beds in contrast to the lenticular channelized sandstones that fill scours/lows in underlying units. The cross-bedded sandstone units are shaded.

graded sandstones and isolated sets of cross-bedded sandstone. The upwards transition into the basin-floor association is marked by the disappearance of the coarse sandy sediment density-flow deposits and an increase in the proportion of mudstone; this change is abrupt.

The association records the introduction of abundant, coarse, fresh volcanic detritus, mainly by gravity-flow processes, into a marine environment. The absence of wave-induced structures indicates deposition below storm-wave base. Pebble imbrication in the sediment density-flow deposits (Fig. 6) indicates derivation of the volcanic detritus from the NW on the subaerial volcanic arc. The cross-bedded sandstones record reworking of this material by NE-flowing bottom currents of probable tidal origin, at right angles to the flow direction of the sediment density-flow deposits. The association forms lenticular isolated bodies along depositional strike, suggesting that the sediment was derived from point sources and did not form a continuous apron at the basin margin. The immaturity of the sediment, the abundance of plant material and the relative scarcity of reworked shallow-marine fossils testify to rapid transport from the subaerial arc into the substorm-wave base setting; little reworking appears to have taken place in shallow coastal environments. In view of such considerations, and the dominance of sediment density-flow deposits, this association is interpreted to represent the toesets of fan deltas that developed in response to localized eruptive events on the arc. In the absence of the proximal, subaerial–shallow-marine, portion of this system, it is not clear whether these deposits represent true fan deltas *sensu* Nemec & Steel (1988), fed by an alluvial fan, or fan-deltoid bodies fed directly by pyroclastic flows. However, sediment density flows formed the dominant mode of sediment transport inviting comparisons with the toeset region of Gilbert-type fan deltas (e.g. Massari & Colella 1988), in which such flows may be derived by periodic collapse of oversteepened, large-scale delta foresets (Postma 1984; Postma & Roep 1985; Postma *et al.* 1988).

A distinctive feature of this association is the presence of thick units of parallel-stratified sandstone, interpreted as being of sediment density-flow origin (see discussion of Facies 3). These were probably the result of deposition from hyperpycnal flows derived directly by fluvial input, further strengthening the fan-delta interpretation. Such high-energy facies are found in a proximal position on deltas (Plink-Björklund & Steel 2004) indicating the fluvial input point was nearby. Such settings also promote the deposition and preservation of plant material (Plink-Björklund & Steel 2004).

In summary, the association is thought to record deposition on the toesets of a marine fan delta. Volcaniclastic sediment originating from localized eruptive centres was deposited from sediment density flows derived either from underflows during flood events or from collapse of oversteepened delta foresets. Tractional bottom currents of probable tidal origin reworked the fan-delta toesets. With the exception of the basal coarsening-upwards unit at locality 1, evidence of progradation of the fan-delta systems is lacking. The association is succeeded at both localities by mud-dominated basin-floor sediments testifying to the demise of the fan-delta systems, perhaps due to the cessation or waning of volcanism at the local eruptive centres on the arc.

Basin-floor association

This association dominates the Hidden Lake Formation over much of its outcrop (Fig. 13). It is characterized by thinly interbedded sandstones and mudstones (Facies 5) and cross-bedded sandstones (Facies 7); evidence of wave activity is absent. A typical section, as illustrated in Figure 18, is overwhelmingly dominated by the above facies; channelized mass-flow

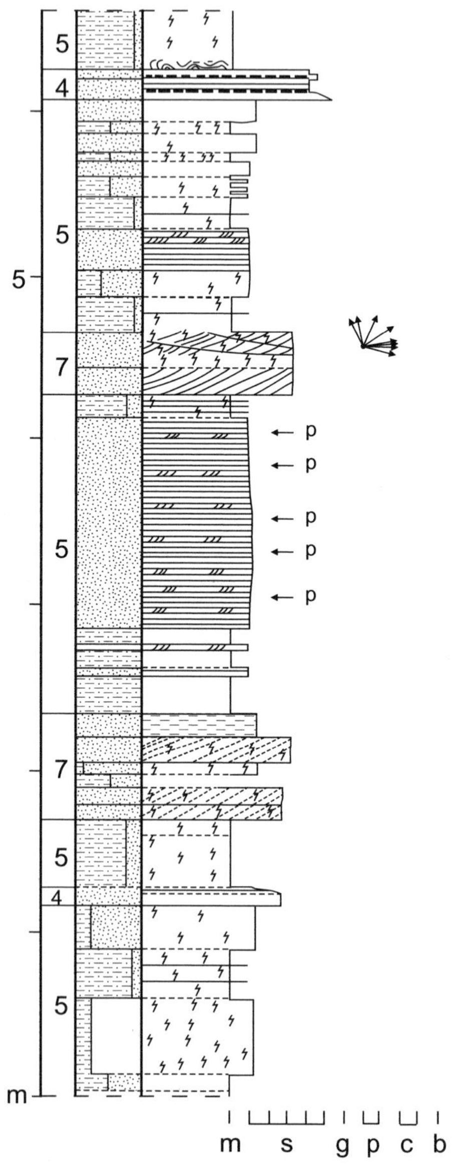

Fig. 18. Sedimentological log through part of the basin-floor association near Hidden Lake (locality 4). For the key, see Figure 14.

conglomerates (Facies 1) and slide deposits (Facies 6) occur only rarely, forming less than 1% of the succession. The relative proportion of facies 5 and 7 varies. Sections dominated by cross-bedded volcaniclastic sandstones record fields of dunes migrating towards the NE, parallel to the basin margin and probably under the influence of tidal currents. Finer grained, mud-rich sections dominated by Facies 5 represent low-energy settings between the dune fields. The rare sediment density flow and slide deposits testify to periodic sediment instability at the margin of, or within, the basin. Slumped and recumbent foresets in the cross-bedded sandstones may also reflect intrabasinal instability. Movement indicators in the slump sheets and mud-rich debris flows (Facies 6) are compatible with derivation from the eastern, rather than from the western, basin margin bordering the volcanic arc. This is consistent with the mud-rich composition of the mass-flow deposits, in contrast to the sand-rich, volcaniclastic sediments derived from the arc. The macrofauna and ichnofauna is of low diversity in most sections, although an increase in the trace-fossil diversity has been recorded in the uppermost levels of the formation (Buatois & López-Angriman 1992*b*; Pirrie *et al.* 2004), heralding the transition into the fossiliferous, intensely bioturbated shelf sediments of the Marambio Group.

A distinctive variant of the association is found at locality 3 (Fig. 1). It consists of mud-rich debris-flow deposits and slide sheets (Facies 6) intimately interbedded with cross-bedded sandstones (Facies 7) (Figs 19 & 20). Although not fully exposed throughout, this unit appears to be about 200 m thick (Fig. 5) with broadly equal proportions of the two facies. A similar, but thinner, unit is also developed at locality 2.

In summary, the association records basin-floor deposition and the absence of wave-related structures suggests deposition below storm-wave base. Furthermore, the presence of abundant sediment density-flow deposits (turbidites, debrites, slides), and the restricted *in situ* macrofauna and ichnofauna, are compatible with these water depths. Sections characterized particularly by debris flow and slide deposits probably reflect basin-floor settings close to intrabasinal slopes and highs. The occurrence of cross-bedded sands of inferred tidal origin may thus appear anomalous, and the association of mud-rich slump/slide and debris-flow deposits with cross-bedded tidal sandstones is striking. Indeed, the latter tidal facies figured prominently in preliminary discussions of this formation (Macdonald *et al.* 1988; Whitham &

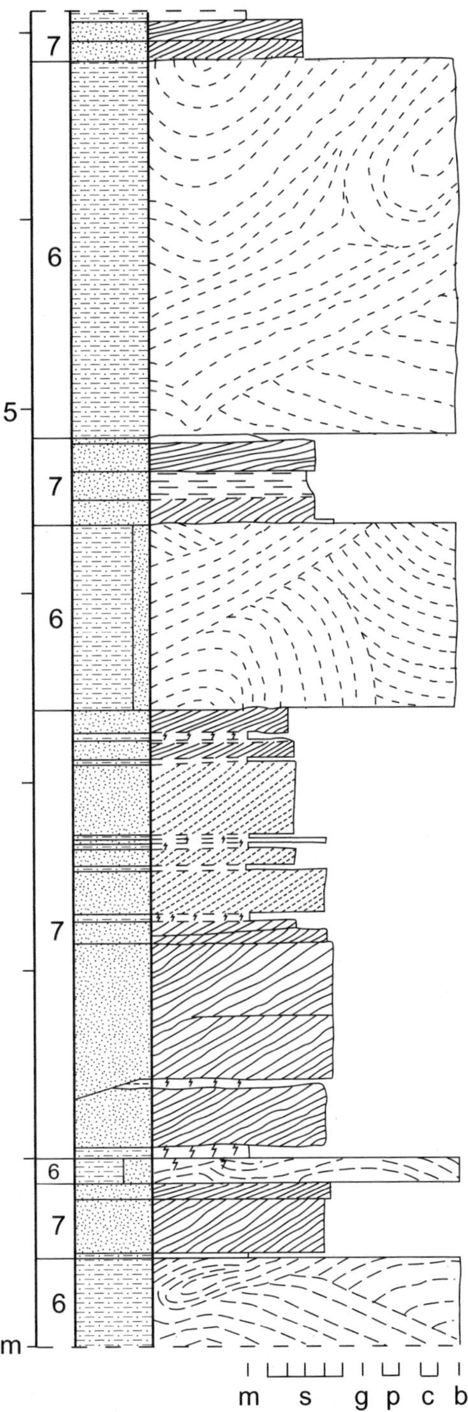

Fig. 19. Sedimentological log through part of the basin-floor association in Whisky Bay (locality 3) where it consists almost entirely of interbedded slide sheets and cross-bedded sandstones. For the key, see Figure 14.

Fig. 20. The basin-floor association in Whisky Bay (locality 3). Planar cross-bedded sandstones are interbedded with mud-rich debris-flow deposits (bottom of the photograph and behind the hammer). Upper mudstone just above the hammer (deposited from suspension, bioturbated) was eroded laterally prior to the deposition of the succeeding cross-bedded sandstone unit. The photograph shows the interval from 0 to 2 m in Figure 19.

Marshall 1988; Ineson 1989) leading to an erroneous shallow-marine, depositional interpretation (see further discussion below).

Depositional setting and stratigraphic evolution

A combination of palaeontological and sedimentological information suggests that the Hidden Lake Formation accumulated in a relatively deep-marine setting below storm-wave base. Sediment density-flow processes were important, particularly in the deposition of the lower part of the succession. A conspicuous feature of the formation, however, is the association of cross-bedded sandstones of inferred tidal origin with facies that otherwise are indicators of a deep-marine environment, and the absence of other shallow-marine indicators. Importantly, cross-bedded sandstones are not found in the underlying deep-marine deposits of the Gustav Group (Ineson 1989) and are uncommon in the shallow-marine deposits of the Marambio Group (Pirrie 1989).

As discussed previously, a sediment gravity-flow origin for the cross-bedding is deemed unlikely based on the orientation of cross-beds relative to sediment density flow, sediment transport indicators and the lateral continuity without change of thickness of the coset units. Further evidence against a sediment gravity-flow origin for the cross-bedding is provided by the fact that cross-bedded sandstones are common not just in the volcaniclastic fan-delta association but also in the basin-floor association. Records of cross-bedded sediment of density-flow origin indicate that such facies are volumetrically insignificant. They occur as isolated sets, or rarely cosets, associated with coarse sediment density-flow deposits that accumulated within or adjacent to submarine channel systems (Winn & Dott 1979; Hein & Walker 1982; Mutti 1992; Broucke *et al.* 2004). In contrast, cross-bedded units are common in the Hidden Lake Formation and are also associated with finer grained lower energy facies.

A tidal origin is favoured here for the cross-bedding given the presence of reactivation surfaces and bioturbated mud-draped foresets reflecting the frequent fluctuations in current energy common in tidal regimes. It should be noted, however, that geostrophic currents cannot be wholly ruled out, in part because they are less well understood. However, it seems unlikely that frequent periods of slack water and current reversal could occur in such currents. Evidence of energetic tidal currents at significant water depths has been presented by Colella & D'Alessandro (1988). These workers documented a succession of marine strata of Pliocene–Pleistocene age with cross-bedded strata of tidal origin, from southern Italy, that was deposited in water depths in excess of 350 m. They proposed that strong tidal currents were produced following the development of a palaeostrait that served to accentuate such currents, in a manner similar to the present-day straits of Messina where tidal sandwaves are reported in water depths in excess of 300 m. The Hidden Lake Formation thus invites comparison with the succession described by Colella & D'Alessandro (1988).

Amplification of marine bottom currents in the proximal zone of the James Ross Basin may have been caused by basinal constriction and shallowing related to partial basin inversion in the early Coniacian. This inversion not only resulted in deformation of the lower Gustav Group along the marginal zone of the basin, and locally the creation of submarine angular unconformities (Whitham & Marshall 1988; Hathway 2000), but may also have led to the development of intrabasinal ridges or highs. Provenance (Browne & Pirrie 1995) and palaeocurrent data (Ineson 1989) from the underlying Whisky Bay Formation also suggest that an irregular basin-floor topography may have controlled sediment dispersal patterns. In the light of evidence for strike-slip tectonics along this margin of the basin (Whitham & Storey 1989; Vaughan & Storey 1997), a complex basin morphology with local basins and highs might be expected, as exemplified by the present-day basinal morphology in the southern California borderland region (Gorsline 1987). Local basin highs may have acted to amplify tidal currents in the manner described by Colella & D'Alessandro (1988). Evidence for an intrabasinal high SE of the present outcrop is provided by the presence of mud-rich slide deposits derived from the east (see above) in the Whisky Bay and Brandy Bay sections. As the basin became tectonically quiescent during the deposition of the Hidden Lake Formation the intrabasinal relief was gradually eliminated by sediment infill, and widespread shallow-marine conditions were established by the end of the Coniacian.

The lower levels of the Hidden Lake Formation testify to a phase of coarse clastic influx in the early Coniacian. A pulse of coarse sediment at this level is evident along the outcrop belt despite the marked lateral variation in facies and inferred environments along depositional strike, parallel to the basin margin. The formation shows a broadly similar overall fining-upwards trend throughout this proximal outcrop belt, probably reflecting a decrease in erosion/transport and sedimentation rates due to the progressive stabilisation of the margin, as tectonism waned and volcanic activity gradually declined. The suggestion of reduced sedimentation rates in the late Coniacian is supported by the recognition of *Glossifungites* 'firm-ground' surfaces and a general increase in trace-fossil diversity in the uppermost Hidden Lake Formation (Buatois & López-Angriman 1992*b*; Buatois 1995; Pirrie *et al.* 2004).

The evolution of the proximal reaches of the basin and the proposed depositional setting of the Hidden Lake Formation are shown schematically in Figure 21. Local eruptive centres on the arc yielded abundant fresh pyroclastic debris creating localized fan deltas that extended out from the basin margin and were significant in the preservation of abundant leaf floras at several levels in the formation (Hayes *et al.* 2006). Facies variations within the succession along depositional strike seem to indicate shallowing of the marginal zone of the basin towards the north-east. Base-of-slope gravity-flow conglomerates and pebbly sandstones dominate the lower levels of the succession in the southern sections, inland of Gin Cove (Fig. 1), whereas cross-bedded tidal sandstones, although present in the upper levels of the formation, are less prominent in this area than further NE, around Whisky Bay and Brandy Bay. A local intrabasinal high (or highs) was responsible for the generation of slide sheets composed of material much finer than would be expected if they had been generated from the direction of the arc.

Discussion

The Coniacian sedimentary record provided by the Hidden Lake Formation bears the imprint of two dominant controlling factors: (1) the tectonic evolution of the basin; and (2) volcanism on the adjacent arc.

Tectonics

Although the detailed tectonic evolution of the Larsen Basin remains poorly understood, both the regional tectonic regime and the local geology suggest that oblique-slip tectonics were influential in the early development of the basin in the late Jurassic–early Cretaceous (Storey & Nell 1988; Whitham & Storey 1989; Hathway 2000). Continuing transpressive tectonic activity throughout the mid-Cretaceous is indicated by the syndepositional deformation of the proximal, deep-water sediments of the Lagrelius Point, Kotick Point and Whisky Bay formations on James Ross Island (Whitham & Marshall 1988), probably due to differential inversion along basin-margin structures accompanying arc rejuvenation. Localized angular unconformities are developed both in the lower Gustav Group (e.g. McArthur *et al.* 2000) and at the base of the Hidden Lake Formation (locality 3).

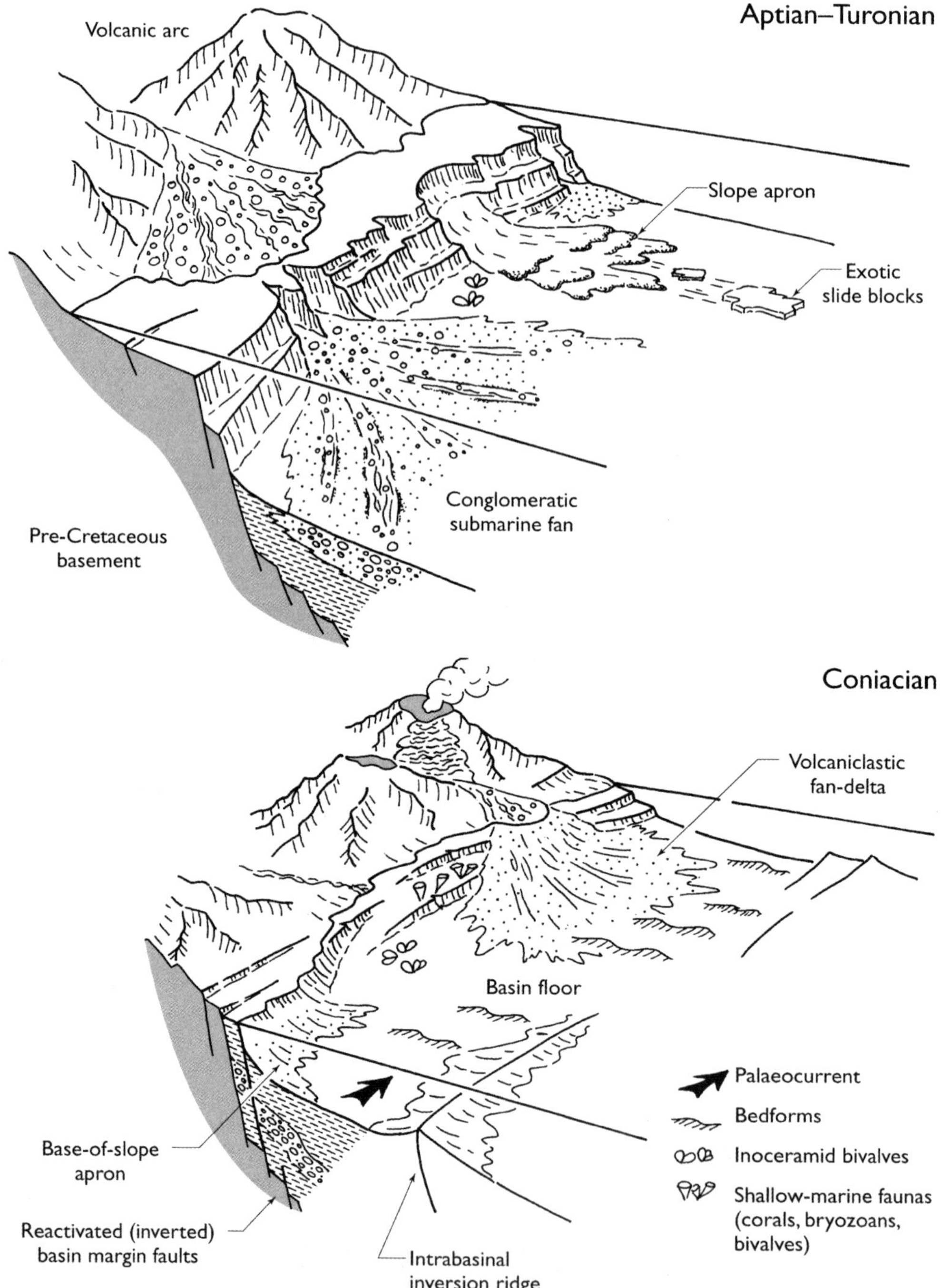

Fig. 21. Schematic representation of the evolution of the western margin of the James Ross Basin in the mid-Cretaceous, viewed towards the north. The Aptian–Turonian period (represented by the lower Gustav Group) was characterized by fault-controlled subsidence and the accumulation of coarse detritus in deep-water fan and slope-apron environments in the proximal reaches of the basin. Marginal and intrabasinal inversion in the Coniacian, coeval with a pulse of arc volcanism, led to radical changes in basin morphology, as recorded by the Hidden Lake Formation (see the text for detailed discussion).

In reviewing the evidence of progressive deformation of the proximal sediment pile, Hathway (2000) suggested that these angular discordances may be analogous to composite syntectonic progressive unconformities described from the Pyrenees (e.g. Riba 1976) that reflect the complex interaction between active basin subsidence and uplift/compression at the basin margin.

Syndepositional formation of the marginal monocline ceased at around the same time as the onset of deposition of the Hidden Lake Formation, as testified by the decrease in vitrinite reflectance values up-section from this level (Whitham & Marshall 1988). However, a number of features of the Hidden Lake Formation suggest that the proximal zone of the James Ross Basin remained tectonically active, albeit to a lesser degree. First, the marked shifts both in depositional environment and formation thickness along depositional strike suggest that lateral variation in differential subsidence across basin-margin faults continued to control sedimentation in the proximal zone. Secondly, the depositional model (Fig. 21) requires the presence of intrabasinal highs or ridges for much of the Coniacian to enhance tidal currents and to shed mud-rich slides; this argues for intrabasinal inversion to maintain the basin configuration. Thirdly, features such as slumped and recumbent foresets in cross-bedded units and intensely loaded, ball-and-pillow structures are consistent with continued seismic activity.

As noted earlier, the Conician record of partial basin inversion is indicative of an element of oblique-slip tectonics in the evolution of the James Ross Basin; that such influences continued to affect basin development is suggested by the description of synsedimentary faults of possible strike-slip origin from lower Maastrichtian strata (Pirrie & Riding 1988).

Volcanism and sediment supply

The Hidden Lake Formation records a major influx of volcaniclastic sediment from the arc into the back-arc basin and is recognized along the length of NW James Ross Island (*c.* 50 km) in a depositional strike section. The presence of pumice, bread-crust bombs and accretionary lapilli, in the volcaniclastic fan-delta association in particular, indicate that the sediment was derived from pyroclastic deposits with only limited reworking. An influx of pyroclastic material is also supported by the fact that the Hidden Lake Formation is compositionally immature in relation to underlying and overlying formations (Pirrie 1991), containing for example abundant unaltered clinopyroxene. In the light of this evidence, it is thought likely that the products of pyroclastic eruptions and the processes associated with their eruption played an important role in the origins of certain facies and in particular the development of the volcaniclastic fan-delta association.

Explosive volcanic eruptions are a common feature of extrusive calc-alkaline volcanism, and the resulting pyroclastic flow and air-fall deposits (Fisher & Schminke 1984) may devastate large areas of vegetation (Collins & Dunne 1986). As a result, the sediment load of fluvial systems draining the vicinity of a volcano following eruptions is commonly dramatically increased, as documented by Schuster (1981) after the Mount St Helens eruption. The increased sediment load leads to rapid growth of deltas at river mouths and the incorporation of abundant woody material in the sediment from vegetation killed by the eruption. Some of this wood is charcoalified, which may be the result of eruption-related wildfire. An excellent example of the marked effect that a pyroclastic eruption may have on the growth of a delta system is illustrated by the 1902 eruption of Santa Maria, Guatamala. The eruption produced an estimated 5.5 km^3 of pyroclastic detritus and caused the Samala river delta to prograde 6.4 km seawards in 20 years (Kuenzi *et al.* 1979). As discussed above, a number of the facies described from the volcaniclastic fan-delta association testify to rapid sedimentation and, at times, a semi-continuous supply of coarse pyroclastic detritus associated with abundant wood, which is commonly charcoalified, and other vegetative matter. A link to contemporaneous subaerial eruptions thus seems highly likely.

In the light of the abundance of pyroclastic material, however, more direct contributions from explosive volcanic eruptions, such as submarine pyroclastic flow deposits, might be expected (e.g. Francis & Howells 1973; Carey & Sigurdsson 1980; Busby-Spera 1988). Such features have not been recognized in the Hidden Lake Formation, suggesting that any contemporaneous pyroclastic eruptions on the arc were either of relatively limited volume or occurred some distance from the basin margin. The only primary pyroclastic beds identified in the sequence are rare thin subaqueous air-fall deposits (Facies 8).

Conclusions

The marine volcaniclastic wedge of the Hidden Lake Formation records a complex depositional history, bearing testament to spatial and temporal variations in syndepositional tectonism and volcanism. The sedimentological data presented here permit the following conclusions to be drawn.

- Three broad depositional systems are identified: base-of-slope, fan delta and basin floor. Depositional processes were dominated by sediment density flows, wave/storm-generated structures are absent and the faunal/ichnofaunal assemblages are of low diversity – this combined evidence suggests deposition in a relatively deep-marine setting. The presence of cross-bedded sandstones of inferred tidal origin thus appears anomalous, but is thought to have been caused by intrabasinal inversion and the creation of a marked basin-floor topography. This topography led to the focusing and amplification of tidal currents.
- The sediment composition testifies to syndepositional andesitic volcanism on the adjacent arc and the resultant transfer of coarse pyroclastic detritus rich in wood/plant debris to the marginal zone of the back-arc basin. This direct link to contemporaneous volcanism is particularly evident within the fan-delta deposits that record rapid sedimentation of coarse pyroclastic detritus on the toes of localized fan deltas, which were probably related to discrete eruptive centres on the arc.
- The Hidden Lake Formation occupies a pivotal position in the genetic evolution of the James Ross Basin, sandwiched between the deep-marine lower Gustav Group and the shelf deposits of the Marambio Group. It records an early Coniacian phase of partial basin inversion that was the culmination of successive pulses of transpressive deformation at the basin margin from the Aptian onwards. This resulted in a shallowing and restructuring of the proximal portion of the basin with the development of intrabasinal inversion ridges or highs that influenced sediment processes and dispersal patterns for much of the Coniacian. The lateral and stratigraphic distribution of facies within the proximal zone of the basin indicates spatial and temporal variation in both the subsidence history along the basin margin and the volcanic evolution of the arc.

The analysis and depositional model presented here provide a broad depositional and genetic framework for a critical phase in the evolution of the James Ross Basin, a basin that is increasingly regarded as a key Cretaceous–Palaeogene reference point in the Southern Hemisphere (e.g. Crame *et al.* 1996; Zinsmeister 1996; Huber 1998). It is hoped that this publication will stimulate further research into this enigmatic and challenging succession.

The authors would like to thank the British Antarctic Survey and S. Bell, D. M. Burkitt, M. P. D. Lewis, B. Newham and M. Sharp for their field support on James Ross Island. D. I. M. Macdonald, K. T. Pickering, F. Ricci-Lucchi and F. Surlyk are thanked for constructive comments on an earlier version of the manuscript. S. Lomas and an anonymous referee significantly improved the paper. D. Pirrie acknowledges NERC-AFI Grant GR3/AFI2/38 to McArthur, Crame, Thirlwall and Kennedy, and the support in the 2002 field season of J. A. Crame and HMS *Endurance*.

References

ALLEN, J.R.L. & BANKS, N.L. 1972. A quantitative analysis of recumbent folded deformed cross-bedding. *Sedimentology*, **31**, 257–283.

ALLEN, P.A. & HOMEWOOD, P. 1984. Evolution and dynamics of a Miocene tidal sandwave. *Sedimentology*, **31**, 63–81.

ANDERTON, R. 1976. Tidal-shelf sedimentation: an example from the Scottish Dalradian. *Sedimentology*, **23**, 429–458.

ASHLEY, G.M. 1990. Classification of large-scale subaqueous bedforms: a new look at an old problem. *Journal of Sedimentary Petrology*, **60**, 160–172.

BALDONI, A.M. & MEDINA, F. 1989. Fauna y microflora del Cretácico en Bahia Brandy, Isla James Ross, Antártida. *Serie Científico, Instituto Antártico Chileno*, **39**, 45–58.

BIBBY, J.S. 1966. *The Stratigraphy of Part of North-east Graham Land and the James Ross Island Group*. British Antarctic Survey Scientific Reports, British Antarctic Survey, **53**.

BROUCKE, O., GUILLOCHEAU, F., ROBIN, C., JOSEPH, P. & CALASSOU, S. 2004. The influence of syndepositional basin floor deformation on the geometry of turbiditic sandstones: a reinterpretation of the Côte de l'Ane area (Sanguinière-Restefondes subbasin, Grès d'Annot, Late Eocene, France). *In*: JOSEPH, P. & LOMAS, S.A. (eds) *Deep Water Sedimentation in the Alpine Basin of SE France: New Perspectives on the Grès d'Annot and Related Systems*. Geological Society, London, Special Publications, **221**, 203–222.

BROWNE, J.R. & PIRRIE, D. 1995. Sediment dispersal patterns in a deep marine back-arc basin: evidence from heavy mineral provenance studies. *In*: HARTLEY, A.J. & PROSSER, D.J. (eds) *Characterization of Deep Marine Clastic Systems*. Geological Society, London, Special Publications, **94**, 139–154.

BUATOIS, L.A. 1995. A new ichnospecies of *Fuersichnus* from the Cretaceous of Antarctica and its palaeoecological and stratigraphic implications. *Ichnos*, **3**, 259–263.

BUATOIS, L.A. & LÓPEZ-ANGRIMAN, A.O. 1992*a*. The

ichnology of a submarine braided channel complex: the Whisky Bay Formation, Cretaceous of James Ross Island, Antarctica. *Palaeogeography, Palaeoclimatology, Palaeoecology*, **94**, 119–140.

BUATOIS, L.A. & LÓPEZ-ANGRIMAN, A.O. 1992*b*. Trazas fósiles y sistemas deposicionales, Grupo Gustav, Cretácico de la isla James Ross. *In*: RINALDI, C.A. (ed.) *Geología de la isla James Ross, Antártida*. Instituto Antártico Argentino, Buenos Aires, 239–262.

BUSBY-SPERA, C.J. 1988. Evolution of a Middle Jurassic back arc basin, Cedros Island, Baja California: evidence from a marine volcaniclastic apron. *Bulletin of the Geological Society of America*, **100**, 218–233.

CANTRILL, D.J. & POOLE, I.P. 2002. Cretaceous to Tertiary patterns of diversity change in the Antarctic Peninsula. *In*: OWEN, A.W. & CRAME, J.A. (eds) *Palaeobiogeography and Biodiversity Change: A Comparison of the Ordovician and Mesozoic–Cenozoic Radiations*. Geological Society, London, Special Publications, **194**, 141–152.

CAREY, S.N. & SIGURDSSON, H. 1980. The Roseau ash: deep-sea tephra deposits from a major eruption on the Island of Dominica, Lesser Antilles Arc. *Journal of Volcanology and Geothermal Research*, **7**, 67–86.

COLELLA, A. & D'ALESSANDRO, A. 1988. Sand waves, *Echinocardium* traces and their bathyal depositional setting (Monte Torre Palaeostrait, Plio-Pleistocene, southern Italy. *Sedimentology*, **35**, 219–237.

COLLINS, B.D. & DUNNE, T. 1986. Erosion of tephra from the 1980 eruption of Mt St Helens. *Bulletin of the Geological Society of America*, **97**, 896–905.

CRAME, J.A., LOMAS, S.A., PIRRIE, D. & LUTHER, A. 1996. Late Cretaceous extinction patterns in Antarctica. *Journal of the Geological Society, London*, **153**, 503–506.

DEL VALLE, R.A., ELLIOTT, D.H. & MACDONALD, D.I.M. 1992. Sedimentary basins of the east side of the Antarctic Peninsula: proposed nomenclature. *Antarctic Science*, **4**, 477–478.

DE MOWBRAY, T. & VISSER, M.J. 1984. Reactivation surfaces in subtidal channel deposits, Oosterschelde, southwest Netherlands. *Journal of Sedimentary Petrology*, **54**, 811–824.

DICKINSON, W.R. 1985. Interpreting provenance relations from detrital modes of sandstones. *In*: ZUFFA, G.G. *Provenance of Arenites*. D. Reidel, Dordrecht, 333–361.

DOTT, R.H. 1963. Dynamics of subaqueous gravity depositional processes. *AAPG Bulletin*, **47**, 104–128.

ENKEBOLL, R.H., ESMAILA, H. and associates. 1982. Petrology and provenance of sands and gravels from the Middle America trench and trench slope, southwestern Mexico and Guatemala. *Initial Reports of the Deep Sea Drilling Project*, **66**, 521–530.

EYLES, N., CLARK, B.M. & CLAGUE, J.J. 1987. Coarse-grained sediment gravity flow facies in a large supra-glacial lake. *Sedimentology*, **34**, 193–216.

FISHER, R.V. & SCHMINKE, H.U. 1984. *Pyroclastic Rocks*. Springer, Berlin.

FRANCIS, E.H. & HOWELLS, M.F. 1973. Transgressive welded ash-flow tuffs among the Ordovician sediments of NE Snowdonia. *Journal of the Geological Society, London*, **129**, 621–641.

GORSLINE, D.S. 1987 Deposition in active margin basins: with examples from the California continental borderland. *In*: GORSLINE, D.S. (ed.) *Deposition in Active Margin Basins*. Society of Economic Paleontologists and Mineralogists, Los Angeles, 33–51.

HATHWAY, B. 2000. Continental rift to back-arc basin: Jurassic–Cretaceous stratigraphical and structural evolution of the Larsen Basin Antarctic Peninsula. *Journal of the Geological Society, London*, **157**, 417–432.

HAYES, P., FRANCIS, J.E., CANTRILL, D.J. & CRAME, J.A. 2006. Palaeoclimate analysis of Late Cretaceous angiosperm leaf floras, James Ross Island, Antarctica. *In*: FRANCIS, J.E., PIRRIE, D. & CRAME, J.A. (eds) *Cretaceous–Tertiary High-latitude Palaeoenvironments: James Ross Basin, Antarctica*. Geological Society, London, Special Publications, **258**, 49–62.

HEIN, F.J. & WALKER, R.G. 1982. The Cambro-Ordovician Cap Enragé Formation, Québec, Canada: conglomeratic deposits of a braided submarine channel with terraces. *Sedimentology*, **29**, 309–329.

HISCOTT, R.N. 1994. Traction carpet stratification in turbidites – fact or fiction. *Journal of Sedimentary Research*, **A64**, 204–208.

HISCOTT, R.N. & MIDDLETON, G.V. 1979. Depositional mechanics of thick-bedded sandstones at the base of a submarine slope, Tourelle Formation (Lower Ordovician), Quebec, Canada. *In*: DOYLE, L.H. & PILKEY, O.H. (eds) *Geology of Continental Slopes*. SEPM Special Publications, **27**, 307–326.

HISCOTT, R.N. & MIDDLETON, G.V. 1980. Fabric of coarse deep-water sandstones, Tourelle Formation, Quebec, Canada. *Journal of Sedimentary Petrology*, **50**, 703–722.

HUBER, B.T. 1998. Tropical paradise at the Cretaceous poles? *Science*, **282**, 2199–2200.

INESON, J.R. 1985. Submarine glide blocks from the Lower Cretaceous of the Antarctic Peninsula. *Sedimentology*, **32**, 659–670.

INESON, J.R. 1989. Coarse-grained submarine fan and slope apron deposits in a Cretaceous back-arc basin, Antarctica. *Sedimentology*, **36**, 793–819.

INESON, J.R., CRAME, J.A. & THOMSON, M.R.A. 1986. Lithostratigraphy of the Cretaceous strata of west James Ross Island, Antarctica. *Cretaceous Research*, **7**, 141–159.

INGERSOLL, R.V. & SUCZEK, C.A. 1979. Petrology and provenance of Neogene sand from the Nicobar and Bengal fans, DSDP sites 211 and 218. *Journal of Sedimentary Petrology*, **49**, 1217–1228.

KNELLER, B.C. 1995. Beyond the turbidite paradigm: physical models for deposition of turbidites and their implications for reservoir prediction. *In*: HARTLEY, A.J. & PROSSER, D.J. (eds) *Characterization of Deep Marine Clastic Systems*. Geological Society, London, Special Publications, **94**, 31–49.

KNUTZ, P.C. & CARTWRIGHT, J.A. 2004. 3D anatomy of late Neogene contourite drifts and associated mass flows in the Faroe–Shetland Basin. *In*: DAVIS, R.J., CARTWRIGHT, J.A., STEWART, S.A., LAPPIN, M. & UNDERHILL, J.R. (eds) *3D Seismic Technology: Application to the Exploration of Sedimentary Basins*. Geological Society, London, Memoirs, **29**, 63–71.

KUENZI, W.D., HORST, O.H. & MCGEHEE, R.V. 1979. Effect of volcanic activity on fluvial–deltaic sedimentation in a modern arc-trench gap, southwestern Guatemala. *Bulletin of the Geological Society of America*, **90**, 827–838.

LEDBETTER, M.T. & SPARKS, R.S.J. 1979. Duration of large-magnitude explosive eruptions deduced from graded bedding in deep-sea ash layers. *Geology*, **7**, 240–244.

LEEDER, M.R. 1999. *Sedimentology and Sedimentary Basins: From Turbulence to Tectonics*. Blackwell Science, Oxford.

LOWE, D.R. 1979. Sediment gravity flows: their classification and some problems of application to natural flows and deposits. *In*: DOYLE, L.H. & PILKEY, O.H. (eds) *Geology of Continental Slopes*. SEPM Special Publications, **27**, 75–82.

LOWE, D.R. 1982. Sediment gravity flows: II. Depositional models with special reference to the deposits of high density turbidity currents. *Journal of Sedimentary Petrology*, **52**, 279–297.

MACDONALD, D.I.M. & BUTTERWORTH, P.J. 1990. The stratigraphy, setting and hydrocarbon potential of the Mesozoic sedimentary basins of the Antarctic Peninsula. *In*: ST JOHN, B. (ed.) *Antarctica as an Exploration Frontier*. AAPG, Studies in Geology, **31**, 101–125.

MACDONALD, D.I.M., BARKER, P.F. *ET AL.* 1988. A preliminary assessment of the hydrocarbon potential of the Larsen Basin, Antarctica. *Marine and Petroleum Geology*, **5**, 34–53.

MASSARI, F. & COLLELA, A. 1988. Evolution and types of fan-delta systems in some major tectonic settings. *In*: NEMEC, W. & STEEL, R.J. (eds) *Fan Deltas: Sedimentology and Tectonic Settings*. Blackie, Glasgow, 103–122.

MASSON, D.G., WYNN, R.B. & BETT, B.J. 2004. Sedimentary environment of the Faroe–Shetland and Faroe Bank channels, north-east Atlantic, and the use of bedforms as indicators of bottom current velocity in the deep ocean. *Sedimentology*, **51**, 1207–1241.

MCARTHUR, J.M., CRAME, J.A. & THIRLWALL, M.F. 2000. Definition of Late Cretaceous stage boundaries in Antarctica using strontium isotope stratigraphy. *Journal of Geology*, **108**, 623–640.

MCCAVE, I.N., MANIGHETTI, B. & ROBINSON, S.G. 1995. Sortable silt and fine sediment size composition slicing – parameters for palaeocurrent speed and palaeoceanography. *Palaeoceanography*, **10**, 593–610.

MULDER, T. & ALEXANDER, J. 2001. The physical character of subaqueous sedimentary density flows and their deposits. *Sedimentology*, **48**, 269–299.

MUTTI, E. 1992. *Turbidite Sandstones*. Agip/Instituto di Geologia, Università di Parma, Parma, Italy.

NEMEC, W. & STEEL, R.J. 1988. What is a fan delta and how do we recognise it? *In*: NEMEC, W. & STEEL, R.J. (eds) *Fan Deltas: Sedimentology and Tectonic Settings*. Blackie, Glasgow, 3–13.

OLIVERO, E.B. & PALAMARCZUK, S. 1987. Ammonites y dinoflagelados Cenomanianos de la Isla James Ross, Antártida. *Ameghiniana*, **24**, 35–49.

PETTIJOHN, F.J., POTTER, P.E. & SIEVER, R. 1972. *Sand and Sandstone*. Springer, New York.

PICKERING, K.T., STOW, D.A.V., WATSON, M. & HISCOTT, R.N. 1986. Deep-water facies, processes and models: a review and classification scheme for modern and ancient sediments. *Earth Science Reviews*, **23**, 75–174.

PIRRIE, D. 1989. Shallow marine sedimentation within an active margin basin, James Ross Island, Antarctica. *Sedimentary Geology*, **63**, 61–82.

PIRRIE, D. 1991. Controls on the petrographic evolution of an active margin sedimentary sequence: the Larsen Basin, Antarctica. *In*: MORTON, A.C., TODD, S.P. & HAUGHTON, P.D.W. (eds) *Developments in Sedimentary Provenance Studies*. Geological Society, London, Special Publications, **57**, 231–249.

PIRRIE, D. & RIDING, J. 1988. Sedimentology and structure of Humps Island, northern Antarctic Peninsula. *British Antarctic Survey Bulletin*, **80**, 1–19.

PIRRIE, D., CRAME, J.A., LOMAS, S.A. & RIDING, J.B. 1997. Late Cretaceous stratigraphy of the Admiralty Sound region, James Ross Basin, Antarctica. *Cretaceous Research*, **18**, 109–137.

PIRRIE, D., FELDMANN, R.M. & BUATOIS, L.A. 2004. A new decapod trackway from the Upper Cretaceous, James Ross Island, Antarctica. *Palaeontology*, **47**, 1–12.

PIRRIE, D., WHITHAM, A.G. & INESON, J.R. 1991. The role of tectonics and eustasy in the evolution of a marginal basin: Cretaceous–Tertiary Larsen Basin, Antarctica. *In*: MACDONALD, D.I.M. (ed.) *Sea Level Changes at Active Plate Margins: Processes and Products*. International Association for Sedimentologists, Special Publications, **12**, 293–305.

PLINK-BJÖRKLUND, P. & STEEL, R.J. 2004. Initiation of turbidity currents: outcrop evidence for Eocene hyperpycnal flow turbidites. *Sedimentary Geology*, **165**, 29–52.

POSTMA, G. 1984. Mass flow conglomerates in a submarine canyon Abrioja fan-delta, Pliocene, southeast Spain. *In*: KOSTER, E.H. & STEEL, R.J. (eds) *Sedimentology of Gravels and Conglomerates*. Canadian Society of Petroleum Geologists, Memoirs, **10**, 237–258.

POSTMA, G. & ROEP, T.B. 1985. Resedimented conglomerates in the bottomsets of Gilbert-type deltas. *Journal of Sedimentary Petrology*, **55**, 874–885.

POSTMA, G., BABIC, L., ZUPANIC, J. & ROE, S.-L. 1988. Delta-front failure and associated bottomset deformation in a marine, gravelly Gilbert-type fan-delta. *In*: NEMEC, W. & STEEL, R.J. (eds) *Fan Deltas: Sedimentology and Tectonic Settings*. Blackie, Glasgow, 91–102.

RAMSAY, P.J., SMITH, A.M. & MASON, T.R. 1996. Geostrophic sand ridge, dune fields and associated bedforms from the northern Kwazulu–Natal shelf, south-east Africa. *Sedimentology*, **43**, 407–419.

RIBA, O. 1976. Syntectonic unconformities of the Alto

Cardena, Spanish Pyrenees: a genetic interpretation. *Sedimentary Geology*, **15**, 213–233.

RIDING, J.B. 1999. *A compilation and interpretation of palynological data on the Hidden Lake Formation (Gustav Group) from James Ross Island, Antarctic Peninsula.* British Geological Survey, Technical Report, **WH/99/115R**, 1–21.

RIDING, J.B. & CRAME, J.A. 2002. Aptian to Coniacian (Early–Late Cretaceous) palynostratigraphy of the Gustav Group, James Ross Basin, Antarctica. *Cretaceous Research*, **23**, 739–760.

RIDING, J.B., KEATING, J.M., SNAPE, M.G., NEWHAM, S. & PIRRIE, D. 1992. Preliminary Jurassic and Cretaceous dinoflagellate cyst stratigraphy of the James Ross Island area, Antarctic Peninsula. *Newsletters in Stratigraphy*, **26**, 19–39.

ROE, S.-L. 1987. Cross-strata and bedforms of probable transitional dune to upper-stage plane-bed origin from a Late Pre-Cambrian fluvial sandstone, northern Norway. *Sedimentology*, **34**, 89–102.

SCHUSTER, R.L. 1981. Effects of the eruption on civil works and operations in the Pacific northwest. *In*: LIPMAN, P.W. & MULLINEAUX, D.R. (eds) *The 1980 Eruptions of Mt St Helens, Washington.* US Geological Survey, Special Publications, **1250**, 701–718.

SHANMUGAM, G., LEHTONEN, L.R., STRAUME, S.E., HODGKINSON, R.J. & SKIBELI, M. 1994. Slump and debris-flow dominated upper slope facies in the Cretaceous of the Norwegian and northern North seas (61–67°N): implications for sand distribution. *AAPG Bulletin*, **78**, 910–937.

STOREY, B.C. 1991. The crustal blocks of West Antarctica within Gondwana: reconstruction and break-up model. *In*: THOMSON, M.R.A., CRAME, J.A. & THOMSON, J.W. (eds) *Geological Evolution of Antarctica.* Cambridge University Press, Cambridge, 587–592.

STOREY, B.C. & NELL, P.A.R. 1988. Role of strike-slip faulting in the tectonic evolution of the Antarctic Peninsula. *Journal of the Geological Society, London*, **145**, 333–337.

STOREY, B.C., VAUGHAN, A.P.M. & MILLAR, I.L. 1996. Geodynamic evolution of the Antarctic Peninsula during Mesozoic times and its bearing on Weddell Sea history. *In*: STOREY, B.C., KING, E.C. & LIVERMORE, R.A. (eds) *Weddell Sea Tectonics and Gondwana Break-up*. Geological Society, London, Special Publications, **108**, 87–103.

SURLYK, F. 1978. *Submarine Fan Sedimentation Along Fault Scarps on Tilted Fault Blocks (Jurassic–Cretaceous Boundary, East Greenland).* Grønlands Geologiske Undersøgelse, Bulletin, **128**.

SURLYK, F. 1984. Fan-delta to submarine fan conglomerates of the Volgian-Valanginian Wollaston Forland Group, East Greenland. *Canadian Society of Petroleum Geologists, Memoirs*, **10**, 359–382.

VAUGHAN, A.P.M. & STOREY, B.C. 1997. Mesozoic geodynamic evolution of the Antarctic Peninsula. *In*: RICCI, C.A. (ed.) *The Antarctic Region: Geological Evolution and Processes.* Terra Antartica Publication, Siena, 373–382.

VISSER, M.J. 1980. Neap-spring cycles reflected in Holcene subtidal large-scale bedform deposits: a preliminary note. *Geology*, **8**, 543–546.

VOIGT, E. 1962. Frühdiagenetische Deformation der turonen Plänerkalke bei Halle/Westf. Als Folge einer Grossgleitung unter besonderer Berücksichtigung des Phacoid-Problems. *Mitteilungen aus dem Geologischen Staatsinstitut in Hamburg*, **31**, 146–275.

WALKER, R.G. 1975. Generalized models for resedimented conglomerates of turbidite association. *Bulletin of the Geological Society of America*, **86**, 737–748.

WHITHAM, A.G. 1993. Facies and depositional processes in an Upper Jurassic to Lower Cretaceous pelagic sedimentary sequence, Antarctica. *Sedimentology*, **40**, 331–349.

WHITHAM, A.G. & MARSHALL, J.E.A. 1988. Syn-depositional deformation in a Cretaceous succession, James Ross Island, Antarctica. Evidence from vitrinite reflectivity. *Geological Magazine*, **125**, 583–591.

WHITHAM, A.G. & STOREY, B.C. 1989. Evidence for strike-slip deformation during Late Jurassic and Early Cretaceous times from the east coast of northern Graham Land, Antarctic Peninsula. *Antarctic Science*, **1**, 269–278.

WINN, R. & DOTT, R., JR. 1979. Deep water fan-channel conglomerates of Late Cretaceous age, southern Chile. *Sedimentology*, **26**, 203–228.

ZINSMEISTER, W.J. 1996. Late Cretaceous faunal changes in the high southern latitudes: a harbinger of global biotic catastrophe? *In*: MacLeod, N. & Keller, G. (eds) *Cretaceous–Tertiary Mass Extinctions: Biotic and Environmental Changes.* W.W. Norton, New York, 303–325.

Palaeoclimate analysis of Late Cretaceous angiosperm leaf floras, James Ross Island, Antarctica

PETA A. HAYES[1], JANE E. FRANCIS[2], DAVID J. CANTRILL[3] & J. ALISTAIR CRAME[4]

[1]*Department of Palaeontology, The Natural History Museum, Cromwell Road, London SW7 5BD, UK (e-mail: P.Hayes@nhm.ac.uk)*

[2]*School of Earth and Environment, University of Leeds, Leeds LS2 9JT, UK*

[3]*Department of Palaeobotany, Swedish Museum of Natural History, Box 50007, Stockholm S-104 05, Sweden*

[4]*British Antarctic Survey, Natural Environment Research Council, High Cross, Madingley Road, Cambridge CB3 0ET, UK*

Abstract: The fossilized remains of Cretaceous angiosperm leaves are preserved within sandstones and siltstones of the Coniacian Hidden Lake Formation (Gustav Group) and the Santonian–early Campanian Santa Marta Formation (Marambio Group) in the James Ross Basin, Antarctic Peninsula region. The leaves represent the remains of vegetation that grew at approximately 65°S on an emergent volcanic arc, now represented by the Antarctic Peninsula, and were subsequently transported and buried in marine sediments in the adjacent back-arc basin. Some of the angiosperm leaf morphotypes show similarities to leaves of living families such as Sterculiaceae, Lauraceae, Winteraceae, Cunoniaceae and Myrtaceae. Palaeoclimate analysis based on physiognomic aspects of the leaves, such as leaf-margin analysis, indicates that the mean annual temperatures for the Hidden Lake and Santa Marta formations were 13–21 °C (mean 17 °C) and 15–23 °C (mean 19 °C), respectively. The fossil plants are indicative of warm climates without extended periods of winter temperatures below freezing and with adequate moisture for growth. This period of Cretaceous warmth in Antarctica corresponds with the Cretaceous thermal maximum, an interval of peak global warmth from the Turonian to the early Campanian.

Vegetation was able to thrive on the Antarctic continent during the Cretaceous, even though it was situated at palaeolatitudes above 60°S, because the Cretaceous greenhouse climate system provided the polar regions with annual warmth (Skelton *et al.* 2003). Studies of the fossilized remains of this vegetation, preserved as leaves, petrified wood, pollen and spores, and reproductive organs (e.g. Askin 1989; Césari *et al.* 2001; Cantrill & Poole 2002; Eklund *et al.* 2004; Poole & Cantrill 2006), are now providing details of the composition and biodiversity of these floras. In addition, the plant fossils contain a record of past climatic conditions that provide clues to the nature of high-latitude Cretaceous climates in the southern polar regions (e.g. Francis & Poole 2002; Howe & Francis 2005; Poole *et al.* 2005). Angiosperm leaf fossils are an important part of the Antarctic fossil record from the mid-Cretaceous onwards. They are particularly valuable because they can provide an indication of past climate, both through Nearest Living Relative (NLR) comparisons and through statistical analyses that are based on the relationship of climate to physiognomic characters of the leaves.

The earliest occurrence of dicotyledonous angiosperm leaf fossils in Antarctica is of late Albian leaves from Alexander Island on the western side of the Antarctic Peninsula (Cantrill & Nichols 1996). From the James Ross Basin, to the east of the Antarctic Peninsula, the oldest known angiosperm leaf fossils are of Coniacian and Santonian age. This particular interval appears to be within a period of peak warming in high latitudes (Huber 1998), so these plant fossils may provide clues about the warmest climates and potentially the most diverse high-latitude floras in the Cretaceous greenhouse. The aim of this paper is to present new information about Late Cretaceous palaeoclimates at high latitudes deduced from two new angiosperm leaf floras.

From: Francis, J. E., Pirrie, D. & Crame, J. A. (eds) 2006. *Cretaceous–Tertiary High-Latitude Palaeoenvironments, James Ross Basin, Antarctica.* Geological Society, London, Special Publications, **258**, 49–62. 0305–8719/06/$15

Geological setting

The fossil leaves are preserved within proximal marine sediments in the James Ross Basin, part of the larger Larsen Basin (Macdonald & Butterworth 1990; del Valle *et al.* 1992; Hathway 2000). This basin contains more than 6 km of sediments that were deposited in a back-arc setting during Late Jurassic–Cenozoic times. The basin formed on the eastern side of an emergent volcanic arc that was constructed during subduction of the Pacific Ocean crust beneath Gondwana. This emergent arc (the eroded roots of which now form the Antarctic Peninsula, Fig. 1) is the likely location of the forests that yielded the fossil plant material (Francis 1986; Hathway 2000).

The basin infill consists of siltstones, sandstones and conglomerates that were deposited in progressively shallowing conditions in proximal submarine-fan and slope-apron settings, shelf settings and deltaic environments (Elliot 1988; Ineson 1989). The succession is divided into three main units: the Gustav Group (Aptian–Coniacian); the Marambio Group (late Coniacian–Palaeocene); and the Seymour Island Group (Palaeocene–latest Eocene/earliest Oligocene) (Pirrie *et al.* 1991; Riding *et al.* 1998; McArthur *et al.* 2000). The palaeolatitude of this region was approximately 65°S during the Cretaceous (Lawver *et al.* 1992).

The oldest flora in this study, the Hidden Lake flora, is preserved within the Hidden Lake

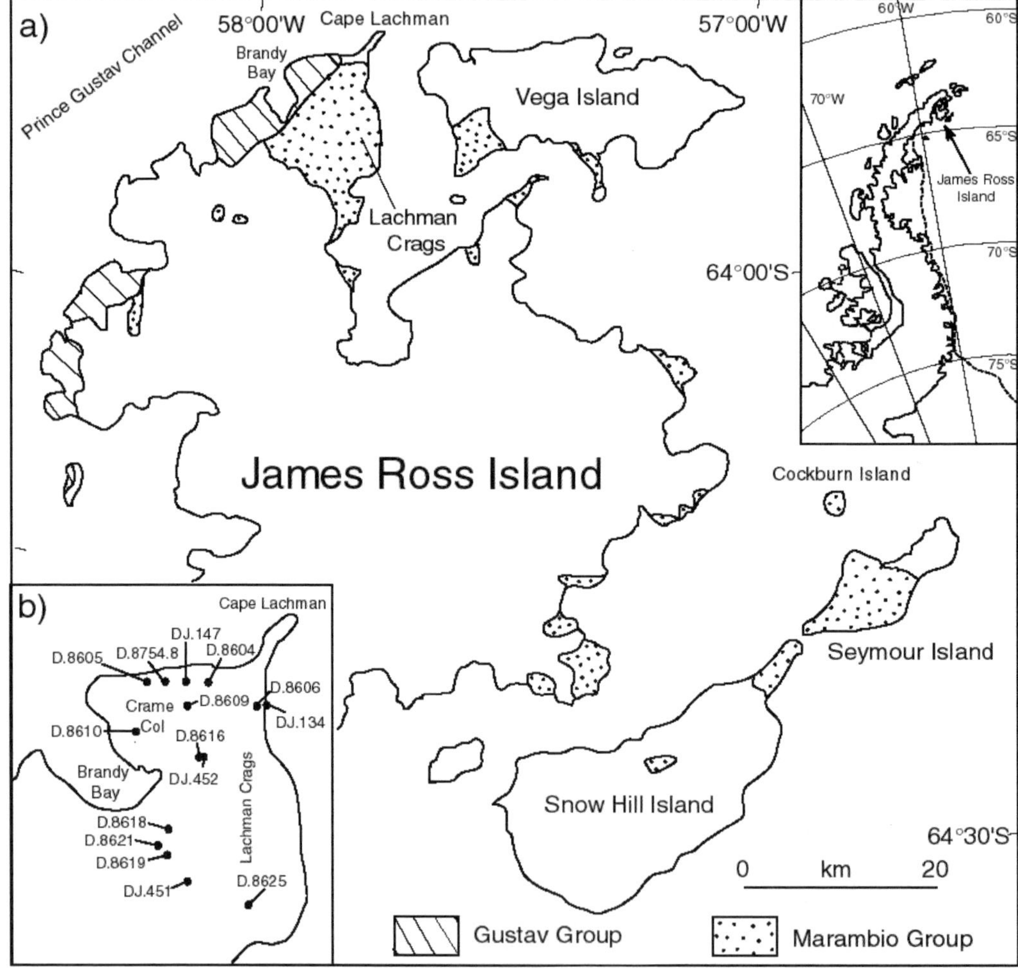

Fig. 1. (**a**) Location map showing the position of the James Ross Island to the east of the Antarctic Peninsula and location of the main geological groups from which the fossil leaves were recovered. (**b**) The leaves come from various localities in the NW part of James Ross Island; numbers indicate location of samples.

Formation of the Gustav Group. This formation consists of 300–400 m of coarse-grained volcaniclastic conglomerates, sandstones and mudstones, representing base-of-slope, fan-delta and basin-floor environments of deposition in a relatively deep marine environment below wave base (Ineson *et al.* 1986; Pirrie *et al.* 1991; Whitham *et al.* 2006). Macrofaunal and palynological studies suggested a probable age range of Coniacian–Santonian, but Sr isotope data confirm an entirely Coniacian age (88.7–86.4 Ma) for this formation (McArthur *et al.* 2000; Riding & Crame 2002). Petrographical studies have shown that the primarily volcaniclastic sediment source was nearby (Pirrie 1991). The fossil leaves in this formation therefore most probably originated on the volcanic arc and were washed only a short distance into the marine basin.

The younger flora in this study is the Santa Marta flora from the Santa Marta Formation of the Marambio Group. This formation conformably overlies the Hidden Lake Formation and is exposed on NW James Ross Island (Fig. 1). Silty and muddy sandstones and marls with concretionary beds were deposited in shallow-marine shelf environments (Pirrie 1989; Crame *et al.* 1991). Petrographical and mineralogical studies indicate that the sediment was sourced from erosion of volcanic and plutonic elements of the Antarctic Peninsula (Browne & Pirrie 1995; Dingle & Lavelle 1998). The flora is found within the Lachman Crags Member, a sequence of sandstones and bioturbated silty sandstones and mudstones, from a mid- to outer-shelf setting (Pirrie 1989; Crame *et al.* 1991). This member has been dated using molluscan faunas and dinoflagellate cysts, which provided an early Santonian–early Campanian age (Crame *et al.* 1991; Keating 1992). However, more recent studies suggest that the Coniacian–Santonian boundary is within the lower 150 m of the Lachman Crags Member (McArthur *et al.* 2000); hence, a late Coniacian–early Campanian age is likely.

Preservation of the floras

Over 200 dicotyledonous angiosperm leaf specimens were included in a study by Hayes (1999). The fossils examined were collected by several Antarctic geologists from 1985 to 1990 from various localities within NW James Ross Island (Fig. 1). All specimens are held at the British Antarctic Survey (BAS), Cambridge, and are numbered with BAS locality and specimen numbers.

Most of the leaf fossils are preserved scattered on uneven bedding planes as impressions, frequently with brown carbonaceous residues retained on the surfaces (Fig. 2). Some of the Santa Marta Formation fossils are also found within fine-grained carbonate concretions and many of these leaf impressions are preserved with a mineral coating of calcite. Approximately 25% of the leaves of both floras are almost whole, some with both part and counterpart preserved, but the majority are fragmentary. Although there is no cuticle present, architectural features of the leaves, such as leaf form, margins and venation patterns of the primary, secondary, tertiary and occasionally higher orders, are clearly visible in most specimens, allowing the isolation of taxonomic groups and palaeoclimatic analysis.

Angiosperms dominate the floras, with some ferns, and rare conifer and bennettitalean remains. Only the angiosperms are considered in this paper as at present the palaeoclimate significance of these other plants is less well constrained. This is the first published palaeoclimate analysis of these Cretaceous angiosperm floras. Further collecting and larger sample size is likely to provide more robust palaeoclimate data in the future.

Floral composition

Each of the leaf fossils was drawn and described in detail by Hayes (1999), including features such as leaf form, apical and basal styles, the nature of the leaf margin and venation characteristics. A multivariate statistical approach was developed to utilize many characters simultaneously in the definition of morphotypes (taxa delineated according to morphological similarities: Hayes 1999). A total of 41 morphotypes were described. Of these, 30 are present within the Hidden Lake flora and 31 within the Santa Marta flora, with 20 common to both floras. Detailed descriptions of each morphotype and the analytical technique used to define the morphotype groups can be found in Hayes (1999). Figure 3 illustrates a typical leaf of each morphotype.

Several of the morphotypes, clearly defined by their distinguishing characters, possess features common to disparate angiospermous groups, but many of the leaf forms show similarities to modern angiosperm orders and even families. The dominant morphotype of both floras is considered to show similarity to the Magnoliales and both floras share a strong component of lauralean-like leaf forms. Morphotypes sharing diagnostic features with leaves of the Sterculiaceae, Lauraceae,

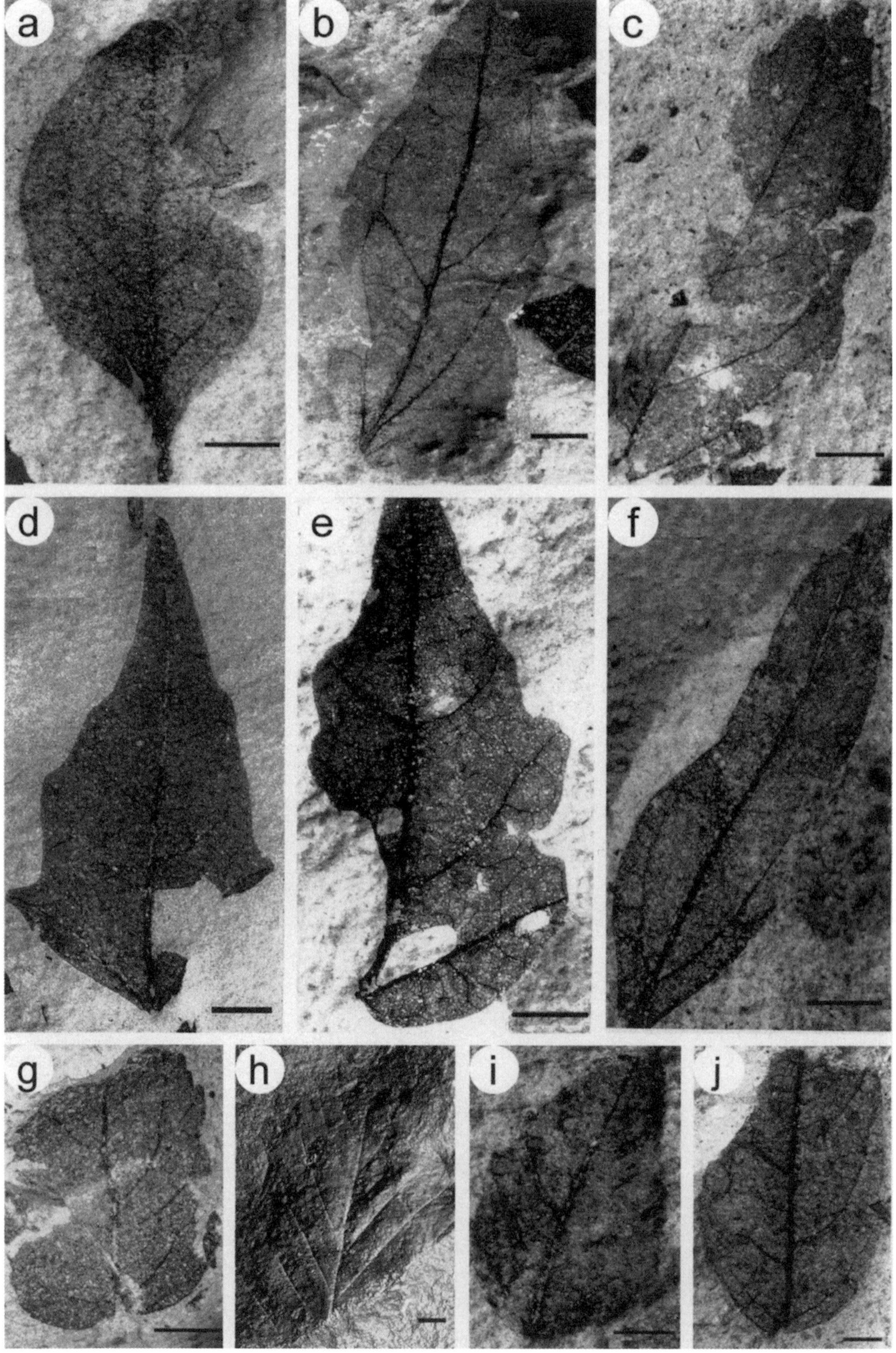
a
b
c
d
e
f
g
h
i
j

Elaeocarpaceae and Winteraceae are common to both floras. A leaf form showing similarities to the Atherospermataceae is present within the Hidden Lake flora only and a morphotype showing similarities to the Myrtaceae is restricted to the Santa Marta flora. Morphotypes with architectural characteristics typical of the Cunoniaceae and Nothofagaceae occur within the Hidden Lake flora but are more abundant within the Santa Marta flora (Hayes 1999).

Palaeoclimate analysis

Techniques

Although fossil leaves from the Hidden Lake and Santa Marta formations are rare and fragmentary, they are a valuable source of information about Late Cretaceous climates in Antarctica. Angiosperm leaf floras can be used to provide qualitative and quantitative assessments of palaeoclimate using a variety of techniques. These include Nearest Living Relative (NLR) approach and analyses of leaf physiognomy, which do not rely on accurate identification of fossil leaf taxa and are thus particularly useful (Spicer 1990).

Nearest Living Relative (NLR) analysis. This approach involves assigning a fossil morphotype to a modern group and extrapolating back the climatic tolerances of the extant taxon (Spicer 1990; Chaloner & McElwain 1997). Qualitative assessments of palaeoclimates can be derived that will provide a first indication of palaeoclimate and the nature of the boundary conditions (e.g. warm, cool, lack of freezing) (Mosbrugger 1999). There are, however, limits to this approach, especially for these Antarctic leaves, as many of the morphotypes possess architectural features that occur across several living families. Application to families or higher groupings rather than genera is problematic because of the breadth of ecological tolerances. In addition, the ecological tolerances of fossil taxa that lived under greenhouse climates not present on Earth today may have been different from those that have suffered geographical restrictions during the Quaternary. Most significantly, angiosperms were rapidly diversifying during the Cretaceous and were not as dominant in global vegetation as they are today (Lidgard & Crane 1990).

Leaf margin analysis (LMA). The analysis of the relationship of non-entire (toothed) leaf margins to mean annual temperature (MAT) is considered one of the most useful and robust techniques in palaeoclimate reconstruction (Wolfe & Upchurch 1987; Burnham 1989). Early work on extant dicotyledonous angiosperm floras from around the world by Bailey & Sinnott (1915, 1916) showed that species with entire margined leaves are dominant in warm tropical environments, while leaves with non-entire margins predominate in cooler temperate regions. Based on extensive studies of living humid to mesic eastern Asian forests, Wolfe (1971, 1979) established a linear relationship between MAT and the proportion of entire-margined species. This relationship only breaks down in areas of low moisture availability and in very cold or arid regions where there is an increased proportion of entire margined species (Gregory & McIntosh 1996). The relationship established for the Northern Hemisphere is an increase of 1 °C for every 3% increase in entire margined species, with 60% corresponding to a MAT of 20 °C. In the Southern Hemisphere there is thought to be a higher proportion of evergreen plants reflected in a higher percentage of entire margined species, and so Wolfe (1979) adjusted the relationship so that a 1 °C increase in MAT corresponds to a 4% increase in entire-margined species, with 68–70% entire-margined species corresponding to a MAT of 20 °C.

The use of this relationship for Southern Hemisphere floras has caused some debate because it was thought that the correlation between LMA and MAT was poor in this region, and that MATs for Australian and New Zealand vegetation were overestimated (Jordan 1997). Greenwood *et al.* (2004) showed, however, that there is a statistically significant relationship between leaf margin and MAT for Australian rainforest vegetation. The slope of

Fig. 2. A selection of angiosperm leaf fossils from the Hidden Lake and Santa Marta formations. (**a**) D.8754.8.45a. Hidden Lake flora. Morphotype 3. (**b**) D.8754.8.54a. Hidden Lake flora. Morphotype 11 (Laurales). (**c**) D.8754.8.4a. Hidden Lake flora. Morphotype 17. (**d**) D.8754.8 3a. Hidden Lake flora. Morphotype 2 (Sterculiaceae). (**e**) D.8754.8.1a Hidden Lake Flora. Morphotype 2 (Sterculiaceae). (**f**) D.8754.8.57a. Hidden Lake flora. Morphotype 11 (Laurales). (**g**) D.8754.8.30a. Hidden Lake flora. Morphotype 20 (Elaeocarpaceae). (**h**) D.8621.27a. Santa Marta flora. Morphotype 7 (Nothofagaceae). (**i**) D.8754.8.8a. Hidden Lake flora. Morphotype 25. (**j**) D.8754.8.42a Hidden Lake flora. Morphotype 10. Scale bar 5 mm for all leaves. Names in brackets indicate possible similar living orders or families.

Fig. 3. Line drawings showing the venation features of the morphotypes identified in the floras discussed here. For detailed descriptions of each morphotype see Hayes (1999).

the regression is similar to that of East Asia, North America and Bolivia, but with a different intercept so that fewer toothed species are present at a given temperature. They proposed that palaeotemperature estimates would be a minimum estimate if the Southern Hemisphere regression line was used and a maximum estimate if the Northern Hemisphere regression relationship of Wolfe (1979) was used. This duel approach was used, for example, by Kennedy *et al.* (2002) in a palaeoclimate analysis of Late Cretaceous and Paleocene floras from South Island, New Zealand and by Hunt & Poole (2003) in palaeoclimate analysis of Tertiary floras from King George Island, West Antarctica.

Simple linear regression (SLR) models. Simple linear regression equations have also been derived for the relationship between MAT and leaf-margin character based on the LMA relationship documented by Wolfe (1979). Several SLR equations have been derived by palaeobotanists who have quantified the relationship of percentage of taxa with entire (untoothed) margins in living floras with climate for specific regions. SLR equations have, for example, been derived by Wing & Greenwood (1993), Wiemann *et al.* (1998) and Wilf (1997) for data from East Asian forests, Australian forests, temperate and tropical floras of North and South America, and from the CLAMP data set of Wolfe (1993) (see below).

Some SLR models are used to estimate precipitation, rather than temperature, based on leaf size (e.g. Wilf *et al.* 1998). Studies by Wiemann *et al.* (1998) concluded that leaf size is one of the most important predictors of precipitation. In these fossil floras only about 25% of the specimens are almost whole leaves, the majority being fragments. However, the likely original size of a leaf was reconstructed by Hayes (1999) by first identifying the morphotype from venation characteristics in the fragment, if well enough preserved, and then reconstructing the whole leaf based on comparable whole specimens of that morphotype. In this way estimates of leaf sizes were successfully obtained for many fragmentary specimens. Maximum estimated leaf areas per fossil morphotype, obtained as described above, were then used to predict mean annual precipitation (MAP). Estimates of MAP using this method (see Table 1) have to be treated with some caution because the fossil leaf assemblages must have been subject to considerable taphonomic bias (Greenwood 1992), especially as it is likely that the larger leaves were eradicated during transport (Ferguson 1985; Gastaldo *et al.* 1996). Estimates of MAP are therefore likely to be minimum estimates.

Multivariate analyses. These models use a selection of leaf attributes, rather than just one as in SLR, to interpret climates. Two techniques are presented below.

Climate–Leaf Analysis Multivariate Programme (CLAMP) (Wolfe 1993; Kovack & Spicer 1996; Wolfe & Spicer 1999). This is a multivariate statistical technique that relates the climatic signal preserved in the physiognomy of leaves of woody dicotyledonous angiosperms to meteorological characteristics of the region in which the plants grow. CLAMP uses Canonical Correspondence Analysis (CANOCO) (Ter Braak 1986), a direct ordination method used widely in plant ecology, to analyse the relationship between leaf attributes and climate. CLAMP datasets of leaf attributes and climate parameters have been compiled for 173 predominantly Northern Hemisphere modern vegetation sites, although subsets of data are often used if more relevant to the likely palaeoclimate (e.g. cold climate sites are often removed). For analysis of fossil floras, the fossil leaf data are positioned within the modern data set and the corresponding palaeoclimate determined from the analysis.

Reliable CLAMP analysis requires sample sizes of greater than 20 species for MAT (25 for precipitation analysis), with the majority of leaf characters preserved. The incomplete preservation of the Antarctic leaves and preponderance of missing characters means that full CLAMP analysis was not suitable in this study. However, simplified CLAMP data sets of modern vegetation and climate characters have been used to derive multiple linear regression models that can be applied to these Antarctic floras, as discussed below.

Multiple linear regression (MLR) models. Several MLR models (regression equations) have been derived from characters in the CLAMP data set, but using only a reduced number of characters that are more frequently preserved in fossil floras, more consistently scored by researchers and more useful for palaeoclimate prediction. For example, Wing & Greenwood (1993) obtained regression equations using a subset of the CLAMP data set with sites experiencing extreme cold removed, thus creating a data set more representative of environments in past greenhouse climates. Wiemann *et al.* (1998) derived their own MLR models from the CLAMP data set for specific sites that included temperate and tropical America and Japan. Their study of correlation coefficients between the CLAMP leaf characters and climate parameters demonstrated that leaf morphology has a stronger relationship with temperature than precipitation. MAT was most highly correlated with leaf-margin character, and growing season precipitation (GSP) with length/width ratio, leaf shape, the presence of an attenuate apex and leaf size. The MLR methods used are shown in Table 1.

Results

Palaeoclimate from NLR analysis

Although the Antarctic fossil leaves cannot be confidently assigned to modern orders or families, comparable modern plant groups that share similar architectural characteristics can provide some indication of the climate window in which the fossils lived. Most living types comparable to the fossil leaves can be found in warm temperate or subtropical zones of the Southern Hemisphere today. For example, the Winteraceae are characteristic of wet tropical montane to cool temperate rainforests of Tasmania and New Zealand. Sterculiaceae is a tropical to subtropical family extending throughout Australia, south Asia, Africa and northern South and Central America. Laurales are trees, shrubs and woody vines that live today in tropical or warm temperate regions with a moist equable climate. Elaeocarpaceae is a widespread family of tropical and subtropical trees and shrubs present in eastern Asia, Australasia, Indomalaysia, the Pacific area, South America and the West Indies. The Cunoniaceae occur as trees, shrubs and woody climbers in tropical and warm temperate Australia, New Guinea and New Caledonia. From the list above it is apparent that examples of plant groups with which the fossil leaves share characteristics are typical of tropical, subtropical or warm temperate regions today. There is little evidence for cool and cold climates from the fossil plants or their possible modern equivalents.

Palaeoclimate analyses of leaf physiognomy

Palaeoclimate data for the Coniacian and late Coniacian–early Santonian, obtained from the Hidden Lake and Santa Marta floras, are shown in Table 1. Leaf margin, SLR and MLR analyses give estimates of mean annual palaeotemperatures that range from 15.2 (±2) to 18.6 (±1.9) °C for the Coniacian, and 17.1 (±2) to 21.2 (±1.9) °C for the late Coniacian–early Santonian. The mean values of MAT, averaging all methods, are 16.9 °C for the Coniacian and 19.1 °C for the late Coniacian–early Santonian.

The lowest temperature estimates were obtained from the SLR analyses using the complete CLAMP data set derived by Wilf (1997); this produced MATs of 15.5 (±3.4) °C for the Hidden Lake flora and 17.8 (±3.4) °C for the Santa Marta flora with large standard errors.

Table 1. *Predicted palaeoclimate data for the Coniacian and late Coniacian–early Campanian, based on LMA, SLR and MLR analyses of the Hidden Lake and Santa Marta floras. (MAT, mean annual temperature; MAP, mean annual precipitation; GSP, growing season precipitation.) Refer to source for details of specific models*

Model	Hidden Lake flora			Santa Marta flora		
	MAT (°C)	MAP (mm)	GSP (mm)	MAT (°C)	MAP (mm)	GSP (mm)
LMA – Northern Hemisphere (Wolfe 1979)	18.0 ± 1.8			20.7 ± 2.1		
LMA – Southern Hemisphere (Wolfe 1979)	16.3 ± 1.6			18.3 ± 1.8		
SLR (Wolfe 1979; Wing & Greenwood 1993)	17.7 ± 0.8			20.1 ± 0.8		
SLR (Wiemann *et al.* 1998)	16.3			18.0		
SLR (Wilf 1997)	17.7 ± 2.0			20.0 ± 2.0		
SLR–CLAMP (Wilf 1997)	15.5 ± 3.4			17.8 ± 3.4		
SLR–CLAMP with cold sites excluded (Wilf 1997)	16.4 ± 2.1			18.4 ± 2.1		
SLR–CLAMP large leaves (Wilf *et al.* 1998)		1093			673	
SLR–leaf-area analysis (Wilf *et al.* 1998)		772 (−233, +333)			956 (−288, +413)	
SLR – large leaves (Wilf *et al.* 1998)		594			591	
MLR – Wing & Greenwood (1993)	15.2 ± 2	2142 ± 580		17.1 ± 2	1991 ± 580	
MLR – Wiemann *et al.* (1998)	18.6 ± 1.9		2630 ± 482	21.2 ± 1.9		2450 ± 482

The inclusion of extremely cold sites with winter freezing in the CLAMP data set is likely to have produced these cooler temperature estimates. However, Wing & Greenwood (1993) and Wilf (1997) derived MLR and SLR models from the CLAMP data set with the cold sites removed, as the exclusion of cold sites has been shown to produce significantly better correlation between leaves and climate in studies of modern floras (Gregory & McIntosh 1996). These produced MATs of 15.2 (±0.2) and 16.4 (±2.1) °C for the Hidden Lake flora, and 17.1 (±0.2) and 18.4 (±2.1) °C for the Santa Marta flora, using the methods of Wing & Greenwood (1993) and Wilf (1997), respectively.

The warmest estimates of MAT for these floras were produced by the Wiemann *et al.* (1998) MLR equation; in testing various SLR and MLR methods on modern vegetation in Florida, Wiemann *et al.* (1998) ground this model to be the closest match to the actual MAT. This method produced MATs of 18.6 ± 1.9 °C for the Hidden Lake flora and 21.2 ± 1.9 °C for the Santa Marta flora.

It is apparent from these analyses that, although there is some overlap in temperature estimates when the potential errors are taken into account, there is a clear signal of an increase in temperature, of about 2 °C, from the Coniacian to the early Santonian.

Estimates of annual precipitation (Table 1) range from 594 to 2142 (±580) mm for the Hidden Lake flora and 673 to 1991 (±580) mm for the Santa Marta flora. The Wiemann *et al.* (1998) model predicts growing season precipitation of 2630 (±482) mm for the Hidden Lake flora and 2450 (±482) mm for the Santa Marta flora. These estimates, especially those above about 2000 mm year^{-1}, are indicative of high rainfall, comparable to rainfall today of tropical regions such as the Amazon Basin and Indonesia tropical rainforest zones (FAO 2003). However, the MLR and SLR models may produce precipitation estimates that are not reliable for this study because they require information on leaf apex character, a feature that is not well preserved in these fossil assemblages, and on whole leaf size. The methods that produce the lowest estimates of rainfall use only leaf size to predict precipitation; however, leaf size amongst the fossils may have been strongly influenced by taphonomic sorting. The estimates for Cretaceous rainfall must therefore be considered with some caution.

Discussion

Analysis of the fossil leaves from James Ross Island using a range of methods indicate warm climates at high latitudes of about 65°S in the mid-Late Cretaceous. Mean annual temperatures are predicted of about 17 °C in the Coniacian (range of all methods 12–21 °C) and 19 °C (range 14–23 °C) in the late Coniacian–early Campanian.

Comparison with other Antarctic palaeotemperature estimates

Estimates of terrestrial palaeoclimates for this interval and this location have been obtained from various sources. For example, Dingle & Lavelle (1998) produced climate curves for the Late Cretaceous and Cenozoic in the James Ross Basin based on the Chemical Index of Alteration that reflects alteration of clay minerals. By assuming that the climate was relatively humid (based on the belief that forests grew on Antarctica through this period), they derived a temperature history for the Antarctic Peninsula. They found evidence of strong chemical weathering during the Santonian–Campanian interval that was interpreted as representing a warm climate, peaking in the mid-Campanian, the warmest part of the Late Cretaceous.

Climatic signals from fossil plants also point to warm terrestrial climates. Rees & Smellie (1989) derived MATs of 13–20 °C for the Cenomanian–Campanian from leaf-margin and leaf-size analyses of angiosperm fossils (although only from six taxa) from Livingston Island, South Shetland Islands. A range of palaeoclimatic data were also derived by Poole *et al.* (2005) from analyses of fossil wood and other plant fossils. They deduced that the Coniacian–Campanian interval was the warmest part of a trend that involved warm wet Late-Cretaceous climates, cooler and drier climates towards the end Cretaceous and into the Palaeocene, a warm wet phase in the latest Palaeocene, followed by a cooler and drier phase through the Middle–Late Eocene. Their study, however, considered the Coniacian–Campanian as one data point, rather than as separate stage intervals, so direct comparison with the two leaf assemblages here is not possible. Their analyses produced a rather broad range of palaeoclimate estimates for this Cretaceous interval, partly due to lack of samples and also due to problems associated with the application of the coexistence approach to Cretaceous floras. Like NLR the Coexistence Approach relies on the extrapolation of the climate tolerances of nearest living relatives, which is considered as less reliable for floras of Cretaceous age due to the evolutionary state of floras at this time, and the unknown response of floras to a greenhouse climate with atmospheric CO_2 levels much higher than present.

An average mean annual temperature for this interval of 15.3 °C was determined by Poole *et al.* (2005), the average of MATs of 13.5 °C derived from physiognomic analyses of angiosperm wood and 17.1 ± 4.2 °C from Coexistence Approach. Despite concerns with the Coexistence Approach outlined above, this temperature of 17.1 °C corresponds well with the MAT derived here for the Coniacian. A mean annual temperature range (MART) of 3.4 °C was derived from physiognomical analysis of wood, and cold month mean temperatures (CMMT) of 29 and 7.6 °C from wood analysis and Coexistence Analysis, respectively; however, Poole *et al.* (2005) conclude that these data are too imprecise to warrant consideration. An average warm month mean of 24.7 °C was derived from Coexistence Analysis. A mean annual precipitation value of 5620 mm was derived from wood analysis (an exceptionally high value found today only in small areas of equatorial Amazonian Andes and Indonesia; FAO 2003) but only 947 ± 60 mm from Coexistence Approach. Wood analysis predicted seasonal variation in rainfall (Poole *et al.* 2005).

The fossil plant analyses above provide palaeoclimate information for terrestrial climates under which the plants grew. It is important to compare the results with marine palaeotemperatures for the same region; however, marine palaeotemperature data may have been influenced by, for example, ocean currents and depth of water, and so comparisons must be made with care. In the adjacent oceans temperatures were also very high, according to oxygen isotope analyses of Ditchfield *et al.* (1994). They found peak warmth in the Coniacian and Santonian, within cooler climates during the Aptian–Cenomanian and from Campanian to Maastrichtian. Isotopic analysis of oysters gave palaeotemperatures of 18.5 °C for the Hidden Lake Formation and higher temperatures of 19.2 °C from ammonites, but 13 °C from belemnites (Pirrie & Marshall 1990; Marshall *et al.* 1993; Ditchfield *et al.* 1994).

Comparison with Arctic Cretaceous climates

This warm peak in the southern high latitudes is also reflected in the northern polar region. In the Arctic region Parrish & Spicer (1988) derived a MAT of about 13 °C at 80°N for the Alaskan North Slope region during the Coniacian, based on leaf-margin analysis and vegetation physiognomy of fossil floras. Using an estimated Cretaceous latitudinal temperature gradient of 0.3 °C per degree latitude (Wolfe & Upchurch 1987), this suggests MATs of approximately 17.5–19 °C at palaeolatitudes of 65°N, comparable to those of the James Ross region. These temperatures for the Arctic are remarkably similar to those from Antarctica. Parrish & Spicer (1988) also found that climate cooled after the Santonian, a trend similar to that seen in the south. From the Russian Arctic Spicer *et al.* (2002) obtained palaeoclimate data for the uppermost Albian–lower Cenomanian from analysis of the Grebenka flora at a palaeolatitude of about 76°N. CLAMP analysis of angiosperms yielded a MAT of 13 ± 1.8 °C moderately high rainfall and no apparent dry season or winter freezing. Herman & Spicer (1996) derived similar MATs of 12.5 ± 1.8 and 9.0 ± 1.8 °C for the North Slope Alaska and Kamchatka, respectively, for Coniacian angiosperm leaves.

Three floras from NE Russia, studied by Craggs (2005), produced much lower MAT estimates of 8.1 ± 1.2, 7.3 ± 1.2 and 9.4 ± 1.2 °C for the early Coniacian. She found cold month mean temperatures around freezing (–1.5 ± 1.9, –2.7 ± 1.9 and 0.9 ± 1.9 °C) for the Russian region. This region was situated at 78°N palaeolatitude, about 13° poleward of the equivalent for the James Ross floras. The low winter temperatures suggest freezing conditions in these northern high latitudes, at odds with other data for peak Cretaceous warmth.

Global climate warmth

Not only is this interval of warm climates apparent at high latitudes but it has been reported from many sites globally. The Cretaceous 'thermal maximum', about 100–80 Ma, has been identified in many reports as a peak of global warmth. The warmest climates occurred at some point between the Cenomanian–Turonian transition and the Early Campanian (Huber *et al.* 1995, 2002; Clarke & Jenkyns 1999; Wilson *et al.* 2002). Poulsen *et al.* (2003) suggested that the Cretaceous thermal maximum attributed to high atmospheric CO_2 levels was at least partly the climatic expression of a tectonically driven oceanographic event (formation of the Equatorial Atlantic gateway). Unfortunately, Turonian leaf floras have not been discovered in Antarctica, but the palaeoclimate data derived from the James Ross floras suggest that the very warm phase is reflected in the Coniacian and Santonian vegetation record.

This warm peak during the Late-Cretaceous may have also been the trigger for expansion of the angiosperm floras in Antarctica. Angiosperm pollen, which first appears in the early Albian Kotick Point Formation (Dettman & Thomson 1987; Riding & Crame 2002) on James Ross Island, increases in abundance in Turonian strata (Keating *et al.* 1992) and maintains high levels throughout the Marambio Group strata (Coniacian–Danian) (Dettmann & Thomson 1987; Dolding 1992; Keating 1992; Dutra & Batten 2000). The timing of this expansion matches well with a marked Turonian–Campanian diversification of angiosperm taxa in the northern high latitudes (Lupia *et al.* 1999), suggesting that angiosperm radiation occurred in both polar regions at the same time. Although this may well have been driven by this interval of global warmth. Lupia *et al.* (1999) state that the sparse sampling of pre-Turonina sediments produced a spurious jump in trend curves for angiosperm abundance.

Comparison with outputs from climate models

The Cretaceous has been a focus for several computer climate models (General Circulation Models, GCMs) (e.g. Valdes *et al.* 1996; Price *et al.* 1998; DeConto *et al.* 2002). Early models predicted seasonally extreme temperatures with mid continental freezing (e.g. Barron *et al.* 1993, 1995), even if atmospheric CO_2 levels were increased. These climate simulations were at odds with geological data that implied equable climates, at least for the mid-Cretaceous (Barron 1983; Francis & Frakes 1993 suggest winter freezing conditions at high latitudes for the Early Cretaceous).

The most detailed climate simulations relevant to the geological age of the floras described here were produced by Valdes *et al.* (1996) for the late Albian–early Cenomanian, using a GCM model with prescribed ocean conditions. For the Antarctica Peninsula region (at that time still joined to South America, Lawver *et al.* 1992) the model predicted summer surface air temperatures of over 20 °C, and winter temperatures of between 4–8 °C. The winter temperatures were above freezing (probably because the adjacent ocean was prescribed as warm), although further south on the main Antarctic continental freezing conditions were predicted. Soil moisture predictions (that indicate the balance between precipitation and evaporation) indicate dry soils in summer but saturated in winter, implying a seasonal precipitation regime. This agrees with palaeobotanical analyses of Poole *et al.* (2005) for the James Ross Basin region and with evidence from palaeosols for seasonally dry climates in the Alexander Island region on the west side of the peninsula (Howe & Francis 2005).

The climate models (NCAR GENESIS GCM with a mixed layer ocean and prescribed ocean heat transport) of Otto-Bliesner & Upchurch (1997) examined the effect of forest vegetation in high latitudes on climate. Although they worked with Maastrichtian boundary conditions and palaeogeography, considered to be a time of cooler climates compared to the early Late Cretaceous, they discovered that the polar forests had a warming effect on climate of about 2.2 °C. They suggested that the low albedo of the polar vegetation led to warmer land surfaces that then warmed adjacent oceans, preventing sea ice formation and causing higher winter temperatures. In addition, DeConto *et al.* (1999) used Campanian palaeogeography and a GCM that was interactively coupled with a predictive vegetation model (EVE) to assess the conditions required to sustain polar forests. Their results indicated that CO_2 levels of 1500 ppm (over four times present-day levels) and greater poleward oceanic heat transport was required to maintain forests at the poles. They found that high-latitude forests reduced surface albedo and also added more water vapour to the atmosphere through evapotranspiration, which both induced greater warming.

The presence of forest vegetation on the Antarctic Peninsula, such as that now preserved as the Hidden Lake and Santa Marta floras, thus provided a positive feedback effect and created warmer climates that favoured the growth of forest vegetation in southern high latitudes.

Summary

Two fossil leaf floras have been recovered from the Coniacian Hidden Lake and late Coniacian–early Campanian Santa Marta formations in the James Ross Basin, Antarctic Peninsula. The floras represent vegetation that once grew on the adjacent emergent volcanic arc, which now forms the Antarctic Peninsula itself, but were subsequently deposited within sediments of the back-arc basin.

Palaeoclimate analysis of the angiosperm leaf floras, using leaf-margin analysis, and simple and multiple linear analysis methods, suggest that the mean annual temperatures for the Hidden Lake and Santa Marta formations were 13–21 °C (mean 17 °C) and 15–23 °C (mean 19 °C), respectively. Rainfall was high but was probably seasonal (although estimates of precipitation must be considered with care). This geological evidence supports climate simulations from computer climate and vegetation models that suggest warm summer temperatures over 20 °C and winter temperatures above freezing. The presence of these floras at high latitudes probably positively contributed to the past polar warmth through their effect of decreasing albedo and increasing atmospheric water vapour (the most powerful greenhouse gas). These warm climates may also have been part of the Cretaceous thermal maximum, an interval of warm global climates, related to a phase of enhanced tectonic activity and CO_2 outgassing.

This work formed part of the PhD project of P. Hayes, supervised by Francis, Cantrill and Crame, undertaken at the University of Leeds and the British Antarctic Survey. We wish to thank NERC and the British Antarctic Survey for CASE funding through an Antarctic Special Topic programme, for loan of fossil specimens and other support. The specimens were collected by D. Pirrie, M. Thomson, A. Whitham and J. E. Francis on previous BAS-supported expeditions. J. E. Francis thanks the Trans-Antarctic Association for support.

References

ASKIN, R.A. 1989. Endemism and heterochroneity in the Late Cretaceous (Campanian) to Paleocene palynofloras of Seymour Island, Antarctica: implications for origins, dispersal and palaeoclimates of southern floras. *In*: CRAME, J.A. (ed.) *Origins and Evolution of the Antarctic Biota*. Geological Society, London, Special Publications, **47**, 107–119.

BAILEY, I.W. & SINNOT, E.W. 1915. A botanical index of Cretaceous and Tertiary climates. *Science*, **41**, 831–834.

BAILEY, I.W. & SINNOT, E.W. 1916. The climatic distribution of certain kinds of angiosperm leaves. *American Journal of Botany*, **3**, 24–39.

BARRON, E.J. 1983. A warm, equable Cretaceous: the nature of the problem. *Earth Science Reviews*, **19**, 305–338.

BARRON, E.J., FAWCETT, P.J., PETERSON, W.H., POLLARD, D. & THOMPSON, S. 1995. A 'simulation' of mid-Cretaceous climate. *Paleoceanography*, **10**, 953–962.

BARRON, E.J., FAWCETT, P.J., POLLARD, D. & THOMPSON, S. 1993. Model simulations of Cretaceous climates: the role of geography and carbon dioxide. *Philosophical Transactions of the Royal Society London*, **B341**, 307–316.

BROWNE, J.R. & PIRRIE, D. 1995. Sediment dispersal patterns in a deep marine back-arc basin: evidence from heavy mineral provenance studies. *In*: HARTLEY, A.J. & PROSSER, D.J. (eds) *Characterization of Deep Marine Clastic Systems*. Geological Society, London, Special Publications, **94**, 139–154.

BURNHAM, R.J. 1989. Relationships between standing vegetation and leaf litter in a paratropical forest: implications for paleobotany. *Review of Palaeobotany and Palynology*, **58**, 5–32.

CANTRILL, D.J. & NICHOLS, G.J. 1996. Taxonomy and palaeoecology of Early Cretaceous (Late Albian) angiosperm leaves from Alexander Island, Antarctica. *Review of Palaeobotany and Palynology*, **92**, 1–28.

CANTRILL, D.J. & POOLE, I. 2002. Cretaceous patterns of floristic change in the Antarctic Peninsula. *In*: CRAME, J.A. & OWEN, A.W. (eds) *Palaeobiogeography and Biodiversity Change: The Ordovician and Mesozoic–Cenozoic Radiations*. Geological Society, London, Special Publications, **194**, 141–152.

CÉSARI, S.N., MARENSSI, S.A. & SANTILLANA, S.N. 2001. Conifers from the Upper Cretaceous of Cape Lamb, Vega Island, Antarctica. *Cretaceous Research*, **22**, 309–319.

CHALONER, W.G. & MCELWAIN, J. 1997. The fossil plant record and global climatic change. *Review of Palaeobotany and Palynology*, **95**, 73–82.

CLARKE, L.J. & JENKYNS, H.C. 1999. New oxygen isotope evidence for long-term Cretaceous climatic change in the Southern Hemisphere. *Geology*, **27**, 699–702.

CRAGGS, H.J. 2005. Late Cretaceous climate signal of the Northern Pekulney Range Flora of northeastern Russia. *Palaeogeography, Palaeoclimatology, Palaeoecology*, **217**, 25–46.

CRAME, J.A., PIRRIE, D., RIDING, J.B. & THOMSON, M.R.A. 1991. Campanian–Maastrichtian (Cretaceous) stratigraphy of the James Ross Island area, Antarctica. *Journal of the Geological Society, London*, **148**, 1125–1140.

DECONTO, R.M., BRADY, E.C., BERGENGREN, J. & HAY, W.W. 1999. Late Cretaceous climate, vegetation, and ocean interactions. *In*: HUBER, B.T., MACLEOD, K.G. & WING, S.L. (eds) *Warm Climates in Earth History*. Cambridge University Press, Cambridge. 275–296.

DEL VALLE, R.A., ELLIOT, D.H. & MACDONALD, D.I.M. 1992. Sedimentary basins on the east flank of the Antarctic Peninsula: proposed nomenclature. *Antarctic Science*, **4**, 477–478.

DETTMANN, M.E. & THOMSON, M.R.A. 1987. Cretaceous palynomorphs from the James Ross Island area, Antarctica – a pilot study. *British Antarctic Survey, Bulletin*, **77**, 13–59.

DINGLE, R.V. & LAVELLE, M. 1998. Late Cretaceous–Cenozoic climatic variations of the northern Antarctic Peninsula: new geochemical evidence and review. *Palaeogeography, Palaeoclimatology, Palaeoecology*, **141**, 215–232.

DITCHFIELD, P.W., MARSHALL, J.D. & PIRRIE, D. 1994. High latitude palaeotemperature variation: New data from the Tithonian to Eocene of James Ross Island, Antarctica. *Palaeogeography, Palaeoclimatology, Palaeoecology*, **107**, 79–101.

DOLDING, P.J.D. 1992. Palynology of the Marambio Group (Upper Cretaceous) of northern Humps Island. *Antarctic Science*, **4**, 311–326.

DUTRA, T.L. & BATTEN, D.J. 2000. Upper Cretaceous floras of King George Island, West Antarctica, and their palaeoenvironmental and phytogeographic implications. *Cretaceous Research*, **21**, 181–209.

EKLUND, H., CANTRILL, D.J. & FRANCIS, J.E. 2004. Late Cretaceous plant mesofossils from Table Nunatak, Antarctica. *Cretaceous Research*, **25**, 211–228.

ELLIOT, D.H. 1988. *Tectonic Setting and Evolution of*

the James Ross Basin, Northern Antarctic Peninsula. Geological Society of America, Memoirs, **169**, 541–555.

FAO. 2003. *Review of World Water Resources by Country*. FAO Water Reports 23, FAO, Rome.

FERGUSON, D.K. 1985. The origin of leaf assemblages – new light on an old problem. *Review of Palaeobotany and Palynology*, **46**, 117–188.

FRANCIS, J.E. 1986. Growth rings in Cretaceous and Tertiary wood from Antarctica and their palaeoclimatic implications. *Palaeontology*, **29**, 665–684.

FRANCIS, J.E. & FRAKES, L.A. 1993. Cretaceous climates. *Sedimentology Review*, **1**, 17–30.

FRANCIS, J.E. & POOLE, I. 2002. Cretaceous and early Tertiary climates of Antarctica: evidence from fossil wood. *Palaeogeography, Palaeoclimatology, Palaeoecology*, **182**, 47–64.

GASTALDO, R.A., FERGUSON, D.K., WALTHER, H. & RABOLD, J.M. 1996. Criteria to distinguish parautochthonous leaves in Tertiary alluvial channel-fills. *Review of Palaeobotany and Palynology*, **91**, 1–21.

GREENWOOD, D.R. 1992, Taphonomic constraints on foliar physiognomic interpretations of Late Cretaceous and Tertiary palaeoclimates. *Review of Palaeobotany and Palynology*, **71**, 142–196.

GREENWOOD, D.R., WOLF, P., WING, S.L. & CHRISTOPHEL, D.C. 2004. Paleotemperature estimation using leaf-margin analysis: is Australia different? *Palaios*, **19**, 129–142.

GREGORY, K.M. & MCINTOSH, W.C. 1996. Paleoclimate and paleoelevation of the Oligocene Pitch-Pinnacle flora, Sawatch Range, Colorado. *Geological Society of America Bulletin*, **108**, 545–561.

HATHWAY, B. 2000. Continental rift to back-arc basin: Jurassic–Cretaceous stratigraphical and structural evolution of the Larsen Basin Antarctic Peninsula. *Journal of the Geological Society, London*, **157**, 417–432.

HAYES, P.A. 1999. *Cretaceous angiosperm leaf floras from Antarctica*. PhD Thesis, University of Leeds.

HERMAN, A.B. & SPICER, R.A. 1996. Palaeobotanical evidence for a warm Cretaceous Arctic ocean. *Nature*, **380**, 330–333.

HOWE, J. & FRANCIS, J.E. 2005. Metamorphosed palaeosols associated with Cretaceous fossil forests, Alexander Island, Antarctica. *Journal of the Geological Society of London*, **162**, 951–957.

HUBER, B.T. 1998. Tropical paradise at the Cretaceous poles? *Science*, **282**, 2199–2220.

HUBER, B.T., HODELL, D.A. & HAMILTON, C.P. 1995. Middle–late Cretaceous climate of the southern high latitudes: stable isotopic evidence for minimal equator-to-pole thermal gradients. *Geological Society of America Bulletin*, **107**, 1164–1191.

HUBER, B.T., NORRIS, R.D. & MACLEOD, K.G. 2002. Deep-sea paleotemperature record of extreme warmth during the Cretaceous. *Geology*, **30**, 123–126.

HUNT, R.J. & POOLE, I. 2003. Paleogene West Antarctic climate and vegetation history in light of new data from King George Island. *In*: WING, S.L., SCHMITZ, B. & THOMAS, E. (eds) *Causes and Consequences of Globally Warm Climates in the Early Paleogene*. Geological Society of America Special Papers, **369**, 395–412.

INESON, J.R. 1989. Coarse-grained submarine fan and slope apron deposits in a Cretaceous back-arc basin, Antarctica. *Sedimentology*, **36**, 793–819.

INESON, J.R., CRAME, J.A. & THOMSON, M.R.A. 1986. Lithostratigraphy of the Cretaceous strata of James Ross Island, Antarctica. *Cretaceous Research*, **7**, 141–159.

JORDAN, G.J. 1997. Uncertainty in palaeoclimatic reconstructions based on leaf physiognomy. *Australian Journal of Botany*, **45**, 527–547.

KEATING, J.M. 1992. Palynology of the Lachman Crags Member, Santa Marta Formation (Upper Cretaceous) of north-west James Ross Island. *Antarctic Science*, **4**, 293–304.

KEATING, J.M., SPENCER-JONES, M. & NEWHAM, S. 1992. The stratigraphical palynology of the Kotick Point and Whisky Bay formations, Gustav Group (Cretaceous), James Ross Island. *Antarctic Science*, **4**, 279–292.

KENNEDY, E.M., SPICER, R.A. & REES, P.M. 2002. Quantitative palaeoclimate estimates from Late Cretaceous and Paleocene leaf floras in the northwest of the South Island, New Zealand. *Palaeogeography, Palaeoclimatology, Palaeoecology*, **184**, 321–345.

KOVACH, W.L. & SPICER, R.A. 1996. Canonical correspondence analysis of leaf physiognomy: a contribution to the development of a new palaeoclimatological tool. *Palaeoclimates*, **2**, 125–138.

LAWVER, L.A., GAHAGAN, L.M. & COFFIN, M.F. 1992. The development of paleoseaways around Antarctica. *Antarctic Research Series*, **56**, 7–30.

LIDGARD, S. & CRANE, P.R. 1990. Angiosperm diversification and Cretaceous floristic trends: a comparison of palynofloras and leaf macrofloras. *Paleobiology*, **16**, 77–93.

LUPIA, R., LIDGARD, S. & CRANE, P.R. 1999. Comparing palynological abundance and diversity: implications for biotic replacement during the Cretaceous angiosperm radiation. *Paleobiology*, **25**, 305–340.

MACDONALD, D.I.M. & BUTTERWORTH, P.J. 1990. The stratigraphy, setting and hydrocarbon potential of the Mesozoic sedimentary basins of the Antarctic Peninsula. *In*: ST JOHN, B. (ed.) *Antarctica as an Exploration Frontier*. AAPG, Studies in Geology, **31**, 101–125.

MARSHALL, J.D., DITCHFIELD, P.W. & PIRRIE, D. 1993. Stable isotope palaeotemperatures and the evolution of high palaeolatitude climate in the Cretaceous. *In*: HEYWOOD, R.B. (ed.) University Research in Antarctica 1989–1992. Proceedings of the British Antarctic Survey Antarctic Special Topic Award Scheme Round 2 Symposium 30th September–1st October 1992. British Antarctic Survey, Cambridge, 71–79.

MCARTHUR, J.M., CRAME, J.A. & THIRLWALL, M.F. 2000. Definition of Late Cretaceous stage boundaries in Antarctica using strontium isotope stratigraphy. *Journal of Geology*, **108**, 623–640.

MOSBRUGGER, V. 1999. The nearest living relative method. *In:* JONES, T.P. & ROWE, N.P. (eds) *Fossil*

Plants and Spores: Modern Techniques. Geological Society, London, 261–265.

OTTO-BLIESNER, B.L. & UPCHURCH, G.R. JR. 1997. Vegetation-induced warming of high-latitude regions during the Cretaceous period. *Nature*, **385**, 804–807.

PARRISH, J.T. & SPICER, R.A. 1988. Late Cretaceous terrestrial vegetation: a near-polar temperature curve. *Geology*, **16**, 22–25.

PIRRIE, D. 1989. Shallow marine sedimentation within an active margin basin, James Ross Island, Antarctica. *Sedimentary Geology*, **63**, 61–82.

PIRRIE, D. 1991 Controls on the petrographic evolution of an active margin sedimentary sequence: the Larsen Basin, Antarctica. *In*: MORTON, A.C., TODD, S.P., HAUGHTON, P.D.W. (eds) *Developments in Sedimentary Provenance Studies*. Geological Society, London, Special Publications, **57**, 231–249.

PIRRIE, D. & MARSHALL, J.D. 1990. High-paleolatitude Late Cretaceous paleotemperatures: new data from James Ross Island, Antarctica. *Geology*, **18**, 31–34.

PIRRIE, D., WHITHAM, A.G. & INESON, J.R. 1991. The role of tectonics and eustacy in the evolution of a marginal basin: Cretaceous–Tertiary Larsen Basin, Antarctica. *In*: MACDONALD, D.I.M (ed.) *Sea Level Changes at Active Plate Margins: Processes and Products*. International Association of Sedimentology, Special Publications, **12**, 293–305.

POOLE, I. & CANTRILL, D.J. 2006. Cretaceous and Cenozoic vegetation of Antarctica integrating the fossil wood record. *In*: FRANCIS, J.E., PIRRIE, D. & CRAME, J.A. (eds) *Cretaceous–Tertiary High-latitude Palaeoenvironments, James Ross Basin, Antarctica*. Geological Society, London, Special Publications, **258**, 63–81.

POOLE, I., CANTRILL, D. & UTESCHER, T. 2005. A multi-proxy approach to determine Antarctic terrestrial palaeoclimate during the Late Cretaceous and Early Tertiary. *Palaeogeography, Palaeoclimatology, Palaeoecology*, **222**, 95–121.

POULSEN, C.J., GENDASZEK, A.S. & JACOB, R.L. 2003. Did the rifting of the Atlantic Ocean cause the Cretaceous thermal maximum? *Geology*, **31**, 115–118.

PRICE, G.D., VALDES, P.J. & SELLWOOD, B.W. 1998. A comparison of GCM simulated Cretaceous 'greenhouse' and 'icehouse' climates: implications for the sedimentary record. *Palaeogeography, Palaeoclimatology, Palaeoecology*, **142**, 123–138

REES, P.M. & SMELLIE, J.L. 1989. Cretaceous angiosperms from an allegedly Triassic flora at Williams Point, Livingston Island, South Shetland islands. *Antarctic Science*, **1**, 239–248.

RIDING, J.B. & CRAME, J.A. 2002. Aptian to Coniacian (Early–Late Cretaceous) palynostratigraphy of the Gustav Group, James Ross Basin, Antarctica. *Cretaceous Research*, **23**, 739–760.

RIDING, J.B., CRAME, J.A., DETTMANN, M.E. & CANTRILL, D.J. 1998. The age of the base of the Gustav Group in the James Ross Basin, Antarctica. *Cretaceous Research*, **19**, 87–105.

SKELTON, P.W., SPICER, R.A., KELLEY, S.P. & GILMOUR, I. 2003. *The Cretaceous World*. Cambridge University Press, Cambridge.

SPICER, R.A. 1990. Fossils as environmental indicators: climate from plants. *In*: BRIGGS, D.E.G. & CROWTHER, P.R. (eds) *Palaeobiology: A Synthesis*. Blackwell Science, Oxford, 401–403.

SPICER, R.A., AHLBERG, A., HERMAN, A.B., KELLEY, S.P., RAIKEVICH, M.I. & REES, P.M. 2002. Palaeoenvironment and ecology of the middle Cretaceous Grebenka flora of northeastern Asia. *Palaeogeography, Palaeoclimatology, Palaeoecology*, **184**, 65–105.

TER BRAAK, C.F.J. 1986. Canonical Correspondence Analysis: a new eigenvector technique for multivariate direct gradient analysis. *Ecology*, **67**, 1167–1179.

VALDES, P.J., SELLWOOD, B.W. & PRICE, G.D. 1996. Evaluating concepts of Cretaceous equability. *Palaeoclimates*, **2**, 139–158.

WHITHAM, A.G., INESON, J.R. & PIRRIE, D. 2006. Marine volcaniclastics of the Hidden Lake Formation (Coniacian) of James Ross Island, Antarctica: an enigmatic element in the history of a back-arc basin. *In*: FRANCIS, J.E., PIRRIE, D. & CRAME, J.A. (eds) *Cretaceous–Tertiary High-latitude Palaeoenvironments: James Ross Basin, Antarctica*. Geological Society, London, Special Publications, **258**, 21–47.

WIEMANN, M.C., MANCHESTER, S.R., DILCHER, D.L., HINOJOSA, L.F. & WHEELER, E.A. 1998. Estimation of temperature and precipitation from morphological characters of dicotyledonous leaves. *American Journal of Botany*, **85**, 1796–1802.

WILF, P. 1997. When are leaves good thermometers? A new case for Leaf Margin Analysis. *Paleobiology*, **23**, 373–390.

WILF, P., WING, S.L., GREENWOOD, D.R. & GREENWOOD, C.L. 1998. Using fossil leaves as paleoprecipitation indicators: an Eocene example. *Geology*, **26**, 203–206.

WILSON, P.A., NORRIS, R.D. & COOPER, M.J. 2002. Testing the Cretaceous greenhouse hypothesis using glassy foraminiferal calcite from the core of the Turonian tropics on Demerara Rise. *Geology*, **30**, 607–610.

WING, S.L. & GREENWOOD, D.R. 1993. Fossils and fossil climate: the case for equable continental interiors in the Eocene. *Philosophical Transactions of the Royal Society London*, **B341**, 243–252.

WOLFE, J.A. 1971. Tertiary climatic fluctuations and methods of analysis of Tertiary floras. *Palaeogeography, Palaeoclimatology, Palaeoecology*, **9**, 27–57.

WOLFE, J.A. 1979. *Temperature Parameters of Humid to Mesic Forests of Eastern Asia and Relation to Forests of Other Regions of the Northern Hemisphere and Australaisia*. US Geological Survey, Professional Papers, **1106**, 1–37.

WOLFE, J.A. 1993. *A Method of Obtaining Climatic Parameters From Leaf Assemblages*. US Geological Survey Bulletin, **2040**.

WOLFE, J.A. & SPICER, R.A. 1999. Fossil leaf character states: multivariate analysis. *In*: JONES, T.P. & ROWE, N.P. (eds) *Fossil Plants and Spores: Modern Techniques*. Geological Society, London, 233–239.

WOLFE, J.A. & UPCHURCH, G.R., JR. 1987. North American nonmarine climates and vegetation during the Late Cretaceous. *Palaeogeography, Palaeoclimatology, Palaeoecology*, **61**, 33–77.

Cretaceous and Cenozoic vegetation of Antarctica integrating the fossil wood record

IMOGEN POOLE[1,2,3] & DAVID J. CANTRILL[4]

[1]*Wood Anatomy Section, National Herbarium of the Netherlands, University of Utrecht Branch, P.O. Box 80102, 3585 CS Utrecht, The Netherlands (e-mail: i.poole@geo.uu.nl)*

[2]*Palaeontological Museum, Oslo University, P.O. Box 1172 Blidern, N-0318 Oslo, Norway*

[3]*Present address: Faculty of Earth Sciences, Organic Geochemistry Group, University of Utrecht, P.O. Box 80021, 3508 TA, Utrecht, The Netherlands*

[4]*Swedish Museum of Natural History, Department of Palaeobotany, Box 50007, Stockholm 104 05, Sweden*

Abstract: A compilation of data for Cretaceous and Cenozoic Antarctic fossil wood floras, predominantly from the James Ross Island Basin, provides a different perspective on floristic and vegetation change when compared with previous studies that have focused on leaf macrofossils or palynology. The wood record provides a filtered view of tree-forming elements within the vegetation, something that cannot be achieved from studies focusing on regional palynology or leaf floras. Four phases of vegetation development in the overstorey are recognized in the Cretaceous and Cenozoic of the Antarctic Peninsula based on the distribution and taxonomic composition of wood floras: Aptian–Albian coniferous forests; ?Cenomanian–Santonian mixed angiosperm forests; Campanian–Maastrichtian southern temperate forests; and Palaeocene–Eocene reduced diversity *Nothofagus* forests. Comparisons between the wood record and information derived from palynological and leaf floras have important implications for our understanding of the spatial composition of the vegetation. There is no doubt that climate change during the Cretaceous and Tertiary influenced the vegetational composition, but evolving palaeoenvironments in the Antarctic Peninsula region were probably of equal, if not greater, importance.

James Eights (1833) reported the first fossil wood from the Antarctic (South Shetland Islands, Antarctic Peninsula), a discovery that pointed to a once vegetated landmass in what is now an icy world. Today, as questions focus on trying to understand the Earth system more fully, it becomes more ever apparent that the high latitudes played an important role in evolution of biotas, development of Southern Hemisphere biogeography (Drinnan & Crane 1990; Cantrill & Poole 2002) and mediating global climates (Upchurch *et al.* 1998). Indeed, the high latitudes are most sensitive to climate fluctuation, yet there is no modern analogue today with which to compare the evidence from the past. An unparalleled record of life in the high southern latitudes is found on the Antarctic Peninsula, which has, in recent years, furthered our knowledge of the palaeoecology of this region through many systematic and climatological studies. It is therefore timely to bring together the works published over the past decade, and revise and update previous conceptions pertaining to the southern high latitudes such that this unique and dynamic ecosystem can be more fully understood.

Four major types of fossil plant material provide information for past vegetation, these include: leaf compressions and impressions; wood; palynomorphs; and dispersed cuticular material, all of which are present in the floras of the Antarctic Peninsula. The Antarctic record of terrestrial vegetation is derived predominantly from leaf and palynomorph records (e.g. Askin 1988, 1992). Abundant leaf floras have been described (e.g. Dusén 1908; Zastawniak 1981, 1990, 1993, 1994; Li & Shen 1990; Li 1994; Zhou & Li 1994; Zastawniak *et al.* 1995; Hayes 1999; Cantrill 2000; Dutra & Batten 2000; Hunt 2001), but these often lack cuticles making systematic identification more problematic. Flowers and fruits also occur, but these are rare (Gandolfo *et al.* 1998; Eklund 2003) and have not to date been extensively studied. By comparison, wood is abundant and thus makes an important contribution to a picture of biodiversity than would otherwise result predominantly from leaf and microfossil

From: Francis, J. E., Pirrie, D. & Crame, J. A. (eds) 2006. *Cretaceous–Tertiary High-Latitude Palaeoenvironments, James Ross Basin, Antarctica.* Geological Society, London, Special Publications, **258**, 63–81. 0305–8719/06/$15

evidence. However, the paucity of solid taxonomic investigations has severely limited the ability to utilize this source of information. This paper synthesizes the recent data on the Cretaceous and Cenozoic of the Antarctic Peninsula region, based largely on wood records from the James Ross Island Basin (e.g. Gothan 1908; Torres *et al.* 1994a, b; Poole & Francis 1999, 2000; Poole *et al.* 2000*a–c*; Poole 2002). To help complete the picture, the fossil wood record from elsewhere in the Antarctic Peninsula (e.g. Torres & Lemoigne 1988, 1989; Chapman & Smellie 1992; Falcon-Lang & Cantrill 2000, 2001*a*; Poole & Cantrill 2001; Poole *et al.* 2001, 2003) has been used to fill the gaps in the James Ross Island Basin biodiversity record.

Geological setting

The Antarctic Peninsula is the remnant of a continental-margin magmatic arc of Mesozoic–Cenozoic age (Storey & Garrett 1985). Formed as a result of subduction of the palaeo-Pacific Plate beneath the western margin of the Antarctic Peninsula, it has good exposures of magmatic-arc, accretionary complex, and fore- and back-arc regions. These provide information on a diversity of environmental settings, many of which contain records of fossil wood. Forearc deposits exposed on the west side of the Antarctic Peninsula include the famous standing forests and associated leaf floras from Alexander Island (Jefferson 1982; Falcon-Lang & Cantrill 2001*a*). These are distal to the arc, and record braided and meandering fluvial environments on a narrow coastal plain (Cantrill & Nichols 1996; Nichols & Cantrill 2002) (Fig. 1). Further north, in the South Shetland Islands, Early Cretaceous (Aptian) intra-arc deposits record the burial of existing topography through the development of local calc-alkaline volcanic edifices (Hathway 1997) (Fig. 1). Fossil wood from palaeotopographic surfaces and entombed in ignimbrite flows record the interaction between vegetation and environmental processes (Falcon-Lang & Cantrill 2002). Similar volcanic-dominated

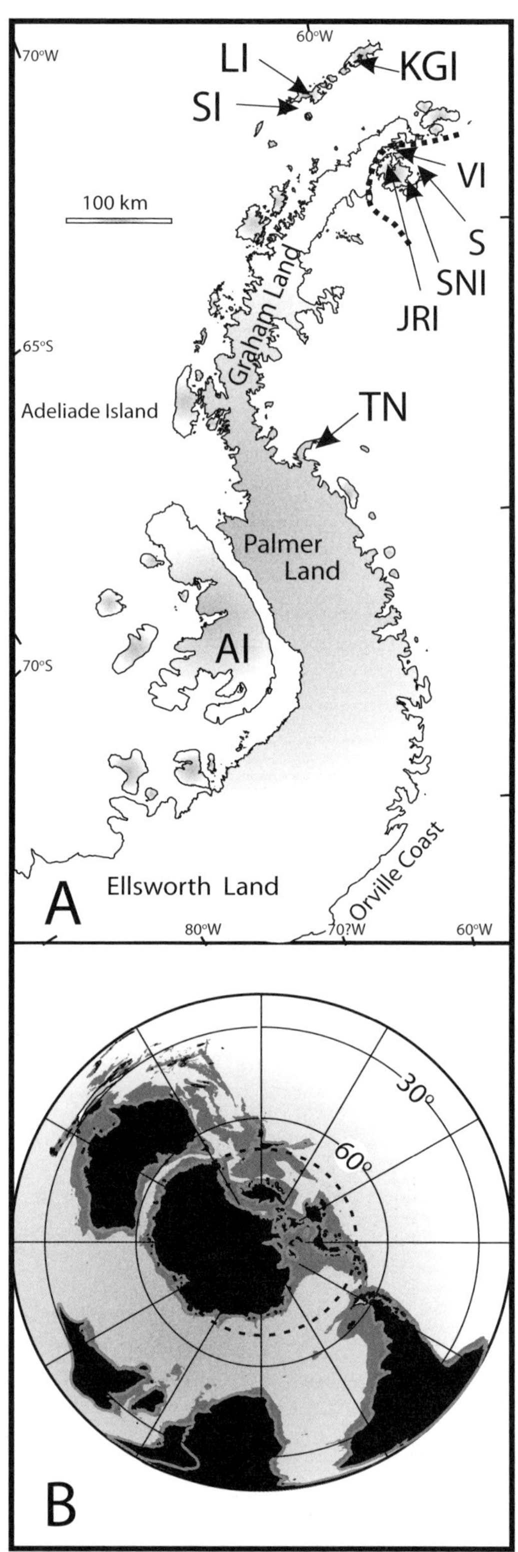

Fig. 1. (**A**) General locality map of the Antarctic Peninsula, with locations of places mentioned in the text. LI, Livingston Island; SI, Snow Island; KGI, King George Island; VI, Vega Island; S, Seymour Island; SNI, Snow Hill Island; JRI, James Ross Island; TN, Table Nunatak; AI, Alexander Island. (**B**) Reconstruction of Late Cretaceous (84 Ma) times showing the high-latitude setting of the palaeofloras, supplied by Dr R. A. Livermore. The dashed line indicates the polar circle.

environments extend from the Late Cretaceous to the ?Oligocene–Miocene in the South Shetland Islands and record the development of vegetation under these conditions. In contrast, the east side of the Antarctic Peninsula was rifted from the margin of Gondwana during the Jurassic. The Early Cretaceous (Aptian)–early Late Cretaceous (Turonian) of the James Ross Basin is characterized by deep submarine fan deposits. A major pulse in arc magmatism in the early Coniacian lead to the development of a complex fan-delta, slope and base-of-slope environment (Hidden Lake Formation: Whitham *et al.* 2006). Plant fossil material becomes increasingly common in the Hidden Lake Formation, and provides a record of vegetation from the eastern side of the Antarctic Peninsula.

This unparalleled record of diverse palaeoenvironments and associated biota from the Early Cretaceous to the Eocene was set in the high southern latitudes. During the Late Cretaceous and early Tertiary the Antarctic Peninsula extended from approximately 68 to 75°S, with Alexander Island lying at about 70°S and King George Island in the South Shetlands at 62°S (Fig. 1). At these palaeolatitudes plant growth was strongly influenced by the unique growing conditions not found on Earth today: a strongly seasonal polar light regime coupled with greenhouse conditions.

Floristic composition and turnover

Basal Cretaceous sedimentary rocks are lacking in the James Ross Island Basin, with the oldest Cretaceous strata being Aptian or slightly older. The oldest strata include the Lagrelius Point (Aptian: Riding *et al.* 1998), Kotick Point (early Albian: Keating *et al.* 1992; Riding & Crame 2002) and Pedersen Nunatak (Aptian: Hathway & Riding 2001) formations (Fig. 2). Unfortunately the record of plant macrofossils from

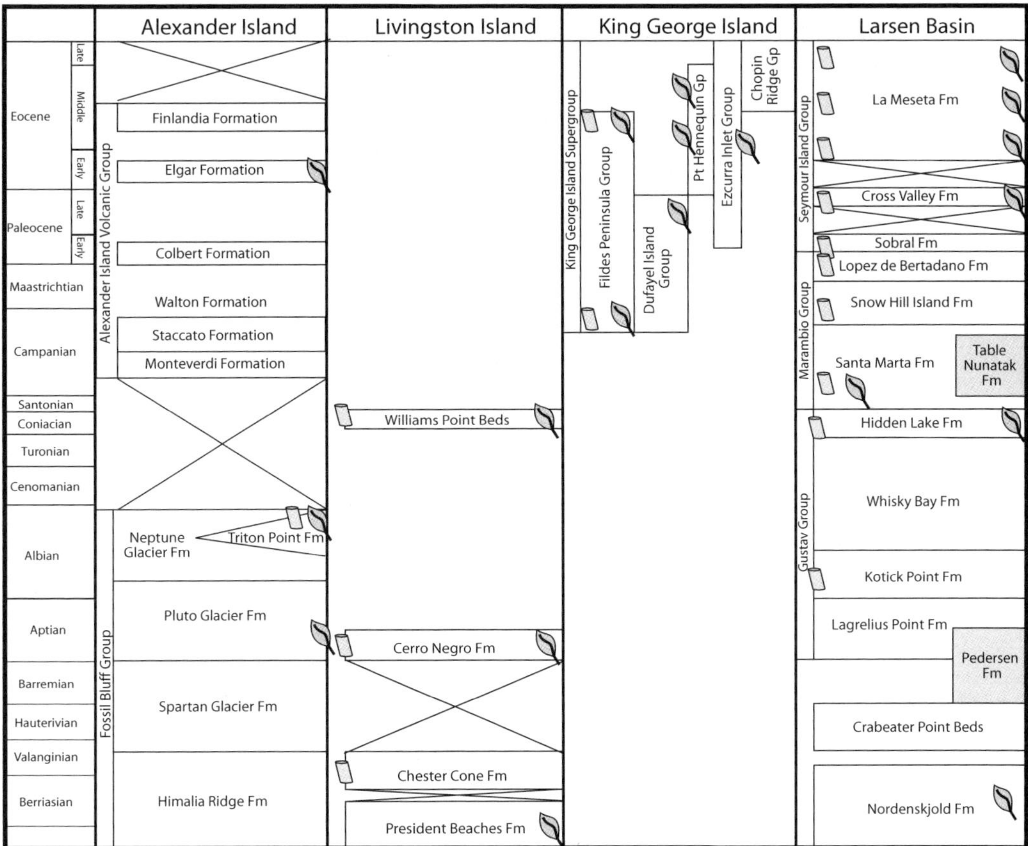

Fig. 2. Stratigraphic summary of key formations and groups for the Cretaceous and Palaeogene of the Antarctic Peninsula (East Antarctica not included). Units with fossil leaf material or wood material are marked. Note that the stratigraphic succession for King George Island is incomplete.

these units is poor. Lower Cretaceous (pre-Aptian) sedimentary rocks occur in the Fossil Bluff Group on Alexander Island (Himalia Ridge and Spartan Glacier formations: Butterworth *et al.* 1988), and on Byers Peninsula, Livingston Island (President Beaches and Chester Cone formations: Crame *et al.* 1993; Duane 1996) (Fig. 2). These strata accumulated in marine settings, generally lack wood, and contain only sparse foliage remains. Consequently, this discussion starts in the Aptian interval when fossil wood becomes more common and terrestrial settings are better documented.

Aptian and Albian interval

The basal Gustav Group, of the James Ross Island Basin, yields sparse fossil wood (Francis 1986; Ottone & Medina 1998; Francis & Poole 2002), and only wood from the early Albian Kotick Point Formation has been formally described (Ottone & Medina 1998). This material was assigned to *Agathoxylon* Hartig, a member of the Araucariaceae. It is unclear if this wood is contemporaneous with deposition, or reworked from older formations (Botany Bay Group). However, most of the wood preserved in the Gustav Group and overlying Marambio and Seymour Island groups is replaced by calcite, whereas Botany Bay Group wood is silicified. The limited number of samples from the Aptian and Albian within the James Ross Island Basin precludes making firm inferences about forest composition, but, fortunately, good Aptian wood floras in the South Shetland Islands (Cerro Negro Formation: Torres *et al.* 1982, 1997*a*; Falcon-Lang & Cantrill 2001*b*), and Albian wood floras from Alexander Island (Falcon-Lang & Cantrill 2001*a*) have been described. Both floras provide a picture of the forest component of the vegetation to both the north and south of the James Ross Island Basin. The general similarity in the composition of these floras enables us to assume that the vegetation within the James Ross Island Basin was comparable. Therefore, in order to set the scene for ensuing vegetational changes, we consider the floras of these two regions during an interval that within the James Ross Island Basin presently lacks data.

The Aptian Cerro Negro Formation in the South Shetland Islands crops out on Livingston Island (on Byers Peninsula) and Snow Island (at President Head) (Hathway 1997). This unit accumulated in a terrestrial intra-arc setting, and was dominated by volcanic processes. A diverse leaf flora occurs in a lacustrine sequence on President Head (Torres *et al.* 1997*a*, *b*: Cantrill 2000). The impression and compression floras from the lacustrine unit are rich in Bennettitales, conifers and other gymnospermous plants (e.g. *Pachypteris* Brongniart), with a minor fern and bryophyte component (Fig. 3C). The coniferous component is dominated by podocarps and taxodiaceous forms with rare araucarian remains (Fig. 3D). This leaf flora is similar in composition to those found within the non-marine units on Byers Peninsula (Hernández & Azcarárte 1971; Cesari *et al.* 1998, 1999), except that the latter are usually less diverse with just one or two taxa. In particular, Bennettitales (*Ptilophyllum* Morris) (Fig. 3C) are encountered more frequently and the floras are richer in araucarian elements in Byers Peninsula localities.

Palynofloras from the same units record only minor differences between the lacustrine unit assemblage (Cantrill 2000) and those in more fully terrestrial environments (Hathway *et al.* 1999). The palynoflora of the lacustrine environment is typically rich in pteridophyte spores, with diversity greater than that seen in the macrofossil record (Cantrill 2000). Localities are often dominated by single taxa, in particular *Cyathidites* Couper and less frequently *Cyatheacidites* Cookson ex Potonié (Lophosoriaceae). This is thought to reflect local abundance or colonization events, as the Lophosoriaceae is a colonizer of disturbed environments (Cantrill 1998). Lycophytes and bryophytes make up only a small component of the flora, whereas conifers tend to be more abundant. Conifer pollen is plentiful and dominated by podocarps, although yields are lower in the lacustrine units. This could be the result of taphonomic bias, as bisaccate grains tend to float and so do not readily become incorporated into lacustrine units.

Relative to the leaf floras, wood is more widespread and locally abundant within the Cerro Negro Formation, which complements and supplements the vegetational picture obtained from the palyno- and leaf floras. Palaeovalleys

Fig. 3. Aptian and Albian floras. (A) and (B) are late Albian Triton Point Formation flora, Alexander Island. (**A**) Standing tree with a pseudomonopodial habit. (**B**) *Tetragleichenites acuta* Nagalingum et Cantrill a gleicheniaceous fern. Scale bar shows 1 mm divisions, KG. 2817.75a/76a. (C) and (D) are Aptian flora from Snow Island, South Shetland Islands. (**C**) *Ptilophyllum Menendez* Cantrill. Scale bar shows 1 cm divisions, P. 2501.1. (**D**) *Elatocladus* sp., a podocarpaceous conifer. Scale bar shows 1 cm, P. 2501.21a.

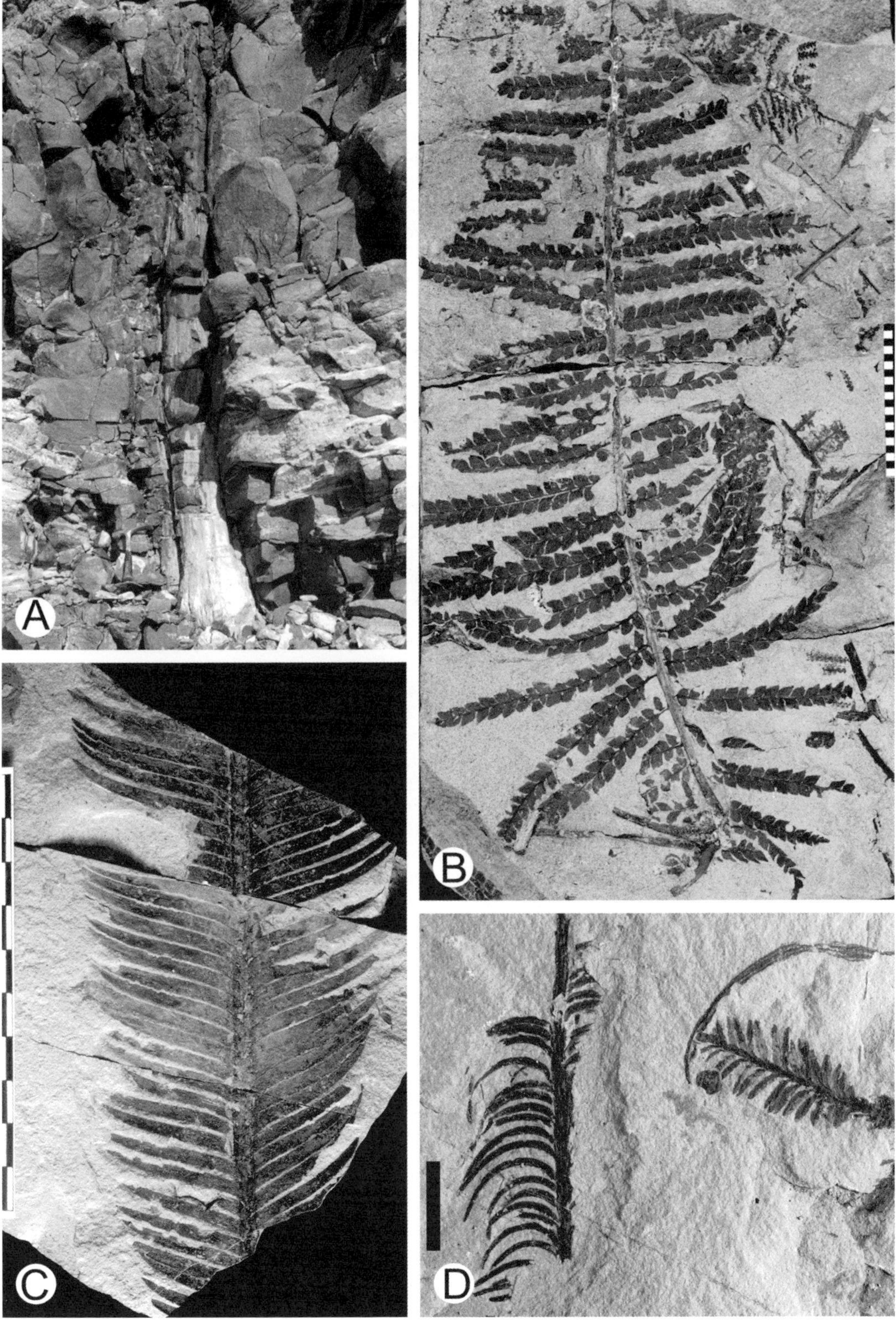
A
B
C
D

contain incipient soils, and trunk wood is scattered across this surface (Hathway 1997; Falcon-Lang & Cantrill 2001*b*). The trees that forested this region were undoubtedly substantial, as evidenced by trunks up to 1.5 m in diameter. Ignimbrite flows on Byers Peninsula contain entrained charred logs up to 5 m in length representing the remains of forest stands that grew on the flanks of a volcanic edifice and which later became entombed during a subsequent eruption. Similar processes have been observed today in New Zealand (Clarkson *et al.* 1988). The leaf floras suggest that the Araucariaceae were important, but the wood record points to forest communities rich in podocarps (61%) with subsidiary araucarians (27%) alongside extinct groups such as *Sahnioxylon* Bose et Sah (12%) (Falcon-Lang & Cantrill 2001*b*; $n = 33$).

Angiosperms first appear as pollen records in early Albian strata on the eastern side of the Antarctic Peninsula (Dettmann & Thomson 1987), but do not appear in the leaf floras until the late Albian (Cantrill & Nichols 1996). The early pollen record is of a low-diversity angiosperm component (mostly *Clavatipollenites* Couper) indicative of a shrubby habit (Dettmann 1989). The late Albian leaf floras are also characterized by herbaceous (e.g. *Hydrocotyllophyllum* Teixeria) and shrubby (e.g. *Dicotylophyllum* Bandulska) forms, but a few leaf types probably represent more substantial plants, perhaps understorey trees (e.g. *Araliaephyllum* Ettingshausen, *Ficophyllum* Fontaine) (Cantrill & Nichols 1996).

The late Albian flora of Alexander Island, situated at a palaeolatitude of 75°S, represents one of the most complete and most southerly forests known to date (Fig. 3A). Within the leaf flora, gymnosperms dominate with arboreal elements such as Araucariaceae, subsidiary Podocarpaceae, minor Taxodiaceae, and an understorey composed of ginkopsids, taeniopterids, bennettitaleans, equisetites, ferns (Fig. 3B) and liverworts (Cantrill 2001*a*; Falcon-Lang *et al.* 2001; Howe & Cantrill 2001). This flora also records the first appearance of angiosperms in the leaf record (Cantrill & Nichols 1996). The famous standing forest horizons contain abundant wood and, although often poorly preserved, are rich in podocarps (85.3%), with fewer araucarians (13.2%) and taxodiaceous (1.5%) conifers (Falcon-Lang & Cantrill 2000; $n = 69$). The discrepancy concerning araucarian dominance may be due to the robust nature of araucarian foliage and reproductive structures (cone scales) having survived destructive taphonomic processes better than podocarps, and thus tending to be over-represented in the leaf record relative to the wood. Alternatively, the best-preserved forest horizons may be in sedimentary environments that lack araucarians and thus account for the deficit of araucarians in the wood flora. Typically, the araucarians are found in more proximal settings, in contrast to the podocarps and taxodiaceous conifers (Cantrill & Falcon-Lang 2001). The lack of angiosperm wood in Albian wood floras from Alexander Island (Falcon-Lang & Cantrill 2000) and the James Ross Island Basin (Ottone & Medina 1998) also supports a more shrubby, herbaceous habit of these first angiosperms.

Inferences pertaining to the physical structure of the vegetation that dominated the high southern latitudes at this time have been drawn from *in situ* 'forests' using stump diameter and density, and growth-ring sequences (e.g. Chaloner & Creber 1989; Falcon-Lang & Cantrill 2000). Alexander Island provides an ideal case study for such a fossil forest and has been studied for many decades (e.g. Jefferson 1981, 1982; Cantrill & Nichols 1996; Falcon-Lang *et al.* 2001; Howe & Cantrill 2001). This forest comprised stumps and standing trees of up to 8 m tall that have been observed in cliff sections (Cantrill 2001*a*). Using the allometric approach of Niklas (1994), it has been suggested that actual heights of 29 m were attained (Falcon-Lang & Cantrill 2000) by these trees which reached ages of more than 180 years (Chapman 1994). The habit of these trees do not appear to be cone-shaped (Brodribb & Hill 2004), as suggested in earlier publications (e.g. Chaloner & Creber 1989), as no evidence of whorled branch insertion have been found at this time (Cantrill 2001*a*). Where branching has been observed it appears to be towards a pseudomonopodial habit and this is more consistent with the habit of extant podocarps. Stand density ranges from a median of 568 stems per hectare (perhaps representing colonization stands: Falcon-Lang *et al.* 2001) to dispersed clumps of individuals similar to those in open woodland today (Cantrill 2001*a*). Such a maximum density would have ensured the minimization of mutual shading due to the low-angle radiation. The productivity of these forests has been thought to be as high as approximately 17.65 m^3 ha^{-1} a^{-1} (Creber & Francis 1999) based on Jefferson's (1981) data, but more recent studies based on revised data and additional fieldwork suggest that this is a drastic over-representation with high productivity probably being closer to 5.62–7.33 m^3 ha^{-1} a^{-1}, similar to the warm temperate araucarian–podocarp stands in North Island, New Zealand today (Falcon-Lang *et al.* 2001).

The question of evergreen v. deciduous

habit for late Albian Antarctic conifers was investigated in detail by Falcon-Lang & Cantrill (2001*b*). Based on five independent techniques, they concluded that the canopy-forming vegetation that included conifers, ginkgos and taeniopterids was predominantly evergreen. Araucarians and podocarps, which dominated the vegetation, held on to their leaves for at least 5–13 years, whereas some of the rarer taxodiaceous conifers were evergreen but with much shorter leaf retention times. Other taxodiaceous conifers, ginkgos and taeniopterids were all deciduous along with the fern and angiosperm components of the understorey (Cantrill & Nichols 1996).

Without preserved *in situ* stumps, estimates of general density, productivity and height of the vegetation cannot be obtained. Therefore, extrapolation from the conclusions drawn by Falcon-Lang *et al.* (2001) suggest that the earlier mid-Cretaceous forests bordering the James Ross Island Basin had an estimated density and productivity similar to the Alexander Island forest. Nevertheless, with the increase in the abundance of the angiosperm component the density, productivity and forest heights would have changed due to different growth rates between angiosperms and conifers, and the influences they exerted on one another. Using growth-ring and palaeotemperature determinations, Specht *et al.* (1992) concluded that the productivity of the angiosperm-dominated vegetation of the Peninsula region during the Late Cretaceous would have remained high.

In summary, plant fossil evidence from wood, leaf and pollen floras indicates the Aptian and Albian overstorey was dominated by conifers, and the understorey dominated by ginkopsids, taeniopterids, bennettitaleans, equisetites, ferns and liverworts with a very minor angiospermous component. Such open-canopied, evergreen araucarian–podocapaceous conifer forests were characteristic of the mid-Cretaceous Pacific margins of the Gondwanan continent (Falcon-Lang & Cantrill 2000) extending from Alexander Island, probably across the James Ross Island Basin, to the South Shetland Islands and as far north as Patagonia and as far east as New Zealand.

Cenomanian–Santonian interval

The plant fossil record from the Cenomanian and Turonian are poorly known in the Antarctic Peninsula. The Whisky Bay Formation (late Albian–latest Turonian: Riding & Crame 2002) is the only unit where strata of this age occur in the James Ross Basin, where it is divided into three members in each of the two main areas of outcrop (Brandy Bay–Whisky Bay area and Gin Cove–Rum Cove area). However, the complex lateral variation makes it difficult to correlate between outcrops and between the threefold subdivision in different parts of the basin (Riding & Crame 2002). Furthermore, considerable problems still exist in identifying and defining Cenomanian strata within the Whisky Bay Formation (Riding & Crame 2002). Marine invertebrate assemblages of Cenomanian age co-occur with palynofloras that are best assigned to the late Albian (Riding & Crame 2002). No detailed collecting for fossil wood has yet taken place, but a few terrestrial microfloras (e.g. Dettmann & Thomson 1987; D 3057.3) point to low-angiosperm diversity (Dettmann 1989; Askin 1992). In the Cenomanian the angiosperm component of the vegetation probably still comprised largely shrubby, herbaceous forms, as evidenced by the deficit of angiosperm wood in deposits older than approximately 90 Ma (Fig. 4) (Cantrill & Poole 2002).

The overlying Coniacian Hidden Lake Formation is rich in angiosperm leaf material (Hayes 1996, 1999; Hayes *et al.* 2006) and contains angiospermous fossil wood, attesting to increasing dominance and radiation into canopy niches by this time. The exact timing of the radiation into the canopy is problematic because of the paucity of data from the Cenomanian and Turonian. Nevertheless, the fossil record documents an increase in angiosperm floral diversity throughout the mid–Late Cretaceous at the expense of ferns and lycophytes initially, and subsequently bryophytes/hepatophytes, bennettites and other gymnosperms (Cantrill & Poole 2002). Evidence suggests that initially these early angiosperms were colonizers occupying under- and middle-storey niches and only later expanding to become more arborescent and invading the overstorey. This invasion would have drastically changed the prevailing landscape and thus the ecosystem. For example, this change enabled ferns to colonize the new niches created by the angiosperms and thus began a fern rediversification (Cantrill & Poole 2002). Important palaeobotanical information on the vegetation during the Coniacian is derived from the Williams Point Beds on Livingston Island, dated as 88 Ma (R. Hunt pers. comm. 1999), lying at a palaeolatitude of approximately 62°S (Grunow *et al.* 1991). The wood, leaves and palynofloras are well preserved, unlike many sites around the Antarctic Peninsula where the palynomorphs are highly degraded or destroyed and the leaves are merely impressions and lack cuticle (Chapman & Smellie 1992). Both macrofossil and microfossil

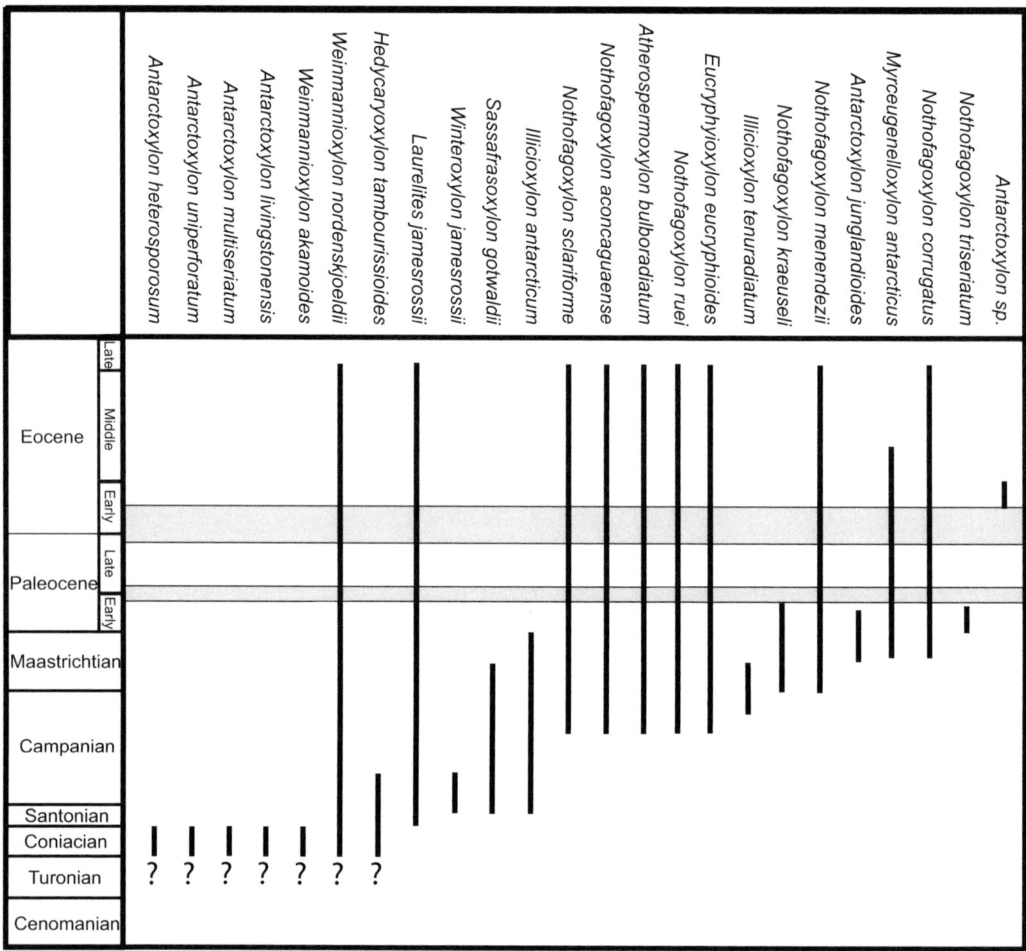

Fig. 4. Biostratigraphic range chart for angiosperm wood taxa (adapted from Cantrill & Poole 2005).

evidence from the Williams Point Beds suggest that the vegetation comprised conifer forest with a diverse angiosperm component. It still remains uncertain as to which group of plants dominated the canopy (Chapman & Smellie 1992; Poole & Cantrill 2001). Preliminary estimates of wood abundance from this locality suggest that between 67 and 81% of the samples are coniferous (Philippe *et al.* 1993; $n = 270$). Ferns, represented by leaves and petrified rachides, are abundant and diverse with large tree ferns and ?bennettites forming part of the arboreal component (Chapman & Smellie 1992; Poole & Cantrill 2001). Understorey fern taxa and Equisetites (Lacey & Lucas 1981) probably inhabited the moist forest floor and banks of streams. Thalloid liverworts (Lacey & Lucas 1981) and epiphytic ferns would have occupied moist shady banks, damp rocks, and trunks and branches of trees (Poole & Cantrill 2001). Unfortunately, this flora is still in need of a thorough revision as it was initially assigned a Triassic age (e.g. Orlando 1968), and the diverse leaf flora, representing many groups of non-woody plants, were assigned to Triassic organ taxa. A revision of the conifer and angiosperm wood component of the flora has already been undertaken by Poole & Cantrill (2001). They found that the size of the material indicates that the woody angiosperm taxa were from large trees. The coniferous element comprised of Araucariaceae and Podocarpaceae, whereas the angiosperms include two species assigned to the Monimiaceae and Cunoniaceae and four other taxa with affinities that lie with the 'Magnoliidae, Hamamelidae and Rosidae' and were assigned to *Antarctoxylon* Poole et Cantrill, an organ genus for fossil woods of equivocal taxonomic affinity (Poole & Cantrill 2001). The early woody–arboreal angiosperm component shows no evidence of distinct growth rings, suggesting that these plants may have been evergreen. With an increase in the abundance of distinct growth rings after this time, it is probable that the angiosperms only later adapted to the seasonal Antarctic environment in becoming predominantly deciduous (Cantrill & Poole 2005). Another slightly younger Coniacian leaf flora occurs in the Hidden Lake Formation on James Ross Island (Fig. 5E–G). Although no cuticles are preserved, thus hindering systematic placement, Hayes (1996, 1999) concluded that the dominant leaf form in this flora shows great similarity to the Magnoliales with a strong component of sterculiaceous (Fig. 5E) and lauralean (Fig. 5F) forms, indicating that angiosperm diversity was well underway by this time.

Indeed, angiosperms had become dominant in the Antarctic macrofloras by the Coniacian (Cantrill & Poole 2002). The increased abundance of wood fragments suggests that the angiosperms were no longer herbaceous, shrubby understorey elements but had become a more important component of the canopy. The Santonian leaf floras still suggest a strong sterculiaceous and lauralean component (Hayes 1996, 1999), alongside woods with lauraceous (Poole *et al.* 2000*c*), cunoniaceous (Poole *et al.* 2000*a*), illiciaceous (Poole *et al.* 2000*b*), atherospermataceous (Poole & Francis 1999; Poole & Gottwald 2001) and winteraceous (Poole & Francis 2000) affinity (Fig. 4). All these elements, with the exception of the illiciaceous, lauraceous and *Sassafrasoxylon* Poole *et al.*, have adisjunct distribution between North America and Asia, and are characteristic of cooler temperate biomes. The presence of the illiciaceous and sassafrasaceous elements could suggest a somewhat warmer temperate biome, or perhaps a more recent adaptation of these plants to warmer latitudes.

Precursors to the changes in vegetation seen in the Campanian–Maastrichtian are present in the Coniacian and Santonian. The first records of *Nothofagidites* Erdtman ex Potonié (Baldoni & Medina 1989; Keating 1992), *Proteacidites* Cookson ex Couper (Dettmann & Thomson 1987; Baldoni & Medina 1989; Barrera *et al.* 1999) and Myrtaceae (Dettmann & Thomson 1987; Baldoni & Medina 1989) occur, although the rare occurrence of these grains indicates the plants were of minor importance in the vegetation. This is supported by the lack of wood from these groups.

Campanian–Maastrichtian interval

Between the Santonian and Campanian there is an unmistakable turnover in pollen taxa that is also reflected in the wood (Fig. 4) and other macro-floras (Dettmann 1989; Askin 1992). Within the angiosperm wood flora the more 'primitive' wood types with affinities to the 'Magnoliidae, Hamamelidae and Rosidae' (Poole & Cantrill 2001) do not continue through to the Campanian. The only primitive types of wood occurring in the Coniacian flora, that also appear in the Campanian, are the cunoniaceous *Weinmannioxylon* Petriella and the monimiaceous *Hedycaryoxylon* Süss (but these xylotypes, *W. ackamoides* Poole et Cantrill and *H. tambourissoides* Poole et Gottwald, both disappear during the Campanian). *Hedycaryoxylon* is replaced by other anatomically different, monimiaceous taxa, whereas the *Weinmannioxylon nordenskjoeldii* Poole *et al.* xylotype continues through until the Eocene (Fig. 4).

Perhaps the most dramatic change is the increase in importance of the Nothfagaceae. The Nothofagaceae is a typical, and one of the most important elements, of the relictual Southern Hemisphere temperate floras today (Manos 1997). Nothofagaceae in the Antarctic pollen floras is first represented by the ancestral *Nothofagidites senectus* Dettmann et Playford, notably distinct from extant *Nothofagus* Blume pollen (Dettmann *et al.* 1990). Importantly, *Nothofagus* pollen is extremely common and its absence can be regarded as evidence for the lack of the genus in areas where it has not been found (Swenson & Hill 2001). This family is only rarely recorded prior to the Campanian, as pollen analyses from Santonian strata (Baldoni & Medina 1989; Keating 1992). Despite the sparse occurrences in the Santonian no wood is known until the Campanian when this group becomes important in both the fossil wood and pollen record (Poole 2002). These changes are supported by an increase in abundance and/or diversity of other taxa including pollen of *Dacrydium* Sol. Ex G. Forst., Proteaceae and Myrtaceae, and a variety of other angiosperms also evidence the changes occurring at, or just prior to, the mid-Campanian (Dettmann & Thomson 1987).

In the Maastrichtian of Vega and Seymour islands on the Antarctic Peninsula the oldest known occurrences of *Nothofagus* subgenera *Fuscospora*, *Lophozonia* and *Brassospora* have been found in the pollen record (Dettmann *et*

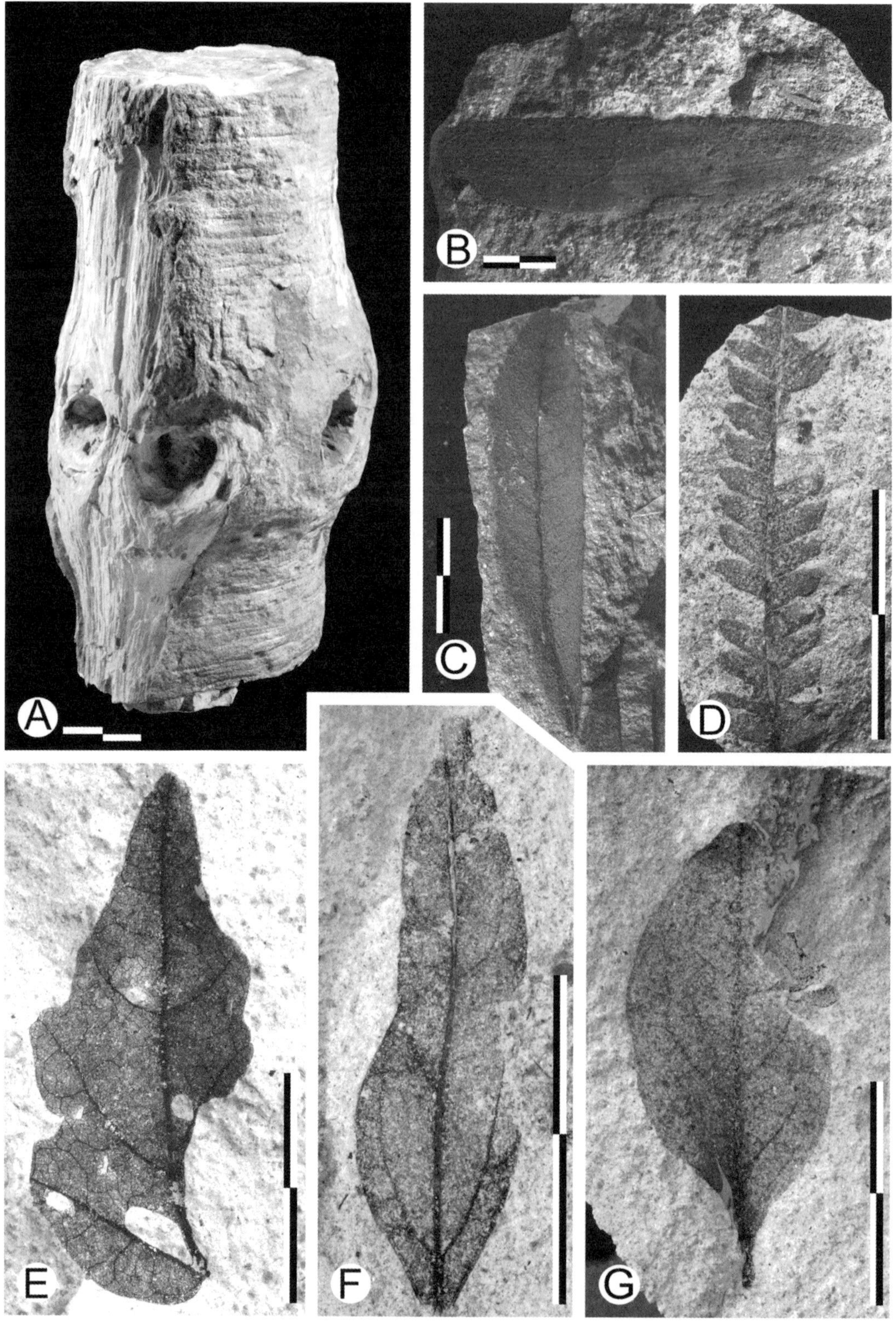
A
B
C
D
E
F
G

al. 1990) alongside wood of the *Fuscospora* type (Poole 2002) (Fig. 4). Manos (1997) and Swenson *et al.* (2000) recognize the subgenus *Lophozonia* as the most basal clade within Nothofagaceae, with the pollen record extending back to the late Campanian in Antarctica (Dettmann *et al.* 1990) and appearing only slightly later in the wood flora (Poole 2002). By the Maastrichtian all but the *Brassospora* subgenus are represented in the wood flora, whereas all four subgenera are represented in the pollen flora. The lack of wood allied to *Brassospora* is not surprising as this group, although a copious pollen producer today, is often confined to montane areas. We can confidently assume that by the mid-Campanian the Nothofagaceae had now become an important and diverse component of the Antarctic ecosystem having substantially changed the face of the flora relative to the ancestral Coniacian–Santonian vegetation.

By the Maastrichtian the wood floras continued to increase in abundance and diversity, with other angiospermous taxa appearing and Nothofagaceae diversifying further (Poole *et al.* 2003). Wood anatomical characters, coupled with good preservation of these now prolific angiosperms, provide us with important indications pertaining to the local environment, as well as the vegetational composition and dynamics. The Antarctic material is comprised predominantly of aerial organs (branches or trunks – only one specimen could be confidently identified as having a root origin), possibly dominated by sap-wood remains (cf. Wheeler & Manchester 2002 and their inferences drawn from the Eocene Clarno Formation woods). The dominance of sap wood is evidenced from preserved fungal hyphae associated with the ray parenchyma and vessel lumens or tracheids, suggesting that these fungi are sap-staining fungi restricted to sap wood. The detachment from the parent plant may have occurred during the Antarctic polar winter or at least during the dormant phase of the year as many of the angiosperm woods have vessels filled with well-developed tyloses, but with no evidence of bud-like outgrowths of parenchyma into the vessel lumen, which mark the beginning of tylose formation (cf. Chattaway 1949; Wheeler & Manchester 2002). The woods probably resided, for a time at least, in a damp, but not waterlogged, aerobic environment. Moist conditions probably prevailed on the magmatic arc adjacent to the James Ross Island Basin, encouraging fungal growth within the woody debris covering the forest floor. This woody debris was probably then transported, by streams or rivers during times of storms or flash floods, to the coast where the material was finally deposited offshore in the shallow-marine environment.

At this time podocarps and angiosperms were the main canopy elements of the perhumid, tall open forests (Askin 1988; Specht *et al.* 1992). Palynofloras suggest the Myrtaceae were present along with diverse species of *Nothofagus*, *Gunnera* L., Proteaceae, Aquifoliaceae, Olacaceae, Loranthaceae and Sapindaceae accompanied by a rich fern component including the Osmundaceae and Gleicheniaceae (e.g. Askin 1989, 1990). At the end of the Maastrichtian the warmer temperate elements, *Sassafrasoxylon* and Illiciaceae, disappeared (Fig. 4), whilst the abundance of fern taxa also decreased (Askin 1988, 1990), possibly in response to a cooling in the climate, leaving a floral composition very similar to that which occurs today in the cool temperate rainforests of South America. In previous publications (Poole *et al.* 2001, 2003) the vegetational composition of the Antarctic Peninsula during the Late Cretaceous and early Tertiary has been likened to the extant low- to mid-altitudinal Valdivian rainforests under the 'Valdivian model'. Indeed the similarities in terms of environmental dynamics in addition to vegetational composition are remarkably similar. Along the Andean margin of South America stratovolcanic activity is a major source of disturbance along with associated events such as landslides, earthquakes and the flooding of lake systems. Glacial erosion and deposition contribute to the general disturbance especially at altitude. The Antarctic Peninsula region during the Cretaceous and Tertiary was a vegetated active volcanic arc and thus would have been subject to similar ecological disturbances. There is no doubt that climate change influenced vegetational composition, although evolving palaeoenvironments in the Antarctic

Fig. 5. Late Cretaceous and Cenozoic floras. (A)–(D) Palaeogene plant fossils from the James Ross Island Basin. (**A**) *Araucaria marenssi* Cantrill et Poole, a petrified araucarian trunk from the Eocene La Meseta Formation, DJ.1057.53. (B)–(D) Leaf fossil from the Palaeocene Cross Valley Formation. (**B**) *Araucaria nathorstii*, DJ.1111.14. (**C**) Angiosperm leaf, D.523.1. (**D**) Fern frond, DJ.1113.134. (E)–(G) Coniacian leaf fossil from the Hidden Lake Formation. (**E**) sterculiaceous taxon, D.8754.8.1a. (**F**) lauraceous taxon, D.8754.8.57a. (**G**) Unidentified angiosperm, D.8754.45a. Scale bar divisions are 1 cm.

Peninsula region probably contributed in equal, if not greater, importance. One good example of environmental dynamics, rather than climate, accounting for vegetation change can be seen in Late Cretaceous–Palaeocene strata on King George Island lying to the north of the James Ross Island Basin. The Late Cretaceous–Early Palaeocene strata is lava-dominated, suggesting that stratovolcanic activity was high and the centre of activity was proximal (Smellie *et al.* 1984; Shen 1994, 1999; Orton 1996). By the Palaeocene epiclastic deposits including tuffaceous rocks indicate that the volcanic dynamics had changed, lowering the rate of disturbance. Late Cretaceous floras of this region also document ecological disturbances: low-density vegetation is dominated by ferns such as *Thyrsopteris* Kunze and *Dacrydium*, and a relatively small percentage of angiosperms with nothofagaceous, sterculiaceous, lauraceous and myrtaceous affinity (Dutra & Batten 2000; Poole *et al.* 2001), whereas post-disturbance vegetation in proximal volcanic settings is characterized by gymnosperms such as cycads and podocarps (Cao 1992; Dutra & Batten 2000; Poole *et al.* 2001).

Palaeocene–Eocene interval

The Valdivian-type ecosystem continues over the Cretaceous–Tertiary boundary, with no suggestion of catastrophic environmental devastation, into the Eocene where the last evidence for this Valdivian-type ecosystem can be found in the palynological record of Seymour Island (Askin 1997) and the wood record of King George Island (Poole *et al.* 2001). Palaeocene and Eocene volcanic activity was still prevalent across the Antarctic Peninsula with terrestrially deposited fossil floras (Fig. 5B, D) providing evidence for ecological disturbance similar to that observed in the Valdivian region today (Hunt 2001; Poole *et al.* 2001 and references therein; Hunt & Poole 2003). Frequent disturbance in the lower part of the sequence on King George Island is characterized by low-diversity flora consisting only of *Nothofagus* and podocarps overlying a coarse volcanic debris flow. Floras more distal from the centre of volcanic activity probably experienced lower levels of disturbance and this is again supported by the floral composition. The presence of myrtaceous, eucryphiaceous and nothofagaceous angiosperms alongside podocarps and Cupressaceae suggest that the flora was relatively diverse and had experienced low or relatively moderate levels of disturbance (Askin 1997; Poole *et al.* 2001). Higher up the sequence and in more distal locations floral diversity increases with pteridophytes (such as Cyatheaceae, Dicksoniaceae and Osmundaceae), conifers (including Araucariaceae, Cupressaceae and Podocarpaceae) and angiosperms (e.g. Nothofagaceae, Proteaceae, Myrtaceae, ?Araliaceae, Anacardiaceae and Cunoniaceae) representing low levels of disturbance, i.e. climax or pre-eruption vegetation. Interestingly, there are no records of *Chusquea* Kunth. (bamboo), a characteristic understorey element of the Valdivian ecosystem today. Poole *et al.* (2001) have suggested that ferns, ubiquitous at this time (Zhou & Li 1994; Mohr 2001), may have filled this niche in these Antarctic forests. By the Late Eocene, angiosperm leaf floras at a palaeolatitude of approximately 62°S show evidence of both deciduous and possible evergreen habits (Hunt & Poole 2003), whereas the wood floras suggest an overriding deciduous habit (Poole *et al.* 2001).

Although this Valdivian-type ecosystem may have persisted for longer, into the Late Eocene, the sudden widespread glaciation of Antarctica and the associated shift towards cooler temperatures at the Eocene–Oligocene boundary (*c.* 34 Ma) (DeConto & Pollard 2003) would have had a detrimental effect on the prevailing vegetation. The cool temperate rainforests would have become less diverse, possibly becoming more analogous to the extant Magellanic subpolar *Nothofagus*- (evergreen and deciduous) dominated forests mixed with conifers (*Podocarpus* L'Her. Ex Pers., *Pilgerodendron* Florin) growing south of approximately 47°S across the southern Aysén and Magallanes regions of Chile and Tierra del Fuego in Argentina (Moore 1983; Veblen *et al.* 1996). Here permanent snow, ice caps and glaciers are present at altitude, cold temperate conditions (MAT 3–6 °C) prevail with high levels of precipitation (MAP 1000–4000 mm) and strong permanent winds (Hoffmann 1975). Indeed, Late Eocene Antarctic fossil assemblages from Seymour Island and McMurdo Sound consist of podocarpaceous and araucariaceous conifers, *Nothofagus* (both deciduous and evergreen) with at least three other angiosperm types and ferns (Dusén 1908; Case 1988; Doktor *et al.* 1996; Askin 1997, 2000; Cantrill 2001*b*; Francis 2000; Pole *et al.* 2000).

Neogene

The cooling climate across the Eocene–Oligocene boundary (e.g. Zachos *et al.* 2001) is attested to both by a more fragmentary Antarctic plant record and also shifts in leaf size classes

(Cantrill 2001*b*). These changes are also seen in the palynological record, with a dramatic decrease in diversity, occurrence and abundance of terrestrial palynomorphs (Raine 1998; Askin & Raine 2000; Raine & Askin 2001). Within the James Ross Basin post-Eocene floras are lacking, due in part to a lack of strata from the latest Eocene until the Miocene (Hobbs Glacier Formation: Pirrie *et al* 1997; Jonkers *et al.* 2002). The overlying James Ross Island Volcanic Group (late Miocene–Pliocene) and the Pliocene Cockburn Island Formation (Jonkers 1998*a*, *b*; Jonkers & Kelley 1998) also lack plant material, and a terrestrial palynoflora has not been documented. Elsewhere in the Antarctic Peninsula the post-Eocene record of macroplants is contentious. In the South Shetland islands the age of a series of Oligocene glacial and interglacials (Birkenmajer 1987, 1990, 1997) has been questioned (Dingle *et al.* 1997; Dingle & Lavelle 1998; Hunt 2001) so that all the fossil floras are now considered to be Eocene in age (Hunt 2001). What is clear from the Oligocene and Miocene strata on King George Island (e.g. Polonez Cove Formation and Cape Melville Formation), and the Miocene and Pliocene units in the James Ross Basin (Hobbs Glacier Formation, James Ross Island Volcanic Group and Cockburn Island Formation), is that conditions for plant growth were extremely unfavourable and a cyclical glacial–interglacial environment prevailed.

The impact of this climate regime on plant communities is best seen in a recent series of cores from offshore Cape Roberts in the Ross Sea (Raine 1998; Askin & Raine 2000; Raine & Askin 2001). Although sedimentation patterns have influenced the accumulation of palynomorphs, the pattern is one of low diversity and abundance interspersed with periods of higher diversity and abundance presumably in response to cold and warm climate periods. By the Early Oligocene the Antarctic flora had become species depauperate with *Nothofagus* and podocarpaceous conifers probably dominating the canopy, and with lycophytes and some bryophytes, and ferns contributing to the understorey (Askin 2000; Pole *et al.* 2000; Cantrill 2001b; Mohr 2001). The flora has analogies with today's Magellanic tundra (*c.* 48°S and the southern tip of Tierra del Fuego and across the Chilean archipelago) which supports predominantly bog communities (Moore 1983; Ruthsatz & Villagran 1991) under a regime of high precipitation (2000–6000 mm year^{-1}) and low temperatures (MAT 5–6 °C). In some sheltered, less unfavourable areas of this tundra small stands of evergreen forest can be found dominated by *Nothofagus* (*N. antarctica* (G. Forst.) Oerst. and *N. betuloides* (Mirb.) Blume), *Drimys* J.R. Forst. Et G. Forst., *Pilgerodendron* and occasional *Maytenus* Molina, elements that have been associated with past Antarctic vegetation. At higher elevations a belt of low 'krumholz' vegetation comprising *Nothofagus* occurs (Moore 1983; Veblen *et al.* 1996). Changes to the Antarctic Neogene vegetation due to cooling of the continent probably resulted in a transition of the vegetation from low forest to krumholz forms and ultimately to tundra. The exact timing of these transitions is unclear, but the terrestrial succession from Cape Roberts (Raine 1998; Askin & Raine 2000; Raine & Askin 2001) shows a general decline in diversity and abundance through the Oligocene and into the Miocene, with fewer abundance spikes in Nothofagaceae. This suggests that, although the vegetation survived successive glaciations presumably in refugia, the ability to recover progressively diminished. The Late Pliocene Meyer Desert Formation (Sirius Group) from East Antarctica contains prostrate, gnarled deciduous *Nothofagus* (as evidenced from leaves, wood and pollen from glacial sediments of the Sirius Formation *c.* 1.8–5.3 Ma: Carlquist 1987; Francis & Hill 1996). Some of the fossil woods show evidence of traumatic events and scarring (Francis & Hill 1996), which is also a common feature in other prostrate species, such as the magellanic *Nothofagus*, to which the habit of these *Nothofagus* fossils can be likened. *Nothofagus* probably had a deciduous habit with single seasonal leaf falls attested to by the strong plicate vernation of fossil *Nothofagus* leaves coupled with dense accumulation of leaves within a single layer (Hill *et al.* 1996). More recent investigations of this flora reveal a surprising floral diversity including conifers, cushion plants, ranunculids, possible grasses and chenopods, and a variety of moss taxa (some cushion forming) growing in a subglacial tundra-like environment (Hill & Truswell 1993; Ashworth & Cantrill 2004). Fossil plant material suggests mean annual temperatures of approximately –20 °C, but possibly colder without snow cover to protect the dormant plants, and growing seasons of up to 3 months with temperatures reaching only about 5 °C in the growing season (Francis & Hill 1996; Hill & Jordan 1996). This evidence lends further support to a magellanic subpolar forest–tundra-type ecosystem.

The change from true magellanic subpolar forest to true tundra in Antarctica at the end of the Tertiary would probably have been punctuated rather than gradual with the flora responding to pulses of glaciations, surviving in refugia during the severest of climates and only making a transitory comeback during the interglacials. Finally, these last remaining plants also succumbed to the deteriorating climates brought about by declining atmospheric CO_2, continental separation and widescale glaciation (Exon *et al.* 2000; DeConto & Pollard 2003) such that today only two species of vascular plant, *Colobanthus quietensis* (Kunth) Bartl. and *Deschampsia antarctica* Desv., along with a few species of lichens and mosses, survive in today's extreme southern high-latitude environments.

Summary

Four stages of vegetation development associated with taxonomic turnover and canopy development are recognized from the Aptian through to the Tertiary across much of the Antarctic Peninsula.

- Early–mid-Cretaceous vegetation was conifer dominated, ranging from forests with lower storeys comprising ginkopsids, spehnopsids, ferns, liverworts alongside extinct taxa, such as bennettites and taeniopterids, to dispersed clumps of individuals more akin to open woodland. The forests had stand densities of approximately 600 stems per hectare, with productivity estimates and taxon composition similar to the warm temperate araucarian–podocarp vegetation of North Island, New Zealand (Falcon-Lang *et al.* 2001; Falcon-Lang & Cantrill 2001*a*). The Alexander Island floras are the only known *in situ* fossil forests to be found in or near the James Ross Island Basin and therefore are taken as a general guide for the ensuing angiosperm-dominated vegetation of the Peninsula region.
- Middle–early Late Cretaceous floras reflect the arrival of the angiosperms at the expense of the bryophytes/hepatophytes, gymnosperms such as bennettites, and, initially, ferns and lycophytes. The initially herbaceous angiosperms had become arboreal by the Coniacian and invaded the conifer-dominated canopy, giving rise to new understorey niches that ferns and liverworts subsequently exploited. By the Santonian angiosperms were an important component of the overstorey, and families including the Sterculiaceae, Lauraceae, Cunoniaceae, Monimiaceae, Atherospermataceae and Winteraceae progressively replaced more 'primitive' forms. Taxonomic similarity with the Vadivian rainforests of Chile was beginning to develop, although the Antarctic vegetation retained some more warm temperate taxa such as the Illiciaceae and *Sassafras*-like plants. The expansion of *Nothofagus* and the replacement of the warm temperate elements in the Campanian changed the vegetation once more. It became similar to the cool temperate rainforests of Valdivia in South America today (Poole *et al.* 2001, 2003).
- The similarities to the Valdivian rainforests strengthened through the Maastrichtian to become dominant up until the Eocene. Vegetation changes reflected environmental disturbances due to volcanic activity that were occurring at this time. Climate deterioration associated with the onset of glaciation in Antarctica led to a further change in the vegetation.
- Diversity decreased towards the Eocene–Oligocene boundary until the forests probably became more similar to the Magellanic subpolar forests of southern South America today where communities are dominated by *Nothofagus* mixed with conifers. The flora became more depauperate during the Oligocene and Miocene, such that by the Miocene–Pliocene it was probably more similar to the Magellanic tundra of southern South America, supporting local stands of *Nothofagus* and a few other angiosperm and conifer taxa. *Nothofagus* decreased in stature as a result of the deteriorating climate conditions. It is likely that the vegetation went through a krumholtz phase in habit prior to becoming low prostrate shrubs in the Pliocene tundra. By the end of the Tertiary, with glaciation widespread and temperatures plummeting, widespread vegetation had been wiped from Antarctica such that only two species of vascular plants occur there today.

We are indebted to BAS for the loan of material, and the opportunity for I. Poole and D. J. Cantrill to undertake fieldwork in Antarctica. They thank Dr R. Hunt and Professor J. Francis for their help with collecting the fossil material; Dr P. Rudell and Professor D. Cutler for continued access to the slide collection at the Jodrell Laboratory, Royal Botanic Gardens Kew; D. Makaham for sectioning modern material for comparisons; and M. Tabecki (BAS) and O. Stiekima (Utrecht University) for sectioning the fossil material used in this study. Much of original work presented here was made possible through NERC funding (grant number GR3/11088) and continued with funding from NWO (grant number ALW/809.32.004).

References

ASHWORTH, A. & CANTRILL, D.J. 2004. Neogene vegetation of the Meyer Desert Formation (Sirius Group) Transantarctic Mountains, Antarctica. *Palaeogeography, Palaeoclimatology, Palaeoecology*, **213**, 65–82.

ASKIN, R.A. 1988. Campanian to Paleocene palynological succession of Seymour Island and adjacent islands, northeastern Antarctic Peninsula. *In*: FELDMANN, R.M. & WOODBURNE, M.O. (eds) *Geology and Paleontology of Seymour Island, Antarctica Peninsula*. Geological Society of America, Memoirs, **169**, 131–153.

ASKIN, R.A. 1989. Endemism and heterochroneity in the Late Cretaceous (Campanian) to Paleocene palynofloras of Seymour Island, Antarctica: implications for origins, dispersal and palaeoclimates of southern floras. *In*: CRAME, J.A. (ed.) *Origins and Evolution of the Antarctic Biota*. Geological Society, London, Special Publications, **47**, 107–119.

ASKIN, R.A. 1990. Campanian to Paleocene spore and pollen assemblages from the upper Campanian and Maastrichtian of Seymour Island. *Review of Palaeobotany and Palynology*, **65**, 105–113.

ASKIN, R.A. 1992. Late Cretaceous-early Tertiary Antarctic outcrop evidence for past vegetation and climates. *In*: KENNETT, J.P. & WARNKE, D.A. (eds) *The Antarctic Palaeoenvironment: A Perspective on Global Change*. American Geophysical Union, Antarctic Research Series, **56**, 61–73.

ASKIN, R.A. 1997. Eocene–?earliest Oligocene terrestrial palynology of Seymour Island. *In*: RICCI, C.A. (ed.) *The Antarctic Region: Geological Evolution and Processes*. Terra Antartica Publication, Siena, 993–996.

ASKIN, R.A. 2000. Spores and pollen from the McMurdo Sound erratics, Antarctica *In*: STILWELL, J.D. & FELDMANN, R.M. (eds) *Palaeobiology and Palaeoenvironments of Eocene Fossiliferous Erratics, McMurdo Sound, East Antarctica*. American Geophysical Union, Antarctic Research Series, **76**, 161–181.

ASKIN, R.A. & RAINE, J.I. 2000. Oligocene and Early Miocene terrestrial palynologyof Cape Roberts Drillhole CRP-2/2A, Victoria Land basin, Antarctica. *Terra Antarctica*, **7**, 493–501.

BALDONI, A.M. & MEDINA, F. 1989. Fauna y microflora del Cretácico, en bahía Brandy, isla James Ross, Antártida. *Instituto Antártico Chileno Serie Científica*, **39**, 43–58.

BARRERA, V., PALAMARCZUK, S. & MEDINA, F. 1999. Palinología de la formación Hidden Lake (Coniaciano–Santoniano), isla James Ross, Antártida. *Revista Espanõla de Micropaleontología*, **31**, 53–72.

BIRKENMAJER, K. 1987. Oligocene–Miocene glaciomarine sequences of King George Island (South Shetland Islands) Antarctica. *Palaeontologia Polonica*, **49**, 9–36.

BIRKENMAJER, K. 1990. Geochronology and climatostratigraphy of Tertiary glacial and interglacial successions on King George Island, South Shetland Islands (West Antarctica). *Zentralblatt für Geologie und Paläontologie*, **1**, 141–151.

BIRKENMAJER, K. 1997. Tertiary glacial/interglacial palaeoenvrionments and sea level changes, King George Island, West Antarctica. An overview. *Bulletin of the Polish Academy of Sciences, Earth Sciences*, **44**, 157–181.

BRODRIBB, T. & HILL, R.S. 2004. The rise and fall of the Podocarpaceae in Australia – a physiological explanation. *In*: HEMSLEY, A.R. & POOLE, I. (eds) *The Evolution of Plant Physiology*. Harcourt, London, 381–399.

BUTTERWORTH, P.J., CRAME, J.A., HOWLETT, P.J. & MACDONALD, D.I.M. 1988. Lithostratigraphy of Upper Jurassic–Lower Cretaceous strata of eastern Alexander Island, Antarctica. *Cretaceous Research*, **9**, 249–264.

CANTRILL, D.J. 1998. Early Cretaceous fern foliage from President Head, Snow Island, Antarctica. *Alcheringa*, **22**, 241–258.

CANTRILL, D.J. 2000. A Cretaceous macroflora from a freshwater lake deposit, President Head, Snow Island, Antarctica. *Palaeontographica Abteilung B*, **253**, 153–191.

CANTRILL, D.J. 2001*a*. *Cretaceous High-latitude Terrestrial Ecosystems: An Example From Alexander Island, Antarctica*. Asociación Paleontológica Argentina, Publicación Especial, **7**, 39–44.

CANTRILL, D.J. 2001*b*. Early Oligocene *Nothofagus* from CRP-3, Antarctica: Implications for the vegetation history. *Terra Antarctica*, **8**, 401–406.

CANTRILL, D.J. & FALCON-LANG, H.J. 2001. Cretaceous (Late Albian) Coniferales of Alexander Island, Antarctica. 2: Leaves, reproductive structures and roots. *Review of Palaeobotany and Palynology*, **115**, 119–145.

CANTRILL, D.J. & NICHOLS, G.J. 1996. Taxonomy and palaeoecology of Early Cretaceous (Late Albian) angiosperm leaves from Alexander Island, Antarctica. *Review of Palaeobotany and Palynology*, **92**, 1–28.

CANTRILL, D.J. & POOLE, I. 2002. Cretaceous to Tertiary patterns of diversity change in the Antarctic Peninsula. *In*: OWEN, A.W. & CRAME, J.A. (eds) *Palaeobiogeography and Biodiversity Change: A Comparison of the Ordovician and Mesozoic–Cenozoic Radiations*. Geological Society, London, Special Publications, **194**, 141–152.

CANTRILL D.J. & POOLE, I. 2005. Taxonomic turnover in wood floras from the Cretaceous and Tertiary of Antarctica: implications for changes in forest ecology. *Palaeogeography, Palaeoecology, Palaeoclimatology*, **215**, 205–219.

CAO, L. 1992. Late Cretaceous and Eocene palynofloras from Fildes Peninsula, King George Island (South Shetland Islands), Antarctica. *In*: YOSHIDA, Y. (ed.) *Recent Progress in Antarctic Earth Science*. Terra Scientific (Terrapus), Tokyo, 363–369.

CASE J. 1988. Paleogene floras from Seymour Island, Antarctic Peninsula. *In*: FELDMANN, R.M. & WOODBURNE, M.O. (eds) *Geology and Paleontology of Seymour Island, Antarctica Peninsula*. Geological Society of America, Memoirs, **169**, 523–530.

CARLQUIST, S. 1987. Pliocene *Nothofagus* wood from the Transantarctic Mountains. *Aliso*, **11**, 571–583.

CESARI, S.N., PARCIA, C.A., REMESAL, M.B. & SALANI, F.N. 1999. First evidence of Pentoxylales in Antarctica. *Cretaceous Research*, **19**, 733–743.

CESARI, S.N., PARCIA, C.A., REMESAL, M.B. & SALANI, F.N. 1999. Paleoflora del Cretácico Inferior de la península Byers, islas Shetland del Sur, Antártida. *Ameghiniana*, **36**, 3–22.

CHALONER, W.G. & CREBER, G.T. 1989. The phenomenon of forest growth in Antarctica: a review. *In*: CRAME, J.A. (ed.) *Origins and Evolution of the Antarctic Biota*. Geological Society, London, Special Publications, **47**, 85–88.

CHAPMAN, J.L. 1994. Distinguishing internal developmental characteristics from external palaeoenvironmental effects in fossil wood. *Review of Palaeobotany and Palynology*, **92**, 1–28.

CHAPMAN, J.L. & SMELLIE, J.L. 1992. Cretaceous fossil wood and palynomorphs from Williams Point, Livingston Island, Antarctic Peninsula. *Review of Palaeobotany and Palynology*, **74**, 163–192.

CHATTAWAY, M.M. 1949. The development of tyloses and secretion of gum in heartwood formation. *Australian Journal of Science*, **2**, 227–240.

CLARKSON, B.R., PATEL, R.N. & CLARKSON, B.D. 1988. Composition and structure of a forest overwhelmed at Pureora, central North Island, New Zealand, during the Taupo eruption (*c*. AD 130). *Journal of the Royal Society of New Zealand*, **18**, 417–436.

CRAME, J.A., PIRRIE, D., CRAMPTON, J.S. & DUANE, A.M. 1993. Stratigraphy and regional significance of the Upper Jurassic–Lower Cretaceous Byers Group, Livingston Island, Antarctica. *Journal of the Geological Society, London*, **150**, 1075–1087.

CREBER, G.T. & FRANCIS, J.E. 1999. Fossil tree-ring analysis: palaeodendrology. *In*: JONES, T.P. & ROWE, N.P. (eds) *Fossil Plants and Spores Modern Techniques*. Geological Society, London, 241–244.

DECONTO, R.M. & POLLARD, D. 2003. Rapid Cenozoic glaciation of Antarctica induced by declining atmospheric CO_2. *Nature*, **421**, 245–249.

DETTMAN, M.E. 1989. Antarctica: Cretaceous cradle of austral temperate rainforests? *In*: CRAME, J.A. (ed.) *Origins and Evolution of Antarctic Biota*. Geological Society, London, Special Publications, **47**, 89–105.

DETTMANN, M.E. & THOMSON, M.R.A. 1987. Cretaceous palynomorphs from the James Ross Island area, Antarctica – a pilot study. *British Antarctic Survey Bulletin*, **77**, 13–59.

DETTMANN, M.E., POCKNALL, D.T., ROMERO, E.J. & ZAMOLA, M.D.C. 1990. *Nothofagidites* Erdtman ex Potonie, 1960; a catalogue of species with notes on the palaeogeographic distribution of *Nothofagus* Bl. (Southern Beech). *New Zealand Geological Survey Bulletin*, **60**, 1–79.

DINGLE, R.V. & LAVELLE, M. 1998. Antarctic Peninsula cryosphere: Early Oligocene (*c*. 30 Ma) initiation and revised glacial chronology. *Journal of the Geological Society, London*, **155**, 433–437.

DINGLE, R.V., MCARTHUR, J.M. & VROON, P. 1997. Oligocene and Pliocene interglacial events in the Antarctic Peninsula dated using strontium isotope stratigraphy. *Journal of the Geological Society, London*, **154**, 257–264.

DOKTOR, M., GAZDZICKI, A., JERMANSKA, A., POREBSKI, S.J. & ZASTAWNIAK, E. 1996. A plant and fish assemblage from the Eocene La Meseta Formation of Seymour Island (Antarctic Peninsula) and its environmental implications. *Palaeontologia Polonica*, **55**, 127–146.

DUANE, A.M. 1996. Palynology of the Byers Group (Late Jurassic–Early Cretaceous) of Livingston and Snow islands, Antarctic Peninsula: its biostratigraphical and palaeoenvironmental significance. *Review of Palaeobotany and Palynology*, **91**, 241–281.

DUSÉN, P. 1908. Über die tertiäre Flora der Seymour-Insel. *Wissenschaftliche ergebnisse Der Schwedischen Südpolar-expedition 1901–1903*, **3**, (3), 1–27.

DRINNAN, A.N. & CRANE, P.R. 1990. Cretaceous Paleobotany and its bearing on the biogeography of austral Angiosperms. *In*: TAYLOR, T.N. & TAYLOR, E.L. (eds) *Antarctic Paleobiology: Its Role in the Reconstruction of Gondwana*. Springer, New York, 192–219.

DUTRA, T. & BATTEN, D. 2000. Upper Cretaceous floras of King George Island, West Antarctica and their palaeoenvironmental and phytogeographic implications. *Cretaceous Research*, **21**, 181–209.

EIGHTS, J. 1833. Description of new crustaceous animal found on the shores of South Shetland Islands, with remarks on their natural history. *Transactions of the Albany Institute*, **2**, 53–69.

EKLUND, H. 2003. First Cretaceous flowers from Antarctica. *Review of Palaeobotany and Palynology*, **256**, 1–31.

EXON, N., KENNETT, J., MALONE, M. & THE LEG 189 SHIPBOARD SCIENTIFIC PARTY. 2000. The opening of the Tasmanian gateway drove global Cenozoic paleoclimatic and paleoceanographic changes: Results of Leg 189. *JOIDES Journal*, **26**, 11–17.

FALCON-LANG, H.J. & CANTRILL, D.J. 2000. Cretaceous (Late Albian) Coniferales of Alexander Island, Antarctica. 1: Wood taxonomy: a quantitative approach. *Review of Palaeobotany and Palynology*, **111**, 1–17.

FALCON-LANG, H.J. & CANTRILL, D.J. 2001*a*. Gymnosperm woods from the Cretaceous (mid-Aptian) Cerro Negro Formation, Byers Peninsula, Livingston Island, Antarctica: the aborescent vegetation of a volcanic arc. *Cretaceous Research*, **22**, 227–293.

FALCON-LANG, H.J. & CANTRILL, D.J. 2001*b*. Leaf phenology of some mid-Cretaceous polar forests, Alexander Island, Antarctica. *Geological Magazine*, **138**, 39–52.

FALCON-LANG, H.J. & CANTRILL, D.J. 2002. Terrestrial palaeoecology of an Early Cretaceous volcanic archipelago, Byers Peninsula and President Head, South Shetlands Islands, Antarctica. *Palaios*, **17**, 535–549.

FALCON-LANG, H.J. CANTRILL, D.J. & NICHOLS, G.J. 2001. Biodiversity and terrestrial ecology of a mid-Cretaceous, high latitude floodplain, Alexander Island, Antarctica. *Journal of the Geological Society, London*, **158**, 709–725.

FRANCIS, J.E. 1986. Growth rings in Cretaceous and Tertiary wood from Antarctica and their palaeoclimatic implications. *Palaeontology*, **29**, 665–684.

FRANCIS, J.E. 2000. Fossil wood from Eocene high latitude forests McMurdo Sound Antarctica. *In*: STILWELL, J.D. & FELDMANN, R.M. (eds) *Palaeobiology and Palaeoenvironments of Eocene Fossiliferous Erratics, McMurdo Sound, East Antarctica*. American Geophysical Union, Antarctic Research Series, **76**, 253–260.

FRANCIS, J.E. & HILL, R.S. 1996. Fossil plants from the Pliocene Sirius Group, Transantarctic Mountains: evidence for climate from growth rings and fossil leaves. *Palaios*, **11**, 389–396.

FRANCIS, J.E. & POOLE, I. 2002. Cretaceous and Tertiary climates of Antarctica: evidence from fossil wood. *Palaeogeography, Palaeoclimatology, Palaeoecology*, **182**, 47–64.

GANDOLFO, M.A., HOC, P., SANTILLANA, S. & MARENSSI, S. 1998. *Una flor fósil morpholoicamente afín a las Grossulariaceae (orden Rosales) de la formación La Meseta (Eoceno Medio) Isla Marambio, Antártida*. Asociación Paleontológica Argentina, Publicación Especial, **5**, 147–153.

GOTHAN, W. 1908. Die fossilen Hölzer von der Seymour und Snow Hill-Insel. *Wissenschaftliche ergebnisse Der Schwedischen Südpolar-expedition 1901–1903*, **3**, (8), 1–33.

GRUNOW, A.M., KENT, D.V. & DALZIEL, I.W.D. 1991. New palaeomagnetic data from Thurston Island: implications for the tectonics of West Antarctica and Weddell Sea opening. *Journal of Geophysical Research*, **96**, 17 935–17 954.

HATHWAY, B. 1997. Nonmarine sedimentation in an Early Cretaceous extensional continental-margin arc, Byers Peninsula, Livingston Island, South Shetland Islands. *Journal of Sedimentary Research*, **67**, 686–697.

HATHWAY, B. & RIDING, J.B. 2001. Stratigraphy and age of the Lower Cretaceous Pederson Formation, northern Antarctic Peninsula. *Antarctic Science*, **13**, 67–74.

HATHWAY, B., DUANE, A.M., CANTRILL, D.J. & KELLEY, S.P. 1999. 40Ar/39Ar geochronology and palynology of the Cerro Negro Formation, South Shetland Islands, Antarctica: a new radiometric tie for Cretaceous terrestrial biostratigraphy in the Southern Hemisphere. *Australian Journal of Earth Sciences*, **46**, 593–606.

HAYES, P. 1996. The Coniacian–Santonian (Late Cretaceous) flora of the Hidden Lake Formation, James Ross Basin, Antarctic Peninsula. *In*: CORSETTII, F. & TIFFNEY, B.H. (eds) *Fifth Quadrennial Conference of the International Organisation of Palaeobotany Abstract Volume*, 41. International Organisation of Palaeobotany and Department of Geological Sciences, University of California, Santa Barbara, USA.

HAYES, P. 1999. *Cretaceous angiosperm floras of Antarctica*. PhD Thesis, University of Leeds.

HAYES, P., FRANCIS, J.E., CANTRILL, D.J. & CRAME, J.A. 2006. Palaeoclimate analysis of the Late Cretaceous angiosperm leaf floras, James Ross Island, Antarctica. *In*: FRANCIS, J.E., PIRRIE, D. & CRAME, J.A. (eds) *Cretaceous–Tertiary High-latitude Palaeoenvironments: James Ross Basin, Antarctica*. Geological Society, London, Special Publications, **258**, 49–62.

HERNÁNDEZ, P.J. & AZCARÁRTE, V. 1971. Estudio paleobotánico preliminar sobre restos de una tafoflora de la Península Byers (Cerro Negro), Isla Livingston, Islas Shetland del Sur, Antártica. *Instituto Antárctico Chileno Serie Científica*, **2**, 15–50.

HILL, R.S. & JORDAN, G.J. 1996. Macrofossils as indicators of Plio-Pleistocene climates in Tasmania and Antartica. *Papers and Proceedings of the Royal Society of Tasmania*, **130**, 9–15.

HILL, R.S. & TRUSWELL, E.M. 1993. *Nothofagus* fossils in the Sirius Group Transantarctic Mountains. *In*: KENNETT, J.P. & WARNKE, D. (eds) *The Antarctic Paleoenvironment: A Perspective on Global Change*. American Geophysical Union, Antarctic Research Series, **60**, 67–73.

HILL, R.S., HARWOOD, D.M. & WEBB, P.N. 1996. *Nothofagus beardmorensis* (Nothofagaceae), a new species based on leaves from the Pliocene Sirius Group, Transantarctic Mountains, Antarctica. *Review of Palaeobotany and Palynology*, **94**, 11–24.

HOFFMANN, J.A.J. 1975. *Atlas climático de América del Sur. Mapas de temperatura y precipitaciones medias*. WMO, UNESCO, Geneve.

HOWE, J. & CANTRILL, D.J. 2001. Palaeoecology and taxonomy of Pentoxylales from the Albian of Antarctica. *Cretaceous Research*, **23**, 779–793.

HUNT, R.H. 2001. *Biodiversity and palaeoecology of Tertiary fossil floras in Antarctica*. PhD Thesis, University of Leeds.

HUNT, R.H. & POOLE, I. 2003. Revising Palaeogene West Antarctic climate and vegetation history in light of new data from King George Island. *In*: WING, S.L., GINGERICH, P.D., SCHMITZ, B. & THOMAS, E. (eds) *Causes and Consequences of Globally Warm Climates in the Early Paleogene*. Geological Society of America, Special Papers, **369**, 395–412.

JEFFERSON, T.H. 1981. *Palaeobotanical contributions to the geology of Alexander Island, Antarctica*. PhD Thesis, University of Cambridge.

JEFFERSON, T.H. 1982. Fossil forests from the Lower Cretaceous of Alexander Island, Antarctica. *Palaeontology*, **25**, 681–708.

JONKERS, H.A. 1998*a*. The Cockburn Island Formation; Late Pliocene interglacial sedimentation in the James Ross Basin, northern Antarctic Peninsula. *Newsletters on Stratigraphy*, **36**, 63–76.

JONKERS, H.A. 1998*b*. Stratigraphy of Antarctic late Cenozoic pectinid-bearing deposits. *Antarctic Science*, **10**, 161–170.

JONKERS, H.A. & KELLEY, S.P. 1998. A reassessment of the age of the Cockburn Island Formation, northern Antarctic Peninsula, and its palaeoclimatic implications. *Journal of the Geological Society, London*, **155**, 737–740.

JONKERS, H.A., LIRIO, J.M., DEL VALLE, R.A. & KELLEY, S.P. 2002. Age and environment of Miocene–Pliocene glaciomarine deposits, James

Ross Island, Antarctica. *Geological Magazine*, **139**, 577–594.

KEATING, J.M. 1992. Palynology of the Lachman Crags Member, Santa Marta Formation (upper Cretaceous) of north-west James Ross Island. *Antarctic Science*, **4**, 293–304.

KEATING, J.M., SPENCER-JONES, M. & NEWHAM, S. 1992. The stratigraphical palynology of the Kotick Point and Whisky Bay formations, Gustav Group (Cretaceous), James Ross Island. *Antarctic Science*, **4**, 279–292.

LACEY, W.S. & LUCAS, R.C. 1981. The Triassic flora of Livingston Island, South Shetland Islands. *British Antarctic Survey Bulletin*, **53**, 157–173.

LI, H. 1994. Early Tertiary Fossil Hill flora from Fildes Peninsula of King George Island, Antarctica. *In*: SHEN, Y. (ed.) *Stratigraphy and Palaeontology of Fildes Peninsula, King George Island, Antarctica.* State Antarctic Committee Monograph, **3**, 133–171.

LI, H. & SHEN, Y. 1990. A primary study of Fossil Hill floras from Fildes Peninsula of King George Island, Antarctica. *Acta Palaeontologica Sinica*, **29**, 147–153.

MANOS, P.S. 1997. Systematics of *Nothofagus* (Nothofagaceae) based on rDNA spacer sequences (ITS): taxonomic congruence with morphologyy and plastid sequences. *American Journal of Botany*, **84**, 1137–1155.

MOHR, B.A.R. 2001. The development of Antarctic fern floras during the Tertiary, and palaeoclimatic and palaeobiogeographic implications. *Palaeontographica Abteilung B*, **259**, 167–208.

MOORE, D.M. 1983. *Flora of Tierra del Fuego*. Anthony Nelson, Shropshire.

NICHOLS, G.J. & CANTRILL, D.J. 2002. Tectonic and climatic controls on a Mesozoic forearc basin succession, Alexander Island, Antarctica. *Geological Magazine*, **139**, 313–330.

NIKLAS, K.J. 1994. Predicting the height of fossil plant remains: an allometric approach to an old problem. *American Journal of Botany*, **81**, 1235–1243.

ORLANDO, H.A. 1968. A new Triassic flora from Livingston Island, South Shetland Islands. *British Antarctic Survey Bulletin*, **16**, 1–13.

ORTON, G.J. 1996. Volcanic environments. *In*: READING, H.G. (ed.) *Sedimentary Environments: Processes, Facies and Stratigraphy*. Blackwell Science, Oxford, 485–567.

OTTONE, E.G. & MEDINA, F.A. 1998. A wood from the Early Cretaceous of James Ross Island. *Ameghiniana*, **35**, 291–298.

PHILIPPE, M., BARALE, G., TORRES, T. & COVACEVICH, V. 1993. First study of *in situ* fossil woods from the Upper Cretaceous of Livingston Island, South Shetland Islands, Antarctica: paaleoecological implications. *Compte Rendu Academie Science Paris Series II*, **317**, 103–108.

PIRRIE, D., CRAME, J.A., RIDING, J.B., BUTCHER, A.R. & TAYLOR, P.D. 1997. Miocene glaciomarine sedimentation in the northern Antarctic Peninsula region: the stratigraphy and sedimentology of the Hobbs Glacier Formation, James Ross Island. *Geological Magazine*, **136**, 745–762.

POLE, M., HILL, R.S. & HARWOOD, D. 2000. Eocene plant macrofossils from erratics, McMurdo Sound, Antarctica. *In*: STILWELL, J.D. & FELDMANN, R.M. (eds) *Palaeobiology and Palaeoenvironments of Eocene Fossiliferous Erratics, McMurdo Sound, East Antarctica*. American Geophysical Union, Antarctic Research Series, **76**, 243–251.

POOLE, I. 2002. Systematics of Cretaceous and Tertiary *Nothofagoxylon*: Implications for Southern Hemisphere biogeography and evolution of the Nothofagaceae. *Australian Systematic Botany*, **15**, 247–276.

POOLE, I. & CANTRILL, D.J. 2001. Fossil woods from Williams Point Beds, Livingston Island, Antarctica: a Late Cretaceous southern high latitude flora. *Palaeontology*, **44**, 1081–1112.

POOLE, I. & FRANCIS, J.E. 1999. The first record of fossil atherospermataceous wood from the upper Cretaceous of Antarctica. *Review of Palaeobotany and Palynology*, **107**, 97–107.

POOLE, I. & FRANCIS, J.E. 2000. The first record of fossil wood of Winteraceae from the Upper Cretaceous of Antarctica. *Annals of Botany*, **85**, 307–315.

POOLE, I. & GOTTWALD, H. 2001. Monimiaceae *sensu lato*, an element of Gondwanan polar forests: Evidence from the Late Cretaceous–early Tertiary wood flora of Antarctica. *Australian Systematic Botany*, **14**, 207–230.

POOLE, I., CANTRILL, D.J., HAYES, P. & FRANCIS, J.E. 2000*a*. The fossil record of Cunoniaceae: new evidence from Late Cretaceous wood of Antarctica. *Review of Palaeobotany and Palynology*, **111**, 127–144.

POOLE, I., GOTTWALD, H. & FRANCIS, J.E. 2000*b*. Illiciaceae, an element of Gondwanan polar forests? Late Cretaceous and Early Tertiary woods of Antarctica. *Annals of Botany*, **86**, 421–432.

POOLE, I., HUNT, R.J. & CANTRILL, D.J. 2001. A fossil wood flora from King George Island: ecological implications for an Antarctic Eocene vegetation. *Annals of Botany*, **88**, 33–54.

POOLE, I., MENNEGA, A.M.W. & CANTRILL, D.J. 2003. Valdivian ecosystems in the late Cretaceous and early Tertiary of Antarctica as evidenced from fossil wood. *Review of Palaeobotany and Palynology*, **124**, 9–27.

POOLE, I., RICHTER, H.G. & FRANCIS, J.E. 2000*c*. Evidence for Gondwanan origins for *Sassafras* (Lauraceae)? Late Cretaceous fossil wood of Antarctica. *International Association of Wood Anatomists Journal*, **21**, 463–475.

RAINE, J.I. 1998. Terrestrial palynomorphs from Cape Roberts Project drillhole CRP-1, Ross Sea, Antarctica. *Terra Antartica*, **5**, 539–548.

RAINE, J.I. & ASKIN, R.A. 2001. Terrestrial palynology of Cape Roberts Project drillhole CRP-3, Victoria Land Basin, Antarctica. *Terra Antartica*, **8**, 389–400.

RIDING, J.B. & CRAME, J.A. 2002. Aptian to Coniacian (Early–Late Cretaceous) palynostratigraphy of the Gustav Group, James Ross Basin, Antarctica. *Cretaceous Research*, **23**, 739–760.

RIDING, J.B., CRAME, J.A., DETTMANN, M.E. & CANTRILL, D.J. 1998. The age of the base of the Gustav Group in the James Ross Basin, Antarctica. *Cretaceous Research*, **19**, 87–105.

RUTHSATZ, B. & VILLAGRAN, C. 1991. Vegetation

pattern and soil nutrients of a magellanic moorland on the Cordillera-de-Piuchue, Chiloe Island, Chile. *Revista Chilena de Historia Natural*, **64**, 461–478.

SHEN, Y. 1994. Subdivision and correlation of Cretaceous to Paleogene volcano-sedimentary sequence from Fildes Peninsula, King George Island, Antarctica. *In*: SHEN Y. (ed.) *Stratigraphy and Palaeontology of Fildes Peninsula, King George Island, Antarctica*. State Antarctic Committee Monograph, **3**, 1–35.

SHEN, Y. 1999. Subdivision and correlation of the Eocene Fossil Hill Formation from King George Island, West Antarctica. *Korean Journal of Polar Research*, **10**, 91–95.

SMELLIE, J.L., PANKHURST, R.J., THOMSON, M.R.A. & DAVIES, R.E.S. 1984. *The Geology of the South Shetland Islands: VI. Stratigraphy, Geochemistry and Evolution*. British Antarctic Survey, Scientific Report, **87**, 1–85.

SPECHT, R.L., DETTMANN, M.E. & JARZEN, D.M. 1992. Community associations and structure in the Late Cretaceous vegetation of southeast Australasia and Antarctica. *Palaeogeography, Palaeoclimatology, Palaeoecology*, **94**, 283–309.

STOREY, B.C. & GARRETT, S.W. 1985. Crustal growth of the Antarctic Penisula by accretion, magmatism and extension. *Geological Magazone*, **122**, 5–14.

SWENSON, U. & HILL, R.S. 2001. Most parsimonious areagrams versus fossils: the case of Nothofagus (Nothofagaceae). *Australian Journal of Botany*, **49**, 367–376.

SWENSON, U., HILL, R.S. & MCLOUGHLIN, S. 2000. Ancestral area analysis of *Nothofagus* (Nothofagaceae) and its congruence with the fossil record. *Australian Systematic Botany*, **13**, 469–478.

TORRES, T. & LEMOIGNE, Y. 1988. Maderas fósiles terciarias de la Formacíon Caleta Arctowski, Isla Rey Jorge, Antártica. *Instituto Antártico Chileno Serie Científica*, **37**, 69–107.

TORRES , T. & LEMOIGNE, Y. 1989. Fossil wood findings of angiosperms and gymnosperms of the Upper Cretaceous at Williams Point, Livingston Island, South Shetland Islands, Antarctica. *Instituto Antártico Chileno Serie Científica*, **39**, 9–26.

TORRES, T., BARALE, G., MÉON, H., PHILIPPE, M. & THÉVENARD, F. 1997*a*. Cretaceous floras from Snow Island (South Shetland Islands, Antarctica) and their biostratigraphic significance. *In*: RICCI, C.A. (ed.) *The Antarctic Region: Geological Evolution and Processes*. Terra Antarctica Publication, Siena, 1023–1028.

TORRES, T., BARALE, G., THÉVERNARD, F., PHILIPPE, M. & GALLEGUILLOS, H. 1997*b*. Morfología y sistemática de la flora del Cretácico Inferior de President Head, isla Snow, archipélago de las Shetland del Sur, Antártica. *Serie Científica del Instituto Antártico Chileno*, **2**, 15–50.

TORRES, T., MARENSSI, S. & SANTILLANA, S. 1994*a*. Maderas fósiles de la isla Seymour, Formacíon La Meseta, Antártica. *Instituto Antártico Chileno Serie Científica*, **39**, 43–58.

TORRES, T., MARRENSSI, S. & SANTILLANA, S. 1994*b*. Fossil wood of La Meseta Formation, isla Seymour, Antártica. *Serie Científica del Instituto Antártico Chileno*, **44**, 17–38.

TORRES, T.G., VALENZUELA, E.A. & GONZALEZ, I.M. 1982. Paleoxilología de Península Byers, Isla Livingston, Antártica. *In*: *Actas 3rd Congresso Geologico Chileno: Concepción, Chile*, A321–A342.

UPCHURCH, G.R., JR, OTTO-BLIESNER, B.L. & SCOTESE, C. 1998. Vegetation-atmosphere interactions and their role in global warming during the latest Cretaceous. *Philosophical Transactions of the Royal Society of London*, **B353**, 97–112.

VEBLEN, T.T., DONOSO, C., KITZBERGER, T. & REBERTUS, A.J. 1996. Ecology of southern Chilean and Argentinian *Nothofagus* forests. *In*: VEBLEN, T.T., HILL, R.S. & READ, J. (eds) *The Ecology and Biogeography of Nothofagus Forests*. Yale University Press, New Haven, CT, 293–353.

WHEELER, E.A. & MANCHESTER, S. 2002. Woods of the Eocene Nut Beds Flora Clarno Formation, Oregon, USA. *International Association of Wood Anatomists Journal Supplement*, **3**, 1–188.

WHITHAM, A.G., INESON, J.R. & PIRRIE, D. 2006. Marine volcaniclastics of the Hidden Lake Formation (Coniacian) of James Ross Island, Antarctica: an enigmatic element in the history of the back-arc basin. *In*: FRANCIS, J.E., PIRRIE, D. & CRAME, J.A. (eds) *Cretaceous–Tertiary High-latitude Palaeoenvironments: James Ross Basin, Antarctica*. Geological Society, London, Special Publications, **258**, 21–47.

ZACHOS, J., PAGANI, M., SLOAN, L., THOMAS, E. & BILLUPS, K. 2006. Trends, rhythms and aberrations in global climate 65 Ma to present. *Science*, **292**, 686–693.

ZASTAWNIAK, E. 1981. Tertiary leaf flora from the Point Hennequin Group of King George Island (South Shetland Islands, Antarctica). Preliminary report. *Studia Geologica Polonica*, **72**, 97–108.

ZASTAWNIAK, E. 1990. Late Cretaceous leaf flora of King George Island, West Antarctica. *In*: KONBLOCH, E. & KVAČEK, Z. (eds) *Proceedings of the Symposium: Palaeofloristic and Palaeoclimatic Changes in the Cretaceous and Tertiary*. Geological Survey, Prague, 81–86.

ZASTAWNIAK, E. 1993. Macroscopic plant remains from Upper Cretaceous and Tertiary of King George Island, (South Shetland Islands, West Antarctica). *Wiadomości Botanicczne*, **37**, 217–219.

ZASTAWNIAK, E. 1994. Upper Cretaceous leaf flora from the Blaszyk Moraine (Zamek Formation), King George Island, South Sheltland Islands, West Antarctica. *Acta Palaeobotanica*, **34**, 119–163.

ZASTAWNIAK, E., WRONA, R., GAZDZICKI, A. & BIRKENMAJER, K. 1995. Plant remains from the top part of the Point Hennequin Group (Upper Oligocene), King George Island (South Shetland Islands, Antarctica). *Studia Geologica Polonica*, **81**, 143–164.

ZHOU, Z. & LI, H. 1994. Early Tertiary ferns from Fildes Peninsula, King George Island, Antarctica. *In*: SHEN, Y. (ed.) *Stratigraphy and Palaeontology of Fildes Peninsula, King George Island*. State Antarctic Committee Monograph, **3**, 173–189.

Late Cretaceous Antarctic fish diversity

J. KRIWET[1], J. M. LIRIO[2], H. J. NUÑEZ[2], E. PUCEAT[3] & C. LÉCUYER[3]

[1]*Museum of Natural History, Palaeontology, Humboldt-University Berlin, Invalidenstr. 43, 10115 Berlin, Germany (e-mail: juergen.kriwet@museum.hu-berlin.de)*

[2]*Instituto Antártico Argentino, Cerrito 1248, Capital Federal (1010), Argentina*

[3]*UMR 5125 – CNRS, UFR Sciences de la Terre, Bat 402 (Géode), Université Claude Bernard Lyon 1, 27–43 Bd du 11 Novembre 1918, 69622 Villeurbanne cédex, France*

Abstract: New material from the Santa Marta Formation (late Coniacian–?early Maastrichtian) of James Ross Island contributes significantly to the current knowledge of Late Cretaceous Antarctic fish diversity. The taxon list for the Santa Marta Formation is extended, and new records of neoselachians and teleosts are reported. The stratigraphic ranges of some previously known taxa are enlarged, and the palaeobiogeography and palaeoecology of Late Cretaceous Antarctic fishes are discussed. Top predators that occupied the higher levels in the food chain along with marine tetrapods dominate the marine faunas from the Santa Marta and López de Bertodano formations. The only fish adapted to crushing hard-shelled invertebrates were the chimeroids. Rays, an important component of marine fish associations, as well as fish from lower trophic levels, remain unknown from the Late Cretaceous of Antarctica.

Fossil fish remains have long been known from the extensive Cretaceous marine deposits of the James Ross Basin on the NE flank of the Antarctic Peninsula (Fig. 1a). The first fish fossils were collected during the 1901–1903 Swedish South Polar Expedition and were subsequently described by Woodward (1908). The James Ross Basin is one of a series of largely back-arc basins that formed in the southernmost South American–Antarctic Peninsula region during the Late Mesozoic–Cenozoic (Riding & Crame 2002), and their origin is closely related to the early stages of Gondwana break-up (Hathway 2000). The islands forming the James Ross Island Group comprise an extensive and remarkably fossiliferous Cretaceous–Palaeogene sedimentary succession, including a well-preserved Cretaceous–Tertiary (K/T) boundary section on Seymour Island (Zinsmeister *et al.* 1989; Zinsmeister 1998; Crame *et al.* 1991). Consequently, the James Ross Island Group is important in reconstructing global Cretaceous and Palaeogene palaeoclimatic and palaeobiological changes (Ditchfield *et al.* 1994; Crame *et al.* 1996; Huber 1998; Riding & Crame 2002).

The regressive Cretaceous megasequence of the James Ross Basin is divided into three principal lithostratigraphic groups: the basal Gustav (Aptian–Coniacian); the Marambio (Coniacian–Danian); and the Seymour Island (Palaeogene) groups (e.g. Crame *et al.* 1991; Riding & Crame 2002). Late Cretaceous fish remains comprising chimeroids, selachians and bony fish have been reported from the Marambio Group of James Ross and Seymour islands (Woodward 1908; Grande & Eastman 1986; Cione & Medina 1987; Grande & Chatterjee 1987; Richter & Ward 1990; Stahl & Chatterjee 1999, 2002). These few published studies indicate that our knowledge of Cretaceous fish faunas in the high southern latitudes, compared to other regions, is still very incomplete. Recent fieldwork by two of the authors (J. M. Lirio, H. J. Nuñez) yielded new fish material from the Late Cretaceous Santa Marta Formation of James Ross Island. Additional unpublished material collected by M. Richter (Rio Grande do Sul, Brazil) and J. J. Hooker (Natural History Museum, London, UK) during the 'James Ross Island Scientific Cruise' of the British Antarctic Survey in 1989, and housed in the Natural History Museum, London, was also examined.

It is the purpose of this paper to review the current state of knowledge of Antarctic Cretaceous fish fossils and to present new material. In addition, this study provides new insights into Antarctic fossil fish communities and biogeography as a basis for future investigations.

Geological and stratigraphic framework

The Marambio Group is exposed in both northern James Ross, Vega and Humps islands in the north and southern James Ross, Snow

From: FRANCIS, J. E., PIRRIE, D. & CRAME, J. A. (eds) 2006. *Cretaceous–Tertiary High-Latitude Palaeoenvironments, James Ross Basin, Antarctica.* Geological Society, London, Special Publications, **258**, 83–100. 0305-8719/06/$15

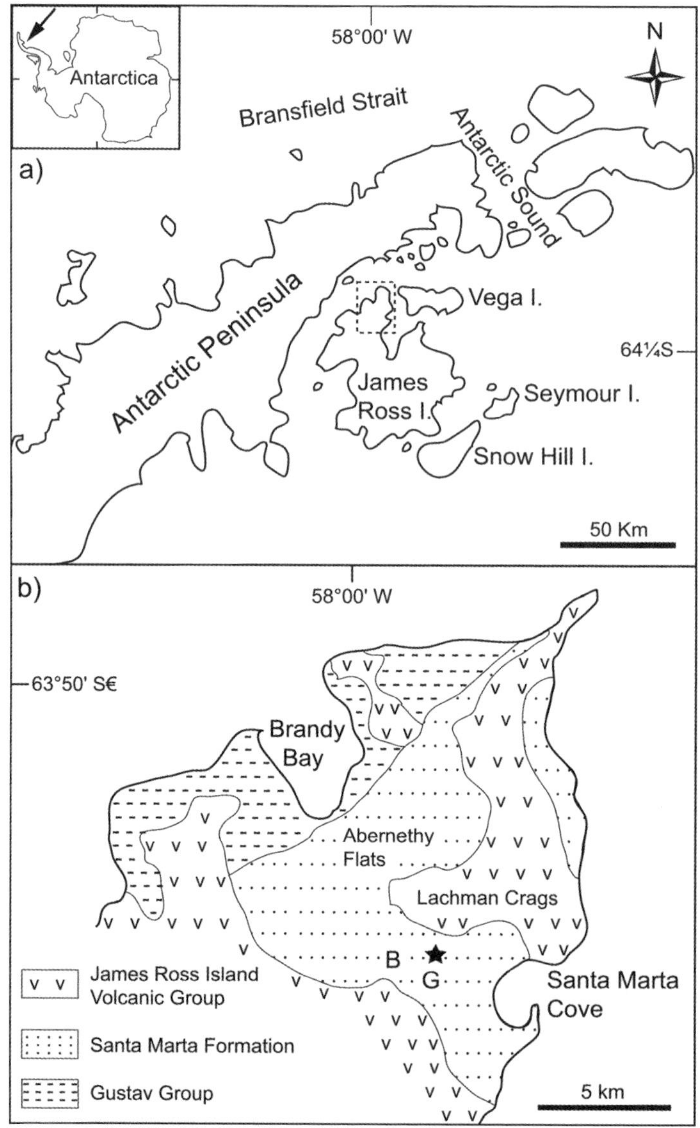

Fig. 1. Sketch maps showing the location of the Antarctic Peninsula (inset: arrow) and study area. (**a**) Location of James Ross Island east of the Antarctic Peninsula. The rectangle indicates the study area. (**b**) Collecting sites in the Santa Marta Formation on northern James Ross Island. B specifies the Argentinean collecting site of Lachman Crags specimens and G the one of Herbert Sound specimens. The asterisk shows the collecting site of the British Antarctic Survey expedition.

Hill, Seymour and Cockburn islands in the south (Olivero *et al.* 1986, 1992; Lirio *et al.* 1989; Pirrie *et al.* 1991, 1997). Both areas are separated by a major fault causing extensive repetition of the stratigraphical sequences. Four formations are included in the Marambio Group (in stratigraphical order): Santa Marta, Snow Hill Island, López de Bertodano and Sobral formations (Pirrie *et al.* 1997). The Santa Marta Formation that crops out on northern James Ross Island between Brandy Bay and Santa Marta Cove (Fig. 1b) represents volcaniclastic, shallow-marine fan and shelf sediments that were deposited adjacent to an active volcanic arc (Fig. 1a). This is the type area for the Santa Marta Formation, as defined by Olivero *et al.* (1986), and the base of the Santa Marta Formation is late Coniacian in age (McArthur *et al.* 2000).

The lower part (*c.* 500 m), which consists of mudstones, siltstones and sandstones together with rare conglomerates, was named the Alpha Member by Olivero *et al.* (1986). The following approximately 350 m-thick Beta Member is characterized by an increased proportion of conglomeratic interbeds; these are up to 4 m thick and poorly sorted. Because of their very similar lithostratigraphic appearance, Crame *et al.* (1991) united the Alpha and Beta members within the Lachman Crags Member. The topmost 250 m-thick Gamma Member of Olivero *et al.* (1986), comprising mainly fine-grained cross-bedded sandstones and shell coquinas, was named the Herbert Sound Member by Crame *et al.* (1991). The lower part of the Lachman Crags Member (Alpha member) is late Coniacian–early Campanian in age, and the upper part (Beta Member) is probably early–late Campanian. A late Campanian–?early Maastrichtian age is assigned to the Herbert Sound (Gamma) Member, based on ammonites, other fossils and strontium isotope dating (Crame *et al.* 1991, 1999; Olivero *et al.* 1992; Olivero & Medina 2000).

Lirio *et al.* (1989) established the 435 m-thick Rabot Formation, which is characterized by less prominent coarse-grained beds and by the absence of coquinas and carbonaceous plant material in the Rabot Point–Hamilton Point area of SE James Ross Island. However, Crame *et al.* (1991) assigned member status to this unit and Pirrie *et al.* (1997) transferred part of it to the Hamilton Point Member. Precise correlation of these members to those of the northern part of the island is difficult, but it would appear that the Rabot Member is the lateral equivalent of both the upper Lachman Crags and lower Herbert Sound members, and probably ranges in age from the early to the late Campanian. The Hamilton Point Member is ?mid–late Campanian in age (Pirrie *et al.* 1997).

Material and methods

The material presented here was collected from siltstones and conglomerates in the uppermost part (Beta Member) of the Lachman Crags Member and from conglomerates in the middle part of the Herbert Sound Member in the northern part of James Ross Island by Argentinean and British field parties (Fig. 1b). The Herbert Sound Member conglomerate that yielded most of the material has an erosive base and increases in thickness southwards up to about 60 cm. It is mainly composed of reworked concretionary sandstones, mudstones, intraclasts and bivalve shells, and is interpreted as a submarine reworked horizon comprising amalgamated beds (Scasso *et al.* 1991). Further published records from the Santa Marta and López de Bertodano formations were scrutinized and used to reconstruct the diversity changes of Antarctic fish in the Late Cretaceous.

As the exact stratigraphic age of the sampled horizons is still unresolved, the faunas were divided into those coming from the (Beta Member) upper part of the Lachman Crags Member and those from the Herbert Sound Member. It is assumed that all this material is Campanian in age. The terminology for sharks used herein follows that of Cappetta (1987); the systematic scheme for sharks is based on Carvalho (1996) and that for teleosts follows Nelson (1994).

Abbreviations: BMNH, Natural History Museum, London. The prefix BAS indicates material collected by the British Antarctic Survey. IAA-IRJ2000-1–IAA-IRJ2001-27 – material housed in the collection of the Instituto Antartico Argentino.

Previous research

The majority of previous studies on Late Cretaceous Antarctic fish concentrated on records from the López de Bertodano Formation of Seymour Island (e.g. Woodward 1908; Grande & Eastman 1986; Grande & Chatterjee 1987). The only account of Late Cretaceous fish from James Ross Island is that of Richter & Ward (1990), who described material from the 'Beta' (Lachman Crags) and 'Gamma' (Herbert Sound) members of the Santa Marta Formation, in the northern part of James Ross Island. The fish association of the Lachman Crags Member is extremely low in taxonomic diversity. Richter & Ward (1990) reported the presence of hexanchiforms (*Chlamydoselachus thomsoni*, *Notidanodon dentatus*) and synechodontiforms (*Sphenodus* sp.), based on three isolated teeth and additional unidentified selachian vertebrae derived from conglomerates and coquinas. The Herbert Sound Member yielded a more diverse fauna of selachians, including *Notidanodon pectinatus* (= *N. dentatus*), *Sphenodus* sp., *Squatina* sp., and undetermined lamniforms, as well as actinopterygians such as *Enchodus* sp. and ?*Sphaeronodus* sp.

Fish from the Santa Marta Formation

Holocephalians

Stahl & Chatterjee (1999, 2002) described the callorhynchid *Ischyodus dolloi* and the chimaerid *Chimaera zangerli* from the Maastrichtian López de Bertodano Formation of

Table. 1. *Occurrence of Late Cretaceous Antarctic neoselachians and teleosts. Asterisks indicate first records. Numbers in brackets denote the number of specimens used in this study*

Santa Marta Formation		López de Bertodano Formation
Lachman Crags Member	Herbert Sound Member	Undifferentiated
Neoselachii		
Chlamydoselachus thomsoni (1)	*Chimaera zangerli** (12)	*Chimaera zangerli* (4)
Notidanodon dentatus (3)	*Chlamydoselachus thomsoni**	*Ischyodus dolloi* (1)
Lamniformes indet. (3)	*Notidanodon dentatus* (8)	*Notidanodon dentatus* (3)
Sphenodus sp. (1)	*Squatina* sp. (11)	Lamniformes indet. (1)
Paraorthacodus sp.* (1)	*Scapanorhynchus* sp.*	*Sphenodus* sp. (1)
	Lamniformes indet. (75)	
	Sphenodus sp. (4)	
	Paraorthacodus sp.* (2)	
Teleostei		
Albuliformes indet.* (2)	Albuliformes indet.* (1)	*Antarctiberyx seymouri* (1)
Ichthyodectiformes indet.* (5)	Ichthyodectiformes* (21)	*Enchodus cf. ferox** (1)
Teleostei indet. (<10)	*Enchodus* sp. (3)	Teleostei indet. (>10)
	Apateodus? sp.* (1)	
	Teleostei indet. (>10)	

Seymour Island. Additional specimens assignable to *Chimaera zangerli* are now recorded in the Herbert Sound Member of James Ross Island. The material consists of two tooth plates that are still embedded in sandy matrix and several tritor fragments. The material will form the focus of a forthcoming taxonomic publication.

Neoselachians

Selachian remains are the main component of the marine vertebrate fauna, and are known from several sites in the Lachman Crags and Herbert Sound members. So far, only neoselachian remains have been discovered. Remains of hybodontoids and rays have not yet been reported from Antarctica. This is almost certainly due to the mode of collection; all material was surface collected and there has been no systematic chemical treatment or screen washing. Sharks include at least 11 species of Hexanchiformes, Lamniformes, Squatiniformes and Synechodontiformes (Table 1).

Hexanchiformes. Hexanchiform sharks are represented by only two species, *Chlamydoselachus thomsoni* Richter & Ward, 1990 (Chlamydoselachoidei) and *Notidanodon dentatus* (Agassiz, 1843) (Hexanchoidei) (see Agassiz, 1833–1844). *Chlamydoselachus thomsoni* is restricted to Antarctica and was based on a single tooth from the 'Beta Member' (= Lachman Crags Member) (Richter & Ward 1990). An additional, smaller specimen (IAA-IRJ2000-1) was recovered from the Herbert Sound Member (Fig. 2a). It is still embedded in an indurate sandy concretion and only the lingual side was prepared due to the very fragile condition of the specimen. The root is slightly shorter than in the specimen described by Richter & Ward (1990); 3.2 mm wide compared to 4.9 mm. The lingual root surface is slightly damaged, and the crown lacks the tips of the mesial and central cusps; all cusps are massive and the apex of the distal cusp is slightly twisted (Fig. 2a). No additional intercalated cusplets between cusps, as in the type-species, are

Fig. 2. Neoselachians from the Herbert Sound Member. (**a**) Tooth of *Chlamydoselachus thomsoni* (IAA-IRJ2000-1) in occlusal view embedded in sandy matrix. The scale bar is 0.5 cm. (**b**) Isolated tooth crown tentatively assigned to *Chlamydoselachus thomsoni* (BAS DJ.172.28) in lateral view. The scale bar is 0.25 cm. (**c**) Split tooth of *Notidanodon dentatus* (IAA-IRJ2000-2) in lateral view exposing modified anaulacorhize root vascularisation pattern. The scale bar is 0.5 cm. (d)–(h) Teeth of *Scapanorhynchus* sp. Scale bars are 1.0 cm. (**d**) Specimen 1 (IAA-IRJ2000-9), labial view. (**e**) Specimen 1 (IAA-IRJ2000–9), labial view. (**f**) Specimen 2 (IAA-IRJ2000-10), labial view. (**g**) Specimen 2 (IAA-IRJ2000-10), mesial view. (**h**) Specimen 2 (IAA-IRJ2000-10), lingual view. (i)–(j) Fragmentary tooth of cf. *Scapanorhynchus* sp. (BAS DJ.172.11). The scale bars are 0.5 cm. (**i**) Labial view. (**j**) Lingual view. (**k**) Tooth of Lamniformes indet. (IAA-IRJ2000-12) in matrix. The scale bar is 0.5 cm. (**l**) Isolated tooth crown of Lamniformes indet. (IAA-IRJ2000-13). The scale bar is 0.5 cm. (**m**) Isolated tooth crown of Lamniformes indet. (IAA-IRJ2000-14). The scale bar is 0.5 cm. (**n**) *Squatina* sp., lingual view. (BAS DJ.172.39). The scale bar is 0.5 cm.

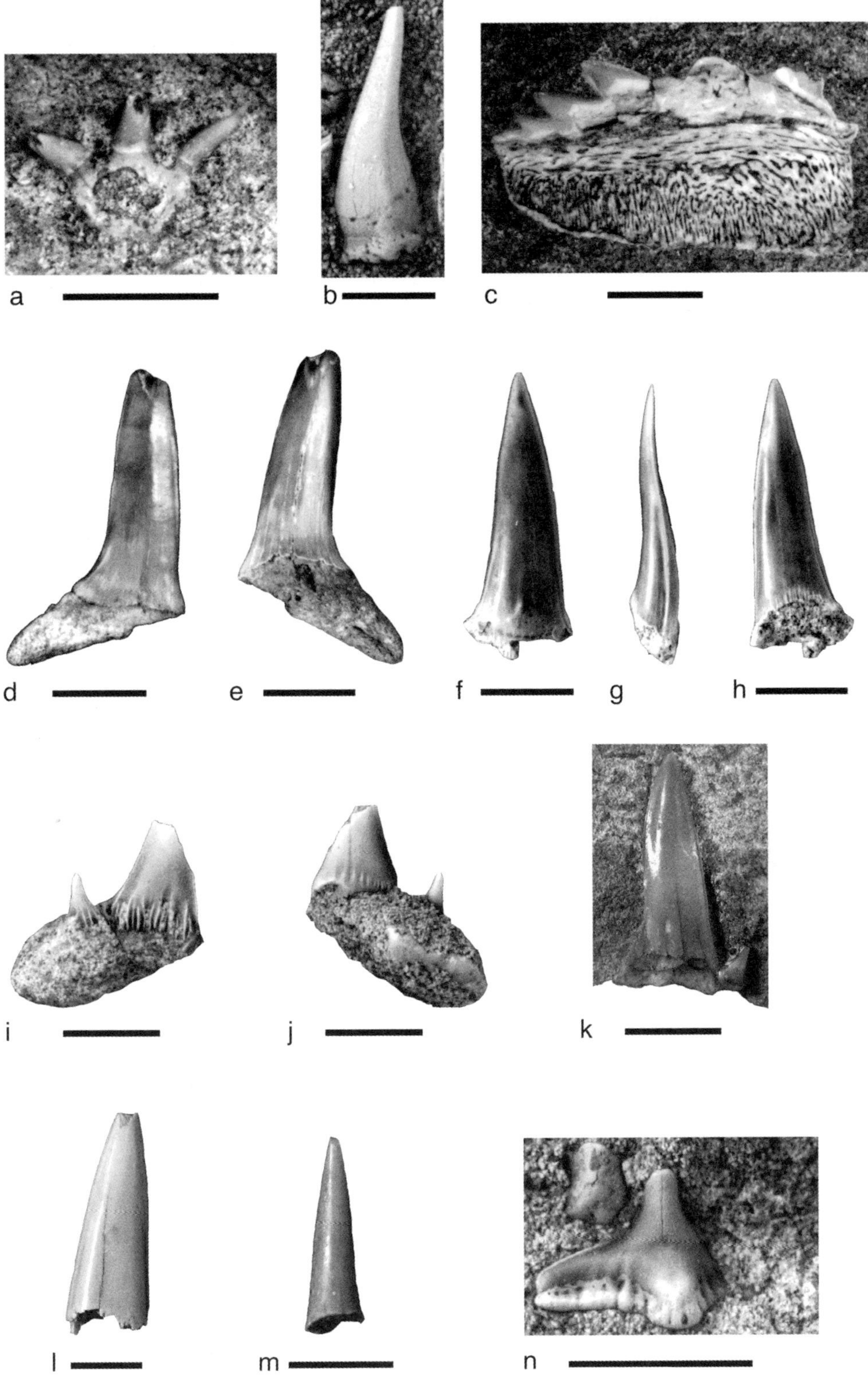
a
b
c
d
e
f
g
h
i
j
k
l
m
n

preserved. The ornamentation is less developed, consisting of fewer and shorter vertical folds than in the Lachman Crags Member specimen and is restricted to basal parts of the cusps. The root is labio-lingually shorter and is less bifid; the crown–root junction is not constricted.

A very fragmentary tooth (BAS DJ.172.28), from the Herbert Sound Member with the central cusp and root remains preserved, displays the general morphology of *Chlamydoselachus* teeth (Fig. 2b). However, this single specimen is about three times larger than the other Herbert Sound Member specimen (IAA-IRJ2000-1), and about twice as large as the Lachman Crags Member specimen described by Richter & Ward (1990).

The differences in the ornamentation and size of the specimens might be related to different positions within the jaws rather than representing different species. Although the attribution of isolated teeth to a definite jaw position is very difficult, general trends might apply to fossil sharks (Welton 1979). The tooth figured by Richter & Ward (1990, fig. 5) was interpreted as coming from an anterior file based on file position reconstructions by Welton (1979) for the extant *C. anguineus*. The morphology of the root, with a smaller median cusp angle in specimen IAA-IRJ2000-1, indicates a lateral to latero-posterior jaw position.

Despite the size of teeth, hexanchoids are relatively uncommon in the Lachman Crags and Herbert Sound members. A single species, *Notidanodon dentatus*, occurs in the Lachman Crags and Herbert Sound members. (NB. The validity of this species is dubious; for a detailed discussion see Cione 1996.) Here, we follow Cione (1996) in regarding *N. pectinatus* as a *nomen dubium* and *N. dentatus* as valid. The presence of *N. dentatus* in the Lachman Crags and Herbert Sound members was indicated by Richter & Ward (1990, fig. 6d) on the basis of a single tooth from each unit (Table 1). Additional new material includes two fragments from the Lachman Crags Member and seven mostly fragmentary teeth from the Herbert Sound Member (IAA-IRJ2000-2–IAA-IRJ2000-8) (Fig. 2c). *N. dentatus* also occurs in the López de Bertodano Formation (Maastrichtian) of Seymour Island (Cione & Medina 1987; Grande & Chatterjee 1987 (as *N. antarcticus*)).

Lamniformes. Lamniform sharks are rare in the Lachman Crags Member (Table 1). Conversely, in the Herbert Sound Member, lamniform teeth are the major component of the selachian association, although most specimens are too fragmentary to be identified below ordinal level. Richter & Ward (1990, fig. 6j, k) figured two tooth crowns that display vertical, slightly flexuous folds on the lingual side. The third specimen assigned to lamniforms by these authors reveals a morphology similar to that of the new material of *Sphenodus* (see below).

At least three additional lamniform teeth (IAA-IRJ2000-9 to IAA-IRJ2000-10, BAS DJ.172.11) of the taxon figured by Richter & Ward (1990, fig. 6j–l) have been recovered (Fig. 2d–k). All these specimens are fragmentary. Specimen IAA-IRJ2000-9 has an awl-shaped, pointed and sigmoidal curved central cusp with a smooth labial face (Fig. 2d), but with numerous weak and parallel vertical folds on the lingual face extending from the crown base to the middle of the crown (Fig. 2e). The labial crown face is flat, lacking a basal labial ledge (Fig. 2d), whereas the lingual one is convex with a faint basal bulge but without a basal crown band (Fig. 2e). The preserved root lobe is slender with pointed terminations and diverges from the crown (Fig. 2d). Lateral cusplets are not present, but this might be due to secondary loss since the tooth is abraded. The central cusp of specimen IAA-IRJ2000-10 is also sigmoidal in profile view (Fig. 2g) and the ornamentation consists of very short lingual folds that are restricted to the crown base (Fig. 2f, h). Specimen BAS DJ.172.11 from the Herbert Sound Member is heavily broken; however, it displays a mesio-distally expanded but short and rounded root lobe (Fig. 2i, j), and a very acute and delicate lateral cusplet that is well separated from the main cusp (Fig. 2i). No basal labial ledge uniting the lateral cusplet with the base of the main cusp is developed (Fig. 2i). Basally, there are numerous short vertical ridges on the labial face (Fig. 2i); the lingual folds are also very short and feeble (Fig. 2j).

The three teeth are more or less damaged and it is not always possible to assign specimens to a specific lamniform; teeth of Cretaceous odontaspidids and mitsukurinids can easily be confused when incomplete. Most specimens share character combinations that occur in several species of *Carcharias* and *Scapanorhynchus*, e.g. cutting edges complete or almost reaching the crown base, acute lateral cusplets and conspicuous ornamentation pattern (specimen IAA-IRJ2000-11). However, the awl-shaped main cusps, absence of a labial basal edge that unites the main cusp and lateral cusplets, and the morphology of the root (pointed and slender root lobes in anterior (Fig. 2d–h) and more spatulate lobes with rounded extremities (Fig. 2i–k) in lateral teeth) are characteristic for teeth of *Scapanorhynchus* rather than odontaspidids. As far as can be ascertained, there is no distinct nutritive groove separating the two root lobes (Fig. 2e) and the protuberance is not shelf-like as it is in most *Carcharias* species. However, the nutritive groove is rather deep and distinct in several specimens of *Scapanorhynchus* and the generally massive lingual protuberance might also be shelf-like in some species (cf. Case & Cappetta 1997). The ornamentation, consisting of rather short and more flexuous lingual folds, is more typical for teeth of odontaspidids. Specimen BAS DJ.172.11 (Fig. 2i) resembles teeth of *Carcharias* in the labial ornamentation and the presence of a labial basal central ridge (Siverson 1996). Labial folds are present in *S.* aff. *praeraphiodon* from the late Cenomanian of Texas (Cappetta & Case 1999) and in *Scapanorhynchus* sp. (= *S. minimus*) from the Cenomanian of France (Landemain 1991). In addition, teeth of *Carcharias* generally have a basal labial ledge that unites the lateral cusplets with the main cusp, which is absent in teeth of *Scapanorhynchus*. Nevertheless, better material is needed to establish the exact systematic position of the Antarctic lamniforms.

Other fragmentary teeth of lamniforms (IAA-IRJ2000-12–IAA-IRJ2000-14) are rather abundant in the Herbert Sound Member (Fig. 2k–m), but are too incomplete for any specific identification. Specimen IAA-IRJ2000-13 (Fig. 2l) is very similar to specimen IAA-IRJ2000-10 (Fig. 2h) and may also belong to *Scapanorhynchus*.

Squatiniformes. Richter & Ward (1990, fig. 6a–c) indicated the presence of *Squatina* in the Herbert Sound Member, with a species similar to *S. hassei* from the Campanian and Maastrichtian of Europe. This identification was based on a single vertebra and two fragmentary teeth. Here we record seven more specimens from the Argentinean collection (IAA-IRJ2000-10–IAA-IRJ2000-16) and two additional specimens from the BAS collections in the Natural History Museum, London. Specimen BAS DJ.172.39 is embedded in matrix and the lingual side is exposed (Fig. 2n); this tooth displays an erect, central cusp and an oblique lateral blade with continuous cutting edge.

The fossil record of *Squatina* extends back to the Middle Jurassic, and the morphology of the tooth crown and root is supposed to be the unifying character of all extant and fossil species. However, a *Squatina*-like morphology is also found in teeth of several extant and fossil orectolobiforms such as *Orectolobus*, *Cretorectolobus* and *Cretascyllium* (Case 1978; Müller & Diedrich 1991; Herman *et al.* 1992; Siverson 1997). In fact, teeth of extant *Orectolobus*, *Eucrossorhinus* and *Squatina* species share some important morphological features such as a root base having a large outer depression and a large inner central protuberance, tooth crown with mesial and distal heels, and a small but well-developed apron and large but narrow uvula (Herman *et al.* 1992). The main feature to distinguish *Squatina* teeth from similar orectolobiform teeth is the root-supported apron. In addition, the lateral blades have denticles or are enlarged at their distal ends in most orectoloboids (D. Long pers. comm. 2004). A single specimen from the Herbert Sound Member (IAA-IRJ2000-10) displays a root-supported apron, and the lateral blades are low, oblique and without denticles or distal enlargements in most Antarctic specimens (e.g. BAS DJ.172.39) (Fig. 2n). Consequently, the Herbert Sound specimen is assigned to *Squatina* without specific identification here.

Synechodontiformes. Antarctic synechodontiform sharks include at least two species. Richter & Ward (1990, fig. 6h, i) indicated the presence of *Sphenodus* sp. (Orthacodontidae) in the Lachman Crags Member on the basis of an isolated and large tooth root, and in the Herbert Sound Member on the basis of an isolated crown with tooth remnants. They stated that these remains belong to a new species resembling *Sphenodus lundgreni* from the Danian of Scandinavia. However, the material was too fragmentary to define this species.

Two additional specimens recovered by the Argentinean expedition (IAA-IRJ2000-17 and IAA-IRJ2000-18) (Fig. 3a–e) and a fragmentary specimen collected by a British Antarctic Survey expedition (BMNH/BAS uncatalogued (uncat.)) provide further morphological information, although both specimens are damaged. The slightly lingually bent tooth crowns (Fig. 3b) are rather broad, almost triangular in labial view, and elongated with pointed apices (Fig. 3a). The cutting edges are well developed and prominent, reaching the base of the crown (Fig. 3b). The labial face is convex in the middle, especially in its central part near the base, with several very short and weak folds along the crown base (Fig. 3a), but does not jut out over the root (Fig. 3b). The lingual face is very convex with numerous stronger vertical folds at the crown base (Fig. 3c). The root is rather narrow and slightly mesio-distally expanded with a horizontal root base and numerous randomly arranged foramina

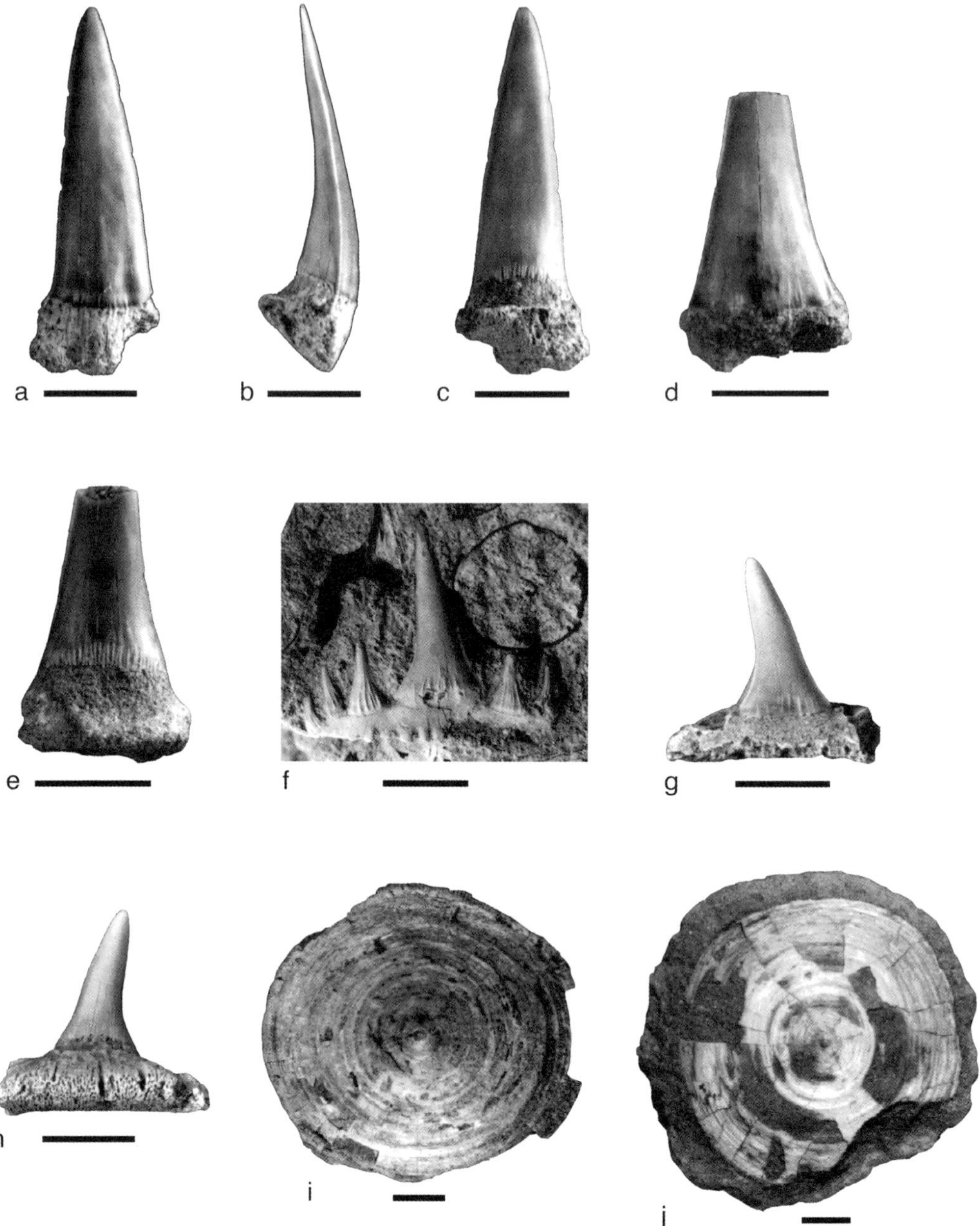

Fig. 3. Neoselachians from the Lachman Crags and Herbert Sound members. (a)–(e) *Sphenodus* sp., Herbert Sound Member. The scale bars are 1.0 cm. (**a**) Specimen 1 (IAA-IRJ2000-17), labial view. (**b**) Specimen 1 (IAA-IRJ2000-17), mesial view. (**c**) Specimen 1 (IAA-IRJ2000-17), lingual view. (**d**) Specimen 2 (IAA-IRJ2000-18), labial view. (**e**) Specimen 2 (IAA-IRJ2000-18), lingual view. (**f**) *Paraorthacodus* sp. (IAA-IRJ2000-19), Lachman Crags Member, labial view. The scale bar is 0.5 cm. (g)–(h) *Paraorthacodus* sp. (BAS DJ.136.2), Herbert Sound Member, lacking lateral cusplets. The scale bars are 0.5 cm. (**g**) Labial view. (**h**) Lingual view. (**i**) Lamniformes indet. (IAA-IRJ2000-20), isolated vertebra, Herbert Sound Member. The scale bar is 1.0 cm. (**j**) Lamniformes indet. (IAA-IRJ2000-21), isolated vertebra, Herbert Sound Member. The scale bar is 1.0 cm.

piercing the surface (Fig. 3a, b, e). The fragmentary crown figured by Richter & Ward (1990, fig. 6l) as Lamniformes indet. displays a very similar morphology and ornamentation to the Argentinean specimens, and is consequently referred to *Sphenodus*.

The fossil record of *Sphenodus* ranges from the Early Jurassic to the Danian (Beaumont 1960). However, the validity of most if not all Late Cretaceous species (e.g. Pictet & Campiche 1858; Priem 1912) is dubious because the material consists mainly of isolated tooth crowns. The Antarctic specimens resemble those from the late Campanian of Angola, especially in the ornamentation (Antunes & Cappetta 2002). Teeth of *S. lundgreni* from the Danian of Scandinavia differ in the higher degree of labial wrinkles that reach far up the crown, and in the more pronounced basal folds.

The specimen identified as *Isurus* sp. from the Maastrichtian López de Bertodano Formation by Grande & Eastman (1986, fig. 3H, I) in fact represents another specimen of this *Sphenodus* species and extends its range from the Campanian to the Maastrichtian in Antarctica. The specimen identified as *Sphenodus*? sp. from the same formation by Grande & Chatterjee (1987, fig. 2F, G) certainly does not belong to a neoselachian shark (see below).

The second synechodontiform species represents the first record of *Paraorthacodus* (Palaeospinacidae) in Antarctica. Two specimens were recovered by the Argentinean (IAA-IRJ2000–19) and BAS expeditions (BAS DJ.136.2), respectively (Fig. 3f–h). Specimen IAA-IRJ2000-19 comes from the Lachman Crags Member and is still embedded in silty matrix that also contains abundant plant remains and a few bivalves (Fig. 3f). The other specimen (BAS DJ.136.2) was recovered from the Herbert Sound Member. The central cusp and the root are well preserved, but the specimen lacks the lateral cusplets (Fig. 3g, h). The teeth are up to 12 mm wide and 11 mm high. The main cusp in both specimens is slender in its upper part with the cutting edges being almost parallel; basally the cusp widens rapidly in its basal third (Fig. 3f–h). The labial face of the central cusp is flat, does not overhang the root, and has folds that differ in length but are restricted to the base (Fig. 3f, g). Conversely, the lingual face is strongly convex so that the margins of the basal part are visible in labial view (Fig. 3g), and has more and tenuous folds ascending halfway up from the base of the cusp (Fig. 3h). There are three pairs of distal and two pairs of mesial cusplets with numerous folds that also ascend half way up in specimen IAA-IRJ2000–19 (Fig. 3f). The root is rather low and displays the typical vascularization pattern of palaeospinacids (Fig. 3f, g). Comparison of the Antarctic specimens with contemporaneous taxa is difficult because of the still insufficient knowledge of Late Cretaceous *Paraorthacodus* species. Six species are regarded as being valid (Siverson 1992): *P. andersoni* (Case 1978) from the ?Cenomanian–Campanian of the USA, France and Sweden; *P. conicus* (Davis 1890) from the Coniacian–Campanian of Kazakhstan, Germany, Belgium and Sweden; *P. patagonicus* (Ameghino 1893) from the Coniacian of Argentina; *P. recurvus* (Trautschold 1877) from the Albian–Cenomanian of Lithuania and Russia; *P. sulcatus* (Davis 1888) from the ?Campanian of New Zealand; and *P. validus* (Chapman 1918) from the Late Cretaceous of New Zealand. The following character combination distinguishes the Antarctic specimens from all other *Paraorthacodus* species: cusp and lateral cusplets labio-lingually compressed; labial crown face very flat; very high, acute and slender main cusp and lateral cusplets; comparably fine and short labial folds; and the basal part of the lingual face of the cusp visible in labial view. This character combination indicates that the Antarctic *Paraorthacodus* represents a hitherto unknown species.

Other remains. Isolated vertebral centra of neoselachians occur sporadically in different horizons of both the Lachman Crags and Herbert Sound members. For example, Richter & Ward (1990, fig. 6a) figured a centrum of typical squatinid appearance from the Herbert Sound Member (cf. Hasse 1882). Some vertebral centra or imprints of centra studied here are of tectospondylic type and can be assigned to squatinids. A few other, more or less circular, vertebral centra and imprints of asterospondylic-type with regular concentric laminae (e.g. IAA-IRJ-2000-20 and IAA-IRJ2000-21) are characteristic of orectolobiforms and lamniforms (Fig. 3i, j). Because no orectolobiform remains have been recovered so far, and because of the size of the centra, these remains are assigned to lamniforms without further identification.

Actinopterygii

The teleostean fauna of the Santa Marta Formation is rather low in diversity compared to contemporaneous fish faunas, and comprises mostly disarticulated fragments. The Lachman Crags Member yielded only isolated scales, and the Herbert Sound Member assemblage is represented by isolated teeth, vertebrae, skull and caudal fin elements.

Elopomorpha. Two single scales from the Lachman Crags Member are assigned to albuliforms, which closely resemble scales of *Osmeroides* (Fig. 4a, b). The better-preserved specimen (IAA-IRJ2000-22) is roughly subrectangular in outline, being only slightly higher than long. There are some deep, slightly convergent anterior radii directed toward the centre of the scale but not actually reaching it (Fig. 4a). The posterior field is marked by faint, subparallel, short grooves obscured by very small granulations. The focus is more or less in the middle of the scale. The posterior margin of the scale is slightly triangular, almost straight (Fig. 4a). Fine circuli occupy the lateral and posterior areas. The second specimen (IAA-IRJ2000-23), although lacking the posterior part, is almost identical to the former (Fig. 4b). The morphology of these scales is very similar to that of *Osmeroides lewesiensis* (BMNH P.10220 and BMNH P.49893) from the Turonian of England, and *Osmeroides* sp. from the late Cenomanian of Germany (BMNH P.306). They also slightly resemble scales assigned to *Osmeroides* from the Turonian of Canada (Fielitz 1996; Wilson & Chalifa 1989). Scales of extant elopomorphs, such as *Megalops atlanticus*, display the same morphology, although they generally have more anterior radii (Roberts 1993, fig. 4B).

Identification of isolated fossil scales is difficult due to poor preservation and insufficient knowledge of scale morphologies in fossil taxa. However, based on comparisons with numerous articulated fish skeletons from the Cretaceous of Europe and North and South America, these Antarctic scales are allocated here to Albuliformes. This is the first record of albuliforms in Antarctica, although positive identification must await the acquisition of better-preserved material. A single, relatively large vertebral centrum from the Herbert Sound Member (IAA-IRJ2000-24) is rostro-caudally compressed and ventrally flattened (Fig. 4c, d). It resembles those described and figured for *Osmeroides* by Loomis (1900).

Ichthyodectiformes. Ichthyodectiform teleosts are represented by numerous scales in the Lachman Crags Member (Fig. 4e–g). The scales are rounded–subtriangular in outline and are wider than long. All have numerous (more than 20) rather deep anterior radii that are directed towards the centre of the scales (Fig. 4f, g). The focus is suboval–oval and is located in the centre of the scales (Fig. 4e). Fine circuli are closely arranged following the outer edge of the scale (Fig. 4e). In addition, punctae decorate the inner parts in a triangular area and randomly around the focus area of the scales (Fig. 4e, f). Bigger scales (e.g. IAA-IRJ2000-25) are more rounded, with the anterior radii being shorter compared to those in the smaller scales and restricted to a small area (Fig. 4e). A few damaged scales from the Herbert Sound Member display the same morphology and are also assigned to ichthyodectiforms (e.g. IAA-IRJ2000-28, Fig. 4h).

The presence of numerous anterior radii, the outer form of the scales and the punctuated inner parts are characteristic of ichthyodectiform scales. Most of these scales resemble closely those of *Ichthyodectes* and *Gillicus* spp. from the Campanian of North America, and might thus be referred to one of these genera. The scales of *Cladocyclus* from the Early Cretaceous of Brazil differ in the general appearance and shorter anterior radii.

Fig. 4. Teleostean remains from the Lachman Crags and Herbert Sound members. (**a**) Isolated scale of Albuliformes indet. (IAA-IRJ2000-22), Lachman Crags Member. The scale bars are 0.5 cm. (**b**) Isolated scale of Albuliformes indet. (IAA-IRJ2000-22, IAA-IRJ2000-23), Lachman Crags Member. The scale bars are 0.5 cm. (c) & (d) Isolated vertebra (IAA-IRJ2000-24), Herbert Sound Member, assigned to Albuliformes. The scale bars are 0.5 cm. (**c**) Anterior view. (**d**) Lateral view. (**e**) Isolated scale of Ichthyodectiformes indet. (IAA-IRJ2000-25), Lachman Crags Member. The scale bar is 0.5 cm. (**f**) Isolated scale of Ichthyodectiformes indet. (IAA-IRJ2000-26), Lachman Crags Member. The scale bar is 0.5 cm. (**g**) Isolated scale of Ichthyodectiformes indet. (IAA-IRJ2000-27), Lachman Crags Member. The scale bar is 0.5 cm. (**h**) Associated scales of Ichthyodectiformes indet. (IAA-IRJ2000-28), Herbert Sound Member. The scale bar is 0.5 cm. (**i**) Isolated tooth of Ichthyodectiformes (IAA-IRJ2000-29), Herbert Sound Member. The scale bar is 0.5 cm. (**j**) Isolated tooth of Ichthyodectiformes (IAA-IRJ2000-30), Herbert Sound Member. The scale bar is 0.25 cm. (**k**) Isolated caudal hypural bone of Ichthyodectiformes indet. (IAA-IRJ2000-31), lateral view, Herbert Sound Member. The scale bar is 0.5 cm. (**l**) Isolated tooth of *Enchodus* sp. (IAA-IRJ2000-32), Herbert Sound Member. The scale bar is 0.5 cm. (**m**) Isolated tooth of *Enchodus* sp. (IAA-IRJ2000-33), Herbert Sound Member. The scale bar is 0.25 cm. (**n**) Isolated scale, Lachman Crags Member tentatively assigned to Ichthyodectiformes indet. (IAA-IRJ2000-34). The scale bar is 0.5 cm. (**o**) Isolated scale of uncertain affinities (IAA-IRJ2000-35), Lachman Crags Member. The scale bar is 0.5 cm. (**p**) Isolated scale of uncertain affinities (?Ichthyodectiformes) (IAA-IRJ2000-36), Lachman Crags Member. The scale bar is 0.5 cm.

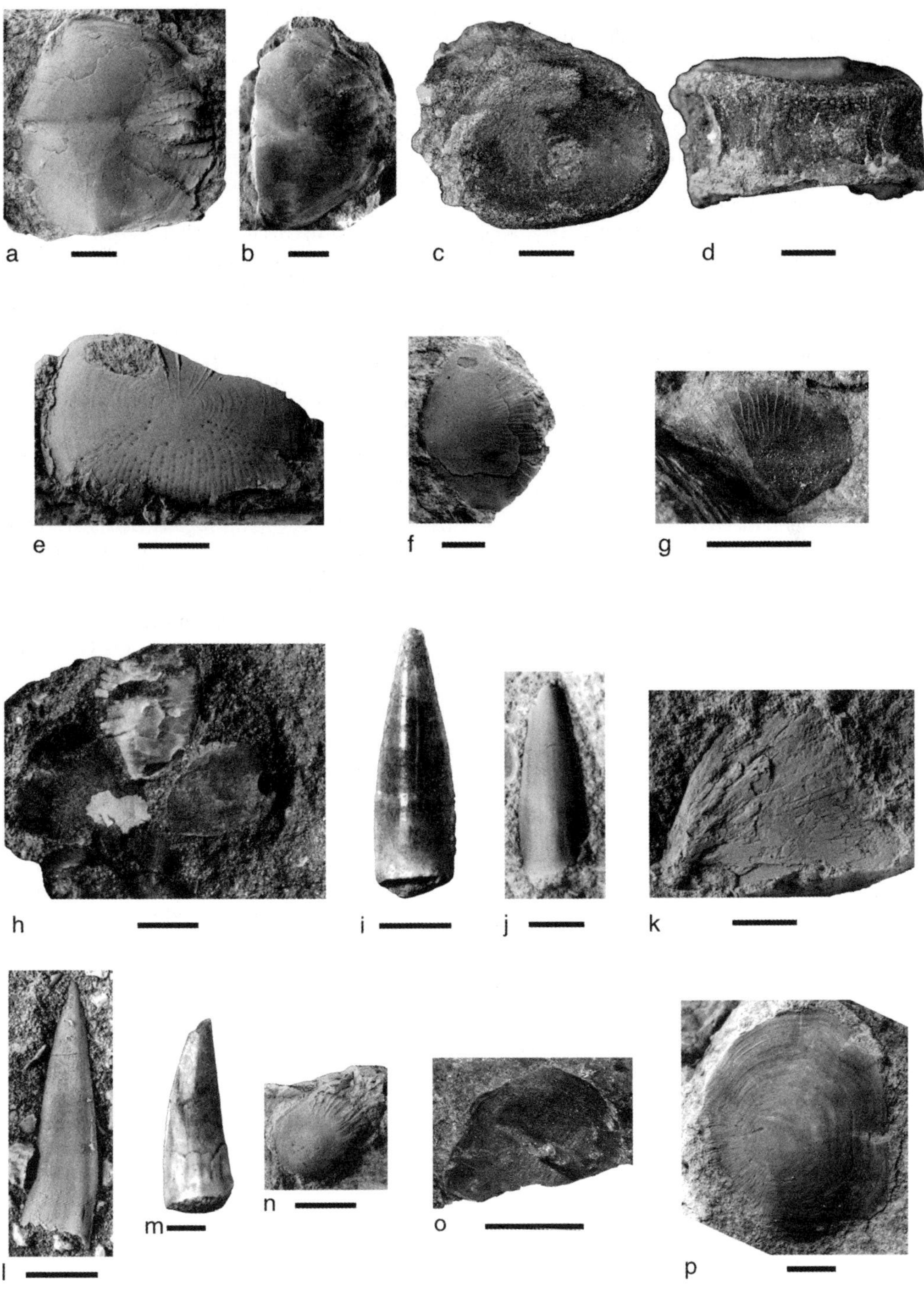
a
b
c
d
e
f
g
h
i
j
k
l
m
n
o
p

Isolated ichthyodectiform teeth are quite common in the Herbert Sound Member. Most teeth are tall, rather slender, laterally compressed and sometimes slightly posteriorly curved but never sigmoidal in lateral view (e.g. IAA-IRJ2000-29 and IAA-IRJ2000-30, Fig. 4i, j). The apices are pointed, the cutting edges distinct and continuous from the tip to the base (Fig. 4i, j). The lateral sides are more or less convex especially in the middle part of the crowns; the base is subcircular–oval. Most teeth are completely smooth; only a few display faint vertical striations along the lateral edge (Fig. 4j), a character that also might be found in larger teeth of some enchodontids. In their general appearance these teeth resemble those of *Enchodus* spp. However, these differ in the sigmoidal curvature of the crown in lateral view, more needle-like appearance, absence of posterior continuous cutting edges, post-apical barb and/or distinct elliptic cross-sections. Some of these characters might be present in ichthyodectiform teeth, but never all together. The most reliable character to distinguish large teeth of enchodontoids and ichthyodectiforms seems to be the sigmoidal curve of the crown in the former group. Because of the apparent differences, these teeth are referred to ichthyodectiforms. However, it is possible that comparable enchodontid teeth have been included with those of ichthyodectiforms. The tooth figured by Richter & Ward (1990, fig. 6e) and referred to as ?*Enchodus* sp. more closely resembles those of ichthyodectiforms. Teeth of *Gillicus* are very similar and most if not all teeth of this type from Antarctica might belong to the same taxon as the scales described above. Whereas teeth of the ichthyodectiform *Cladocyclus* from Brazil differ in having a folded base, those of North American *Xiphactinus* species are more massive, stouter and exhibit striations on the lateral surfaces (cf. Schwimmer *et al.* 1997).

An isolated hypural bone (IAA-IRJ2000-31) that is very high and slightly inclined ventrally is tentatively also assigned to ichthyodectiforms (Fig. 4k). Hypural bones of enchodontids are sometimes also enlarged, but still narrower than the Antarctic specimen. In addition, the enlarged hypurals of enchodontids often bear basal openings that are absent in this specimen.

Alepisauriformes. Alepisauriform remains are extremely rare. Richter & Ward (1990, fig. 6g) assigned an isolated and damaged tooth to *Enchodus* that displays the typical morphology of enchodontid palatine teeth. Two isolated teeth (IAA-IRJ2000-32 and IAA-IRJ2000-33) that are referred here to *Enchodus* were recovered from the Herbert Sound Member (Fig. 4l, m). Both specimens display the typical sigmoidal curvature in lateral view.

A single specimen was erroneously identified as ?*Sphaeronodus* by Richter & Ward (1990, fig. 6f). This specimen resembles teeth of the alepisauriform *Apateodus*. An isolated, rather small scale from the Lachman Crags Member (IAA-IRJ2000-34) displays the general morphology found in several extant alepisauriforms (Fig. 4n). It is roughly rectangular in outline and crenate with a rounded focus and closely arranged, fine circuli.

Teleostei indet. A number of isolated teeth and scales from both members cannot be assigned to any teleostean group because they are too fragmentary or to unspecific (e.g. IAA-IRJ2000-35 and IAA-IRJ2000-36, Fig. 4o, p). A moderately large scale (IAA-IRJ2000-36) with rounded outline and almost straight anterior margin from the Lachman Crags Member is characterized by very fine circuli and short, narrow, posterior radii (Fig. 4p). A very similar scale was figured by Wilson & Chalifa (1989, fig. 7M) from the Turonian of Canada without any taxonomic allocation. This scale resembles those of *Cladocyclus mawsoni* from the Late Cretaceous of Brazil (e.g. BMNH P.3872a and BMNH P.9615) and thus might also represent an ichthyodectiform. Additional isolated bones and small vertebral centra indicate the likely presence of further unidentified bony fishes.

Maastrichtian teleosts

Maastrichtian Antarctic fish are exclusively known from the López de Bertodano Formation of Seymour Island (Table 1). Grande & Chatterjee (1987) described *Antarctiberyx seymouri* and indicated the presence of *Orthacodus* (= *Sphenodus*) sp. in the same formation (Grande & Chatterjee 1987, fig. 2F, G). However, the single specimen of *Orthacodus* differs significantly from teeth of *Sphenodus* (see above). It displays some striations on the lateral sides and a posterior groove that almost reaches the apex. In this respect, the tooth resembles closely those of *Enchodus ferox* from the Campanian and Maastrichtian of the Western Interior Seaway of North America and the Maastrichtian of Morocco. Additional teleostean material includes about 100 isolated bones and vertebral centra, indicating that a probably diverse teleostean fauna waits to be described.

Palaeobiogeographical and palaeoecological implications

The James Ross Basin was a back-arc basin with rather unstable margins; sediment was intermittently transported into it from an active magmatic arc to the NW (Crame *et al.* 1991). After a period of deep-marine sedimentation (Gustav Group) sediments of the Marambio Group were deposited at shelf depth, with the Lachman Crags Member representing a mid–outer shelf setting and the overlying Herbert Sound Member an inner shelf environment with significant shallowing of the basin (Pirrie 1990; Crame *et al.* 1991; McArthur *et al.* 2000).

Oxygen isotope analyses of planktonic foraminifera from the Falkland Plateau (high-latitude South Atlantic) suggest that warm surface waters may have endured in this region from the Turonian through to the early Campanian, before the beginning of a long-term cooling in the late Campanian through to the end of the Maastrichtian (Huber *et al.* 1995). Direct surface and midwater access for South Atlantic equatorial waters into the Weddell Basin was initiated in the Turonian, shortly after the transpolar shallow seaway linking the southern Weddell Sea to the Tasman and eastern Australian margins had been closed (Dingle 1999; Dingle & Lavelle 2000). The mid-Campanian witnessed an enlargement of the trans-equatorial Tethyan seaway, which was at its largest in the mid-Campanian–Maastrichtian. This may well have been a time of enhanced faunal exchanges between the James Ross Basin and lower latitudes, with large predatory fish having cosmopolitan or widespread distributions entering Antarctic waters.

Hexanchiformes

The hexanchoid *Notidanodon* displays a bipolar distribution with most records being from the Northern Hemisphere Late Cretaceous and Palaeogene (Cappetta 1987). The species *N. dentatus* is a typical Southern Hemisphere hexanchoid that is known from Angola (Antunes & Cappetta 2002), New Zealand (Davis 1890) and Antarctica. The stratigraphically youngest records come from the Palaeocene of Seymour Island and possibly the Danian of New Zealand. According to Cione (1996), *N. dentatus* preferred warm temperate to temperate waters.

The feeding behaviour of fossil hexanchoids is still not fully understood. Extant hexanchoids feed on a wide range of marine organisms, including other sharks, rays, chimaeras, bony fish, squids, crabs, shrimps, carrion and even seals (e.g. Bigelow & Schroeder 1984). Fossil hexanchoid teeth are also often associated with plesiosaur or cetacean bones (e.g. Welles 1943) and it is usually assumed that hexanchoids were scavenging on carcasses. The co-occurrence of *N. dentatus* teeth and plesiosaur remains in the López de Bertodano Formation (Maastrichtian) of Seymour Island (Cione & Medina 1987) and in the Herbert Sound Member (Richter & Ward 1990; this study) supports this interpretation.

The Cretaceous and Cenozoic distribution of *Chlamydoselachus* is rather patchy and thus similar to the distribution of the single extant species. Cretaceous species are relatively rare and most records come from the Campanian. Antunes & Cappetta (2002) described *C. goliath* from the late Campanian, and *C. gracilis* and *Chlamydoselachus* sp. from the late Campanian–early Maastrichtian of Angola, respectively. In addition, *Chlamydoselachus* occurs in the Santonian and Maastrichtian of Japan (M. Goto pers. comm. 2001), indicating a distribution of *Chlamydoselachus* in both hemispheres early in its evolutionary history. It displays its greatest diversity in the Cenozoic with all species being restricted to the Northern Hemisphere (Pfeil 1983). The extant *C. anguineus* is bathydemersal, ranging from surface waters to more than 1200 m depth on outer continental and insular shelves, and upper slopes. *C. thomsoni* represents an endemic faunal element in the Santa Marta Formation, and its occurrence in mid–outer (Lachman Crags Member) and inner shelf settings (Herbert Sound Member) is in accordance with the bathymetric distribution of other *Chlamydoselachus* species.

The dentition of fossil and extant species is of clutching type with generally very slender, needle-like cusps, and similar prey is assumed for fossil and extant species (e.g. cephalopods, fish). However, *C. thomsoni* is characterized by more robust cusps than other Campanian–Maastrichtian species; this supports the interpretation by Richter & Ward (1990) of prey with harder skeletal structures such as belemnites and ammonites. Other prey includes heavier bodied fish, such as a variety of bony fish, and also sharks (e.g. Cox & Francis 1997).

Lamniformes

Lamniforms include large predaceous sharks of demersal and mesopelagic forms occurring from surface waters to the deep sea (e.g. Last & Stevens 1994). Cretaceous lamniforms are among the largest known selachians (Siverson 1999). The only undoubtedly identifiable lamniform teeth from the Late Cretaceous of Antarctica belong to the exclusively Late Cretaceous mitsukurinid *Scapanorhynchus*. At least seven species of *Scapanorhynchus* have been described that are widely distributed in the Northern Hemisphere, and range from the Albian to the Maastrichtian. The oldest record (without specific identification) comes from the Aptian of Japan (Goto *et al.* 1993). Southern Hemisphere occurrences are extremely rare: they include the Turonian and Campanian–Maastrichtian of Angola and Maastrichtian of Brazil (Cappetta 1987; Antunes & Cappetta 2002). *Scapanorhynchus* is an epibenthic–eurybathic predator with a cosmopolitan distribution and it is interpreted here as a casual inhabitant of Antarctic waters, probably related to seasonal feeding migrations. The dental morphology implies small fish and soft-bodied invertebrates as prey. The Herbert Sound Member records expand the range of *Scapanorhynchus* to Antarctica.

Squatiniformes

Modern and fossil species of *Squatina* are widely distributed from cold northern boreal waters to the tropics, and occur at intertidal–upper continental slope depths (e.g. Compagno 1984). They are bottom dwellers that ambush their prey from buried positions in mud or sand. Modern angel sharks display rather pronounced patterns of endemism. The wide distribution of several fossil species might be explained by the rather conservative dental morphology that renders the identification of species difficult. Cretaceous Southern Hemisphere occurrences of *Squatina* are extremely rare (e.g. Maastrichtian of Chile; M. Suarez pers. com. 2001). Many *Squatina*-like specimens might belong either to *Cretorectolobus* or other orectolobiforms (e.g. Siverson 1997).

Synechodontiformes

Cretaceous occurrences of *Sphenodus* (Orthacodontidae) are extremely rare compared to its Jurassic distribution. Only two species from the Early Cretaceous (*S. salandianus* and *S. subaudianus*) and two from the Late Cretaceous (*S. planus* (Cenomanian) and *S. sennessi* (Santonian)) have been described so far (Duffin & Ward 1993). Additional unidentified specimens were reported from the Campanian of Angola (Antunes & Cappetta 2002). The Antarctic and Angola specimens represent a third, still unnamed, Late Cretaceous species. Apart from the Antarctic and Angola specimens, which are the youngest Late Cretaceous records, all other material is from the Northern Hemisphere. However, *Sphenodus* had already displayed a Northern and Southern Hemisphere distribution in the Late Jurassic (Arratia *et al.* 2002). It represents a typical pelagic predator that might have followed its prey during seasonal feeding migrations, in a way similar to that of the lamniforms reported from Antarctica.

The distribution of palaeospinacids, especially that of *Paraorthacodus*, is very similar to that of the hexanchoid *Notidanodon*. Although the species diversity of *Paraorthacodus* is rather low in the Cretaceous, it displays a wide, bipolar distribution in warm–cool temperate areas. The occurrence of *Paraorthacodus* from the Late Cretaceous of Antarctica supports this interpretation. Species of *Paraorthacodus* are interpreted as small, slow-swimming predators in shallow-marine environments with a diet similar to extant orectolobiforms of fish and soft-bodied invertebrates (Thies & Reif 1985).

Elopomorpha

Elopomorph teleosts are a highly diversified group with a rich fossil record. Southern Hemisphere records are, for example, from the Early Cretaceous of Brazil (e.g. Maisey 1991), and from the Late Cretaceous of Brazil (*Brannerion*) and Australia (*Istieus*) (Nelson 1994). The Antarctic specimens assigned to albuliforms expand the range of this group into Antarctic waters during the Late Cretaceous.

Ichthyodectiformes

The fossil history of ichthyodectiforms ranges back to the Middle Jurassic and, as early as the Late Jurassic, ichthyodectiforms are known from Antarctica (Arratia *et al.* 2004). In the Cretaceous ichthyodectiforms are widespread, but are most abundant in the Late Cretaceous of the Northern Hemisphere and include some very large taxa (e.g. *Xiphactinus*). Southern Hemisphere occurrences are mainly of Early Cretaceous age, and include *Cooyoo* from Australia and *Cladocyclus* from freshwater and marine deposits of Brazil (Lees & Bartholomai 1987; Maisey 1991). The large ichthyodectiform *Xiphactinus* has also been reported from the Late Cretaceous of Australia (Bardack 1965).

The rare Cretaceous Southern Hemisphere records of ichthyodectiforms might be related to collecting and/or identification biases.

Alepisauriformes

Alepisauriforms display rather wide distribution patterns in the Late Cretaceous (Kriwet 2003*a*). For example, the enchodontid *Enchodus* is reported from 18 Late Cretaceous localities (Chalifa 1996). The only other positive Southern Hemisphere occurrence of *Enchodus* comes from the Late Cretaceous (?Campanian) of Brazil. *Enchodus* is interpreted as an open-water pelagic predator that also ventured into near-shore areas from the open ocean (Goody 1976). *Apateodus* is very rare and most records are confined to the Late Cretaceous of Europe and North Africa. Alepisauriforms preyed on fish and probably soft-bodied cephalopods, but were in turn preyed on by marine tetrapods, such as plesiosaurs (Cicimurri & Everhart 2001).

Diversity and abundance

Unfortunately, patterns of diversity and abundance in the Early Cretaceous fish of Antarctica are still poorly known (Kriwet 2003*b*). The Herbert Sound Member yielded the highest number of taxa in this study with at least seven chondrichthyans (eight if lamniforms are considered) and four teleosts, compared to four (five if lamniforms are considered) chondrichthyans and two teleosts from the Lachman Crags Member, and four (five if lamniforms are considered) chondrichthyans and at least two teleosts from the López de Bertodano Formation (Table 1).

Most common are unidentified lamniform remains including isolated and fragmentary teeth and vertebrae. Teeth and bones assigned to ichthyodectiforms are the second most common group of fish remains. *Enchodus*, which is quite common in the Late Cretaceous of the Northern Hemisphere, is here reported from both the Santa Marta and López de Bertodano formations: remains such as tooth plates and isolated dental tritors of *Chimaera zangerli* are also rather abundant in the Herbert Sound Member and López de Bertodano Formation.

Reviewing all available information it is apparent that the collections of Late Cretaceous Antarctic fish are still extremely incomplete (if compared to contemporaneous collections form the Western Interior of North America, NW Europe, etc.). This is especially so for both the latest Cretaceous and the Palaeogene, and further intensive taxonomic investigations of these intervals are required.

Conclusions

- The main localities that have yielded Late Cretaceous Antarctic fish to date are situated in the northern part of James Ross Island and on Seymour Island. Three faunas in stratigraphic order can be distinguished that differ slightly in taxonomic composition. The fish fauna from the late Campanian part of the Lachman Crags Member comprises eight taxa, five of which are sharks and three are teleosts. The slightly younger Herbert Sound association comprises at least 11 taxa plus some still unidentified lamniforms. The lamniform *Scapanorhynchus*, the synechodontiform *Paraorthacodus*, as well as an elopomorph, ichthyodectiforms and an alepisauriform close to *Apateodus*, are reported from the Campanian of Antarctica for the first time.
- The stratigraphic ranges of *Chlamydoselachus thomsoni*, *Paraorthacodus* sp. and ichthyodectiforms include both members of the Santa Marta Formation. The stratigraphic range of the chimeroid *Chimaera zangerli*, which was previously only known from the Maastrichtian López de Bertodano Formation, is extended back into the Herbert Sound Member. *Enchodus* cf. *ferox*, previously identified as *Sphenodus* sp., is reported from the López de Bertodano Formation for the first time.
- Medium- to large-sized top predators that occupied the higher levels in the food chain, along with marine tetrapods, dominate the Late Cretaceous Antarctic marine vertebrate faunas. Fish occupying lower levels in the trophic chain are extremely rare in the fossil record. The only fish adapted to crushing hard-shelled prey was a chimeroid. Rays, which are quite common in other Late Cretaceous fish associations (e.g., Maastrichtian of Morocco), are still not known from the Late Cretaceous of Antarctica.
- Many fish above species level, and especially the teleosts, belong to cosmopolitan groups or had wide geographical distributions during the Late Cretaceous. The migration of Northern Hemisphere fish was facilitated and supported by wide open-marine seaways between the Weddell Basin and the Tethyan realm.
- Late Cretaceous Antarctic fish diversity seems highest in the Campanian, with a decrease in the Maastrichtian, and might be

assumed to be related to long-term Late Cretaceous climatic trends. However, the collections are too imperfect for reconstructing any diversity patterns at the moment.

- The Lachman Crags and Herbert Sound members are characterized by several species endemic to Antarctica (e.g. *Chlamydoselachus thomsoni* and *Chimaera zangerli*). In addition, *Notidanodon dentatus*, both synechodontiforms, and probably *Squatina* sp. may be species typical of the high latitudes. Consequently, the Late Cretaceous Antarctic fish fauna is assumed to consist of two groups. One is indicative of the Weddellian Province (Antarctica–Patagonia–New Zealand), and is characterized by high-latitude species, while the second (including probably most of the lamniforms, ichthyodectiforms and alepisauriforms) is widespread to cosmopolitan.

Fieldwork of J. M. Lirio and H. J. Nuñez in Antarctica was supported by the Instituto Antártico Argentino. J.A. Crame (Cambridge, UK) is thanked for providing valuable data and for sharing his knowledge about Antarctic Late Cretaceous stratigraphy with us. He also improved the language and reviewed the manuscript. We thank J. E. Martin (Rapid City, USA) and D.J. Long (San Francisco, USA) for their critical reviews and helpful suggestions that improved the manuscript considerably. This research has been supported by a Marie Curie Fellowship of the European Community programme 'Improving Human Research Potential and the Socio-economic Knowledge Base' under contract number HPMF-CT-2001-01310 to J. Kriwet. P. Forey and A. Longbottom (BMNH, London) are thanked for the possibility to study material in the collections under their care. We acknowledge S. Powell (Bristol, UK) and the Photographic Unit of the Natural History Museum for preparing photographs for the figures.

References

AGASSIZ, L. 1833–1844. *Recherches sur les Poissons Fossiles,* 5 volumes. Petitpierre, Neuchâtel et Soleure (with supplements).

AMEGHINO, F. 1893. Sobre la presencia de vertebrados de aspecto Mesozóico, en la formación Santacruceña de la Patagonia austral. *Revista del Jardín Zoológico de Buenos Aires*, **1**, 76–84.

ANTUNES, M.T. & CAPPETTA, H. 2002. Sélaciens du Crétacé (Albien–Maastrichtien) d'Angola. *Palaeontographica A*, **264**, 85–146.

ARRATIA, G., KRIWET, J. & HEINRICH, W.-D. 2002. Selachians and actinopterygians from the Upper Jurassic of Tendaguru, Tanzania. *Mitteilungen aus dem Museum für Naturkunde zu Berlin, Geowissenschaftliche Reihe*, **5**, 207–230.

ARRATIA, G., SCASSO, R. & KIESSLING, W. 2004. Late Jurassic fishes from Longing Gap, Antarctic Peninsula. *Journal of Vertebrate Paleontology*, **24**, 41–55.

BARDACK, D. 1965. Anatomy and evolution of chirocentrid fishes. *University of Kansas Paleontological Contributions*, **10**, 1–88.

BEAUMONT, G., DE. 1960. Contribution à l'étude des genres *Orthacodus* Woodw. et *Notidanus* Cuv. (Selachii). *Mémoires Suisses de Paléontologie*, **77**, 1–46.

BIGELOW, H. & SCHROEDER, W. 1948. *Fishes of the Western Atlantic*, Volume 1. Memoirs of the Sears Foundation for Marine Research, Yale, 59–546.

CAPPETTA, H. 1987. Mesozoic and Cenozoic Elasmobranchii, Chondrichthyes II. *In:* SCHULTZE, H.-P. (ed.) *Handbook of Paleoichthyology 3B.* Gustav Fischer, Stuttgart.

CAPPETTA, H. & CASE, G.R. 1999. Additions aux faunes de sélaciens du Crétacé du Texas (Albien supérieur–Campanian). *Palaeo Ichthyologica*, **9**, 5–111.

CARVALHO, M., DE 1996. Higher-level elasmobranch phylogeny, basal squaleans, and paraphyly. *In:* STIASSNY, M.L., PARENTI, L.R. & JOHNSON, G.D. (eds) *Interrelationships of Fishes.* Academic Press, San Diego, CA, 35–62.

CASE, G.R. 1978. A new selachian fauna from the Judith River Formation (Campanian) of Montana. *Palaeontographica A*, **160**, 176–205.

CASE, G.R. & CAPPETTA, H. 1997. A new selachian fauna from the Late Maastrichtian of Texas (Upper Cretaceus/Navarroan); Kemp Formation. *Münchner Geowissenschaftliche Abhandlungen A*, **34**, 131–189.

CHALIFA, Y. 1996. A new species of *Enchodus* (Aulopiformes: Enchodontidae) from the Northern Negev, Israel with comments on evolutionary trends in the Enchodontoidei. *In:* ARRATIA, G. & VIOHL, G. (eds) *Mesozoic Fishes – Systematics and Paleoecology.* Dr Friedrich Pfeil, Munich, 349–367.

CHAPMAN, F. 1918. Description and revision of the Cretaceous and Tertiary fish remains of New Zealand. *Palaeontological Bulletin*, **7**, 1–46.

CICIMURRI, D.J. & EVERHART, M.J. 2001. An elasmosaur with stomach contents and gastroliths from the Pierre Shale (late Cretaceous) of Kansas. *Transactions of the Kansas Academy of Science*, **104**, 129–143.

CIONE, A.L. 1996. The extinct genus *Notidanodon* (Neoselachii, Hexanchiformes). *In:* ARRATIA, G. & VIOHL, G. (eds) *Mesozoic Fishes – Systematics and Paleoecology.* Dr Friedrich Pfeil, Munich, 63–72.

CIONE, A.L. & MEDINA, F.A. 1987. A record of *Notidanodon pectinatus* (Chondrichthyes, Hexanchiformes) in the Upper Cretaceous of the Antarctic Peninsula. *Mesozoic Research,* **1**, 79–88.

COMPAGNO, L.J.V. 1984. *FAO Species Catalogue. Volume 4. Sharks of the World. An Annotated and Illustrated Catalogue of Shark Species Known to Date. Part 1. Hexanchiformes to Lamniformes.* FAO Fisheries Synopses, **125**, 1–249.

COX, G. & FRANCIS, M. 1997. *Sharks and Rays of New Zealand.* Canterbury University Press, Canterbury.

CRAME, J.A., LOMAS, S.A., PIRRIE, D. & LUTHER, A.

1996. Late Cretaceous extinction patterns in Antarctica. *Journal of the Geological Society, London*, **153**, 503–506.

CRAME, J.A., MCARTHUR, J.M., PIRRIE, D. & RIDING, J.B. 1999. Strontium isotope correlation of the basal Maastrichtian stage in Antarctica to the European and US biostratigraphic schemes. *Journal of the Geological Society, London*, **156**, 957–964.

CRAME, J.A., PIRRIE, D., RIDING, J.B. & THOMSON, M.R.A. 1991. Campanian–Maastrichtian (Cretaceous) stratigraphy of the James Ross Island area, Antarctica. *Journal of the Geological Society, London*, **148**, 1125–1140.

DAVIS, J.W. 1888. On the fossil fish-remains from the Tertiary and Cretaceous–Tertiary formations of New Zealand. *Scientific Transactions of the Royal Dublin Society*, **4**, 1–62.

DAVIS, J.W. 1890. On the fossil fish of the Cretaceous formations of Scandinavia. *Scientific Transactions of the Royal Dublin Society*, **4**, 363–434.

DINGLE, R.V. 1999. Walvis Ridge barrier: its influence on palaeoenvironments and source rock generation deduced from ostracod distribution in the early South Atlantic Ocean. *In:* CAMERON, N., BATE, R.H. & CLURE, R. (eds) *Hydrocarbon Habitats of the South Atlantic.* Geological Society, London, Special Publications, **153**, 293–302.

DINGLE, R.V. & LAVELLE, M. 2000. Antarctic Peninsula Late Cretaceous–Early Cenozoic palaeoenvironments and Gondwana palaeogeographies. *Journal of African Earth Sciences*, **31**, 91–105.

DITCHFIELD, P.W., MARSHALL, J.D. & PIRRIE, D. 1994. High latitude palaeotemperature variations: New data from the Tithonian to Eocene of James Ross Island, Antarctica. *Palaeogeography, Palaeoclimatology, Palaeoecology*, **107**, 79–101.

DUFFIN, C.J. & WARD, D.J. 1993. The Early Jurassic palaeospinacid sharks of Lyme Regis, southern England. *Belgian Geological Survey, Professional Papers*, **264**, 53–102.

FIELITZ, C. 1996. A Late Cretaceous (Turonian) ichthyofauna from Lac de Bois, Northwest Territories, Canada, with palaeobiogeographic comparisons with Turonian ichthyofaunas of the Western Interior Seaway. *Canadian Journal of Earth Sciences*, **33**, 1375–1389.

GOODY, P.C. 1976. *Enchodus* (Teleostei: Enchodontidae) from the Upper Cretaceous Pierre Shale of Wyoming and South Dakota with an evaluation of the North American enchodontid species. *Palaeontographica A*, **152**, 91–112.

GOTO, M., UYENO, T. & YABUMOTO, Y. 1993. Summary of Mesozoic elasmobranch remains from Japan. *In*: ARRATIA, G. & VIOHL, G. (eds) *Mesozoic Fishes – Systematics and Paleoecology.* Dr Friedrich Pfeil, Munich, 71–85.

GRANDE, L. & EASTMAN, J.T. 1986. A review of Antarctic ichthyofaunas in the light of new fossil discoveries. *Palaeontology*, **29**, 113–137.

GRANDE, L. & CHATTERJEE, S. 1987. New Cretaceous fish fossils from Seymour Island, Antarctic Peninsula. *Palaeontology*, **30**, 829–837.

HASSE, C. 1882. *Das natürliche System der Elasmobranchier auf Grundlage des Baues und der Entwicklung ihrer Wirbelsäule. Eine morphologische und paläontologische Studie. Besonderer Theil.* von Gustav Fischer, Jena, 97–179.

HATHWAY, B. 2000. Continental rift to back-arc basin: Jurassic–Cretaceous stratigraphical and structural evolution of the Larsen Basin, Antarctic Peninsula. *Journal of the Geological Society, London*, **157**, 417–432.

HERMAN, J., HOVESTADT-EULER, M. & HOVESTADT, D.C. 1992. Part A: Selachii. No. 4: Order Orectolobiformes – Families: Brachaeluridae, Ginglymostomidae, Hemiscylliidae, Orectolobidae, Parascylliidae, Rhiniodontidae, Stegostomidae. Order Pristiphoriformes – Family: Pristiophoridae. Order Squatiniformes – Family: Squatinidae. *Bulletin de l'Institut Royal des Sciences naturelles de Belgique*, **65**, 193–254.

HUBER, B.T. 1998. Tropical paradise at the Cretaceous poles? *Science*, **282**, 2199–2200.

HUBER, B.T., HODELL, D.A. & HAMILTON, C.P. 1995. Middle-Late Cretaceous climate of the southern high latitudes: stable isotope evidence for minimal equator-to-pole thermal gradients. *Bulletin of the Geological Society of America*, **107**, 1164–1191.

KRIWET, J. 2003*a*. Lancetfish teeth (Neoteleostei, Alepisauroidei) from the Early Cretaceous of Alcaine, NE Spain. *Lethaia*, **12**, 323–332.

KRIWET, J. 2003*b*. First record of Early Cretaceous shark (Chondrichthyes, Neoselachii) from Antarctica. *Antarctic Science*, **15**, 519–523.

LANDEMAIN, O. 1991. Sélaciens nouveaux du Crétacé supérieur du sud-ouest de la France. Quelques apports à la systématique des elasmobranches. *Saga Information, Société Amicale des Géologues Amateurs*, **1**, 1–45.

LAST, P.R. & STEVENS, J.D. 1994. *Sharks and Rays of Australia.* CSIRO, Australia.

LEES, T. & BARTHOLOMAI, A. 1987. Study of Lower Cretaceous actinopterygian (class Pisces) *Cooyoo australis* from Queensland, Australia. *Memoirs of the Queensland Museum*, **25**, 177–192.

LIRIO, J.M., MARENSSI, S.A., SANTILLANA, S.N., MARSHALL, P.A. & RINALDI, C.A. 1989. Marambio Group at the southeastern part of James Ross Island, Antarctica. *Contribución del Instituto Antártico Argentino*, **371**, 1–45.

LOOMIS, F.B. 1900. Die Anatomie und die Verwandtschaft der Ganoid- und Knochenfische aus der Kreide Formation von Kansas. *Palaeontographica*, **46**, 213–283.

MAISEY, J.G. 1991. *Santana Fossils: An Illustrated Atlas.* Tropical Fish Hobbyist, T.F.H. Publications, NJ.

MCARTHUR, J.M., CRAME, J.A. & THIRLWALL, M.F. 2000. Definition of Late Cretaceous stage boundaries in Antarctica using Strontium Isotope Stratigraphy. *Journal of Geology*, **108**, 623–640.

MÜLLER, A. & DIEDRICH, C. 1991. Selachier (Pisces, Chondrichthyes) aus dem Cenomanium von Ascheloh am Teutoburger Wald (Nordrhein-Westfalen, NW-Deutschland). *Geologie und Paläontologie in Westfalen*, **20**, 1–105.

NELSON, J.S. 1994. *Fishes of the World.* Wiley, New York.

OLIVERO, E.B. & MEDINA, F.A. 2000. Patterns of Late Cretaceous ammonite biogeography in southern high latitudes: the family Kossmaticeratidae in Antarctica. *Cretaceous Research*, **21**, 269–279.

OLIVERO, E.B., MARTINIONI, D.R. & MUSSEL, F.J. 1992. Upper Cretaceous sedimentology and biostratigraphy of Western Cape Lamb (Vega Island, Antarctica). Implications on sedimentary cycles and evolution of the basin. *In:* RINALDI, C.A. (ed.) *Geología de la Isla James Ross*. Instituto Antártico Argentino, Buenos Aires, 147–166.

OLIVERO, E.B., SCASSO, R.A. & RINALDI, C.A. 1986. Revision of the Marambio Group, James Ross Island, Antarctica. *Contribución del Instituto Antártico Argentino*, **331**, 1–28.

PFEIL, F.H. 1983. Zahnmorphologische Untersuchungen an rezenten und fossilen Haien der Ordnungen Chlamydoselachiformes und Echinorhiniformes. *Palaeo Ichthyologica*, **1**, 1–315.

PICTET, F.J. & CAMPICHE, G. 1858. Description des fossiles du terrain Crétacé des environs de Sante-Croix. *Matériaux pour la Paléontologie Suisse*, **2**, 1–380.

PIRRIE, D. 1990. A new sedimentological interpretation of the Santa Marta Formation, James Ross Island. *Antarctic Science*, **2**, 77–78.

PIRRIE, D., CRAME, J.A., LOMAS, S.A. & RIDING, J.B. 1997. Late Cretaceous stratigraphy of the Admiralty Sound Region, James Ross Basin, Antarctica. *Cretaceous Research*, **17**, 109–137.

PIRRIE, D., CRAME, J.A. & RIDING, J.B. 1991. Late Cretaceous stratigraphy and sedimentology of Cape Lamb, Vega Island, Antarctica. *Cretaceous Research*, **12**, 227–258.

PRIEM, F. 1912. Sur les poissons fossiles des terrains Tertiaires du sud de la France. *Bulletin de la Société Géologique de France, 4 série*, **12**, 213–245.

RICHTER, M.A. & WARD, D.J. 1990. Fish remains from the Santa Marta Formation (Late Cretaceous) of James Ross Island, Antarctica. *Antarctic Science*, **2**, 67–76.

RIDING, J.B. & CRAME, J.A. 2002. Aptian to Coniacian (Early–Late Cretaceous) palynostratigraphy of the Gustav Group, James Ross Basin, Antarctica. *Cretaceous Research*, **23**, 739–760.

ROBERTS, C.D. 1993. Comparative morphology of spined scales and their phylogenetic significance in the Teleostei. *Bulletin of Marine Science*, **52**, 60–113.

SCASSO, R.A., OLIVERO, E.B. & BUATOIS, L. 1991. Lithofacies, biofacies and ichnoassemblage evolution of a shallow submarine volcanoclastic fan-shelf depositional system (Upper Cretaceous – James Ross Island, Antarctica. *Journal of South American Earth Sciences*, **4**, 239–260.

SCHWIMMER, D.R., STEWART, J.D. & WILLIAMS, G.D. 1997. *Xiphactinus vetus* and the distribution of *Xiphactinus* species in the eastern United States. *Journal of Vertebrate Paleontology*, **66**, 994–1001.

SIVERSON, M. 1992. Late Cretaceous Paraorthacodus (Palaeospinacidae, Neoselachii) from Sweden. *Journal of Paleontology*, **35**, 519–554.

SIVERSON, M. 1996. Lamniform sharks from the mid-Cretaceous Alinga Formation and Beedagong Claystone, Western Australia. *Palaeontology*, **39**, 813–849.

SIVERSON, M. 1997. Sharks from the Mid-Cretaceous Gearle Siltstone, southern Carnarvon Basin, Western Australia. *Journal of Vertebrate Paleontology*, **17**, 543–465.

SIVERSON, M. 1999. A new large lamniform shark from the uppermost Gearle Siltstone (Cenomanian, Late Cretaceous) of Western Australia. *Transactions of the Royal Society of Edinburgh: Earth Sciences*, **90**, 49–66.

STAHL, B.J. & CHATTERJEE, S. 1999. A Late Cretaceous chimaerid (Chondrichthyes, Holocephali) from Seymour Island, Antarctica. *Palaeontology*, **42**, 979–989.

STAHL, B.J. & CHATTERJEE, S. 2002. A Late Cretaceous callorhynchid (Chondrichthyes, Holocephali) from Seymour Island, Antarctica. *Journal of Vertebrate Paleontology*, **22**, 848–850.

THIES, D. & REIF, W.-E. 1985. Phylogeny and evolutionary ecology of Mesozoic Neoselachii. *Neues Jahrbuch für Geologie und Paläontologie Abhandlungen*, **169**, 333–361.

TRAUTSCHOLD, H. 1877. Über Kreidefossilien Russlands. *Bulletin de la Société de Moscou*, **11**, 332–349.

WELLES, S.P. 1943. Elasmosaurid plesiosaurs with description of new material from California and Colorado. *Memoirs of the University of California*, **13**, 125–254.

WELTON, B.J. 1979. *Late Cretaceous and Cenozoic Squalomorphii of the Northwest Pacific region.* PhD thesis, University of California, Berkeley, CA.

WILSON, M.V.H. & CHALIFA, Y. 1989. Fossil marine actinopterygian fishes from the Kaskapau Formation (Upper Cretaceous: Turonian) near Watino, Alberta. *Canadian Journal of Earth Sciences*, **26**, 2604–2620.

WOODWARD, A.S. 1908. On fossil fish-remains from Snow Hill and Seymour Islands. *Wissenschaftliche Ergebnisse der Schwedischen Südpolar-Expedition 1901–1903*, **3**, 1–4.

ZINSMEISTER, W.J. 1998. Discovery of fish mortality horizon at the K–T boundary on Seymour Island: re-evaluation of events at the end of the Cretaceous. *Journal of Paleontology*, **72**, 556–571.

ZINSMEISTER, W.J., FELDMANN, R.M., WOODBURNE, M.O. & ELLIOT, D.H. 1989. Late Cretaceous/ Earliest Tertiary transition on Seymour Island, Antarctica. *Journal of Paleontology*, **63**, 731–738.

Biostratigraphy of the Mosasauridae (Reptilia) from the Cretaceous of Antarctica

JAMES E. MARTIN

Museum of Geology, SD School of Mines and Technology, Rapid City, SD 57701, USA (e-mail: James.Martin@sdsmt.edu)

Abstract: Field expeditions to Seymour, James Ross and Vega islands, Antarctic Peninsula, have resulted in significant specimens of mosasaurs (marine lizards). Enough taxonomic diversity is now known to allow preliminary biostratigraphic assessment. These specimens were collected from the Late Campanian levels of the Santa Marta Formation and the Late Campanian–Late Maastrichtian López de Bertodano Formation. All specimens are from near-shore marine deposits, and an overall shallowing appears in younger successions. Most mosasaur individuals are represented by single specimens, but, in some cases, jaws and cranial material are preserved. Based on these specimens, tylosaurines appear in the Santa Marta Formation and extend into the López de Bertodano Formation, and Plioplatecarpines and mosasaurines occur in the upper López de Bertodano Formation. Therefore, the extent of mosasaur ranges in Antarctica is similar to that elsewhere in the world where the bulk of mosasaur differentiation occurred in the Campanian and Maastrichtian. Not enough specimens are available from the Campanian sedimentary rocks to determine whether the faunal turnover during the Campanian–Maastrichtian in North America exists in Antarctica. Some differences in taxa occur between Antarctica and elsewhere, but these are at generic levels, not subfamilial. Even so, most genera are similar, but the tylosaurines may be endemic, and, if the *Moanasaurus* reference is substantiated, this genus is also known only in New Zealand and Antarctica. Therefore, the ranges and occurrences of taxa suggest a mix of endemic and cosmopolitan genera. Additional field investigations are required to refine biostratigraphic ranges, and the ranges of many taxa such as those of *Mosasaurus* and *Plioplatecarpus* will be extended into older deposits.

Marine reptiles were some of the first vertebrate fossils to be found in the Late Cretaceous of Antarctica, and mosasaurs, the marine squamates, were initially reported by Gasparini & del Valle (1981). However, it was not until recently (Martin *et al.* 1999*a*, *b*, 2002) that systematic descriptions were undertaken. These descriptions included specimens collected by earlier Argentinean expeditions and later joint expeditions between Argentina and the United States during 1998 and 1999. Additional specimens had been collected previously by United States expeditions, and some of these specimens are included herein, thanks to access of the collections at the University of California, Riverside. Martin & Crame (2006) describe mosasaur specimens in the British Antarctic Survey collections, and their ranges are denoted herein. All of the Antarctic mosasaur specimens have been collected from the eastern side of the Antarctic Peninsula in the James Ross Basin from Seymour, James Ross and Vega islands (Fig. 1). Combination of the contextual data of the mosasaurs from various collections provides enough information to result in the first biostratigraphic analysis of Antarctic Mosasauridae.

Biostratigraphy of any taxonomic group is constantly in flux, and the mosasaurs of Antarctica are no exception. Undoubtedly, as fieldwork continues, the ranges and correlations herein will be altered to reflect advances in the knowledge of the stratigraphy and distribution of these ubiquitous marine lizards. The mosasaurs are important biostratigraphically because many genera exhibit a worldwide distribution. They appear relatively late in the Mesozoic during the Late Cretaceous, but diversify and become numerous throughout the world. Owing to their relatively short geological history, diversity and wide dispersal, they are ideal creatures for biostratigraphic correlation. The biostratigraphy of the Antarctic Mosasauridae is part of an overall biostratigraphic zonation begun in the 1970s and continuing as major mosasaur collections are made from the Late Cretaceous of South Dakota and elsewhere. Thus far, a major faunal turnover occurs in the Interior Epicontinental Seaway of North America in the upper Campanian portion of the Pierre Shale (Martin *et al.* 1996). Because Antarctica possesses a complete section through this

From: Francis, J. E., Pirrie, D. & Crame, J. A. (eds) 2006. *Cretaceous–Tertiary High-Latitude Palaeoenvironments, James Ross Basin, Antarctica.* Geological Society, London, Special Publications, **258**, 101–108. 0305–8719/06/$15

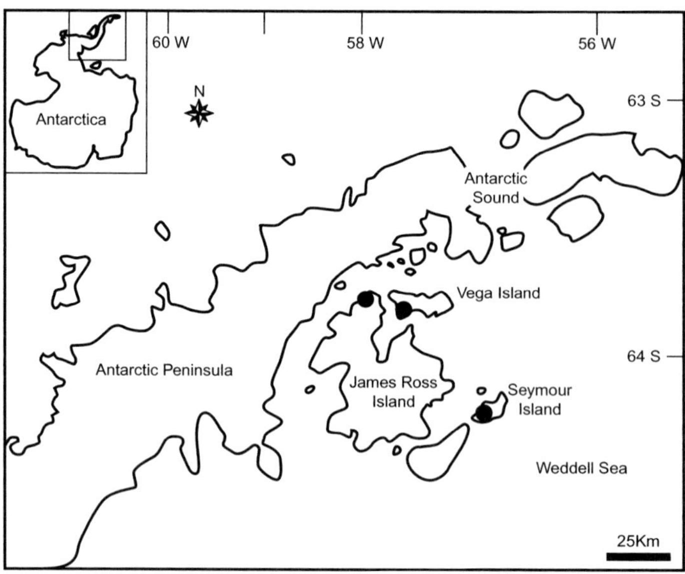

Fig. 1. Sources of mosasaur specimens from Antarctica.

interval, Antarctica will eventually serve as a test of the extent of this turnover.

The methodology employed herein is the recognition of local range zones, normally at a generic level. The local range zones are based on first local appearances (lowest stratigraphic occurrence (LO) of Aubry 1995) and last local occurrences (highest stratigraphic occurrence (HO) of Aubry 1995). Much discussion has revolved around the utility of last appearances, particularly the last appearance datum (LAD), a datum of regional extent (see Woodburne 1996); however, in this case, the last appearances are exceedingly important because many appear to coincide with the Cretaceous–Tertiary boundary and are therefore of global significance and truly represent a last appearance datum. Moreover, disregard of the data *a priori* that may become important with future work seems unwarranted. In order to delineate local range zones, the stratigraphic framework must be established. Much of the stratigraphic delineation has been undertaken by Argentinean and British researchers (see Olivero *et al.* 1986, 1992; Crame *et al.* 1991, 2004; Pierre *et al.* 1991, 1997; Marenssi & Santillana 1998). Currently, there are some differences in the interpretation of Cretaceous units and stratigraphic boundaries related to the various islands upon which the mosasaur specimens were collected. Most stratigraphic questions have been resolved, and the stratigraphic positions have been superimposed on the stratigraphic framework of Crame *et al.* (1991, 2004). This stratigraphic framework is currently being refined, but refinements should not materially affect the local range zones noted herein.

The environment of deposition in which the mosasaur specimens were entombed was on a relatively shallow, marine shelf. The overall stratigraphic succession coarsens upwards and represents deeper marine conditions shallowing up-section. Moreover, the succession indicates shallowing toward the shoreline from east to west towards the Antarctic Peninsula. As a result, the depositional package is thicker on Seymour Island to the east and thinner on Vega Island to the west (Figs 2 & 3). Mosasaur biostratigraphy is based on stratigraphic positions, but correlation of stratigraphic sections is necessarily based on other criteria: radiometric means, physical correlation and invertebrate fossil correlations, among others. Therefore, mosasaur ranges of the Antarctic Peninsula area are expressed in terms of the global marine stages.

Marine reptiles, mosasaurs and plesiosaurs, are relatively well represented in Antarctica. Percentages of mosasaurs v. plesiosaurs in Antarctica are in contrast to those of the Late Cretaceous Interior Epicontinental Seaway of North America. Plesiosaurs are more abundant than mosasaurs in Antarctica. Also, a relatively high number of juvenile mosasaurs and plesiosaurs was observed in the Antarctic collections. Moreover, plesiosaurs are more commonly represented by partial skeletons, whereas mosasaur specimens are normally represented by isolated elements; however,

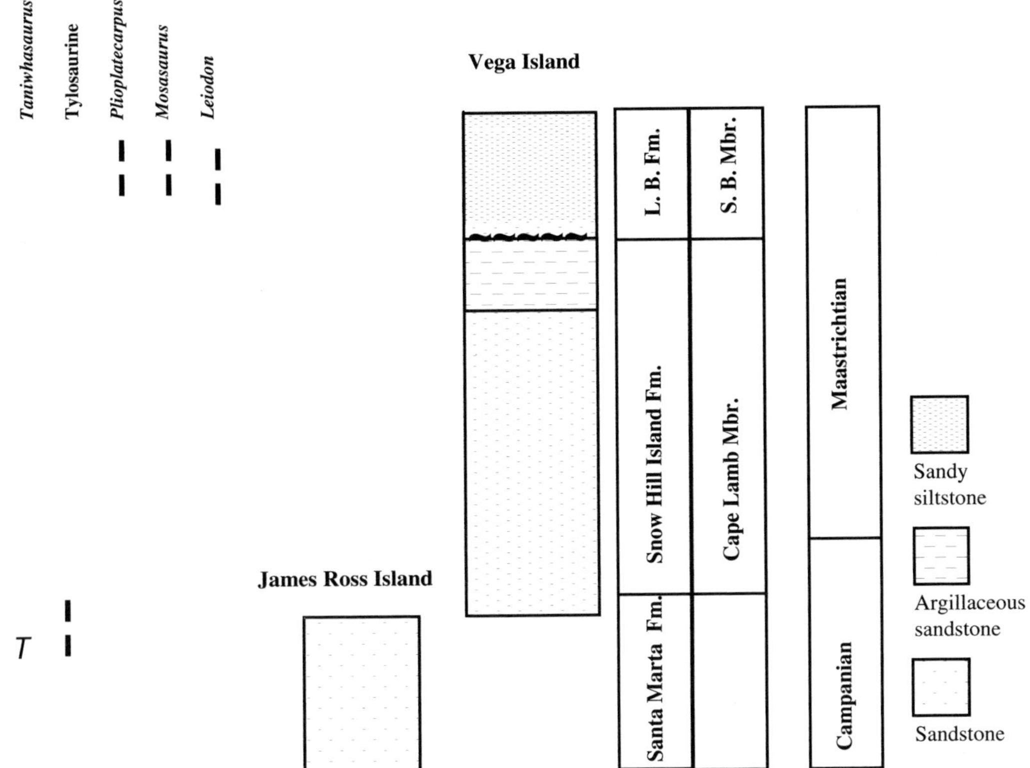

Fig. 2. Stratigraphic occurrences of mosasaurs in the James Ross and Vega island areas (stratigraphy based on Crame *et al.* 1991, 2004). *T* designates the position of *Taniwhasaurus antarcticus*; S.B., Sandwich Bluff; and L.B., López de Bertodano.

cranial material, jaws and associated vertebrae (particularly caudals) occur. Some specimens exhibit evidence of transport and possible reworking. However, most lowest and highest stratigraphic occurrences are represented by little or non-reworked specimens. Representatives of all the major subfamilies of the Mosasauridae (Mosasaurinae, Tylosaurinae + Plioplatecarpinae = Russellosaurina) are represented in Antarctica. Of these, the Mosasaurinae are the most abundantly represented based on current collections. The Antarctic mosasaur specimens are listed systematically below with their stratigraphic positions, and their biostratigraphic ranges are plotted in Figures 2 & 3.

Stratigraphic palaeontology

SQUAMATA

MOSASAURIDAE

Tylosaurinae

Lakumasaurus antarcticus (IAA 2000-JR-FSM-1, where IAA is Instituto Antártico Argentina, Buenos Aires, Argentina), now considered *Taniwhasaurus antarcticus* (Martin & Fernandez 2005) represented by a partial skull (Fig. 4) and post-cranial elements, was found on James Ross Island from the 'upper part of Gamma Member of the Santa Marta Formation (Upper Cretaceous, Latest Campanian–Early Maastrichtian' (Novas *et al.* 2002). The 'Gamma Member' would equate to the Herbert Sound Member, and Crame *et al.* (1991, 2004) indicate that the Santa Marta Formation is, at the youngest, Late Campanian. Therefore, the holotype of this tylosaurine mosasaur appears to have been collected from the Late Campanian (Fig. 2). Another tylosaurine specimen, MLP 86-X-28-7 (where MLP is Museo de La Plata, La Plata, Argentina), an anterior caudal vertebra (Martin *et al.* 2002), may well be referred to this taxon and is likewise from the Santa Marta Formation of James Ross Island (Fig. 2). Therefore, the local range zone for the giant mosasaurs, Tylosaurinae, from James Ross Island is from the Late Campanian.

Two other tylosaurine specimens, MLP

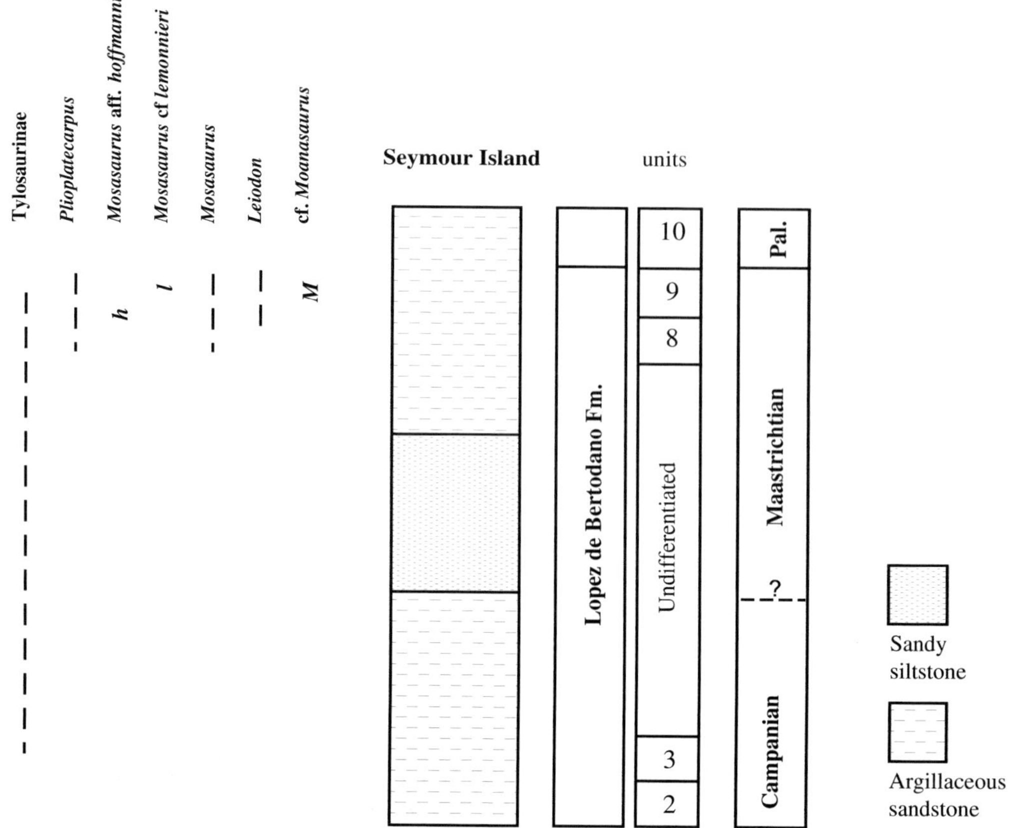

Fig. 3. Stratigraphic occurrences of mosasaurs on Seymour Island (stratigraphy based on Macellari 1988; Crame *et al.* 1991, 2004). *M* designates the position of cf. *Moanasaurus* (MLP 83-I-12); *l* designates the position of *Mosasaurus* sp. cf. *M. lemonnieri* (MLP 92-XII-30); and *h* designates the position of *Mosasaurus* sp. cf. *M. hoffmanni* (DJ.1053.10/MOW 9513).

87-II-7-1 and MLP 79-I-1-12 – large trunk and caudal vertebrae, respectively – are from the Upper Maastrichtian portion of the López de Bertodano Formation of Seymour Island. In addition, a short series of articulated caudal vertebrae, DJ.956.41 (see Martin & Crame 2006), were collected from the basal portion of the Lower Maastrichtian López de Bertodano Formation on Seymour Island. Thus, the local range zone of the tylosaurinae from Seymour Island extends from the Early to Late Maastrichtian López de Bertodano Formation (Fig. 3).

Therefore, the biostratigraphic range of tylosaurine mosasaurs from Antarctica extends from the Late Campanian Santa Marta Formation to the Late Maastrichtian López de Bertodano Formation. Although all of these specimens may represent *Taniwhasaurus*, the indisputable known generic distribution is only from the Late Campanian Santa Marta Formation.

Plioplatecarpinae

Plioplatecarpus is known from Seymour and Vega islands in the James Ross Basin (Fig. 1). The genus is known from vertebrae, MLP 79-I-1 (units 8–9) and MLP 79-I-19 (units 8 and 9), from the Upper Maastrichtian López de Bertodano Formation of Seymour Island (Martin *et al.* 2002). Teeth of this genus occur from the Upper Maastrichtian portion of the López de Bertodano Formation (DJ.952.249, tooth) and from within 1 m of the Cretaceous–Tertiary boundary (DJ.1020.2-C, pterygoid tooth; DJ.1020.2-H, tooth). These teeth are characteristically striated and small; however, *Taniwhasaurus* teeth are also striated (an unusual characteristic of the Tylosaurinae). Consequently, these specimens and others assigned to *Plioplatecarpus* from Antarctica may represent very young individuals of this taxon; however, their number and small size suggest assignment to *Plioplatecarpus*. Therefore, the local range of *Plioplatecarpus*

Fig. 4. *Taniwhasaurus antarcticus* Holotype (IAA 2000-JR-FSM-1). Dorsal view of cranium, and lateral views of lower jaws.

from Seymour Island is from the Late Maastrichtian, with the highest occurrence near the Cretaceous–Tertiary boundary (Fig. 3). The genus is also recorded (Martin *et al.* 2002) from the Late Maastrichtian Sandwich Bluff Member of the López de Bertodano Formation on Vega Island (Fig. 2). Therefore, the known biostratigraphic range of the genus in Antarctica is restricted to the Late Maastrichtian. As the Antarctic collections are enlarged, the biostratigraphic range should extend into older rocks as it does elsewhere.

Mosasaurinae

The most diverse subfamily of mosasaurs known from Antarctica is the Mosasaurinae. At least three genera and four species occur. Most are typical taxa that have been found in the circum-Atlantic, but one, cf. *Moanasaurus* (Martin *et al.* 2002), has only been found in New Zealand. However, as pointed out previously, the vertebral morphology (Fig. 5A, C–E) upon which this reference was made may not be substantiated when additional specimens are collected or some species regarded as *Mosasaurus* from Antarctica may in fact be *Moanasaurus*. Until such time, the referred specimen (MLP 83-I-12) was collected from the range of units 8 and 9 (M. Reguero, pers. comm. 2003) of the López de Bertodano Formation from the Late Maastrichtian (Fig. 3).

A relatively derived mosasaurine appears to be represented by *Leiodon* (Fig. 5F), a taxon characterized by extremely laterally compressed teeth with smooth enamel. Most specimens (MLP 98-I-10-15, MLP 98-I-10-23 and MLP 98-I-10-12) have been collected from Vega Island from the Upper Maastrichtian Sandwich Bluff Member of the López de Bertodano Formation (Martin *et al.* 2002). Another specimen, DJ.952.266, a tooth from the Upper Maastrichtian López de Bertodano Formation of Seymour Island, is described in this volume (Martin & Crame 2006). Essentially, the genus occurs in Late Maastrichtian sedimentary rocks in Antarctica (Figs 2 & 3).

The genus *Mosasaurus* appears taxonomically the most diverse genus of mosasaur in Antarctica. Even so, only two species have been referred to the genus. One is a smaller taxon provisionally referred to *Mosasaurus* sp. cf. *M. lemmonieri* (Martin *et al.* 2002), a species known from the type Maastrictian area of Europe. This specimen (MLP 92-XII-30) is represented by cranial elements collected from the Late Maastrichtian unit 9 of the López de Bertodano Formation on Seymour Island (Fig. 3). The second species, *Mosasaurus* sp. aff. *M. hoffmanni* (DJ.1053.10 and MOW 9513, where MOW is Michael O. Woodburne field number), is represented by a large, fragmentary skull, from the Late Maastrichtian López de Bertodano Formation (lower unit 9), Seymour Island

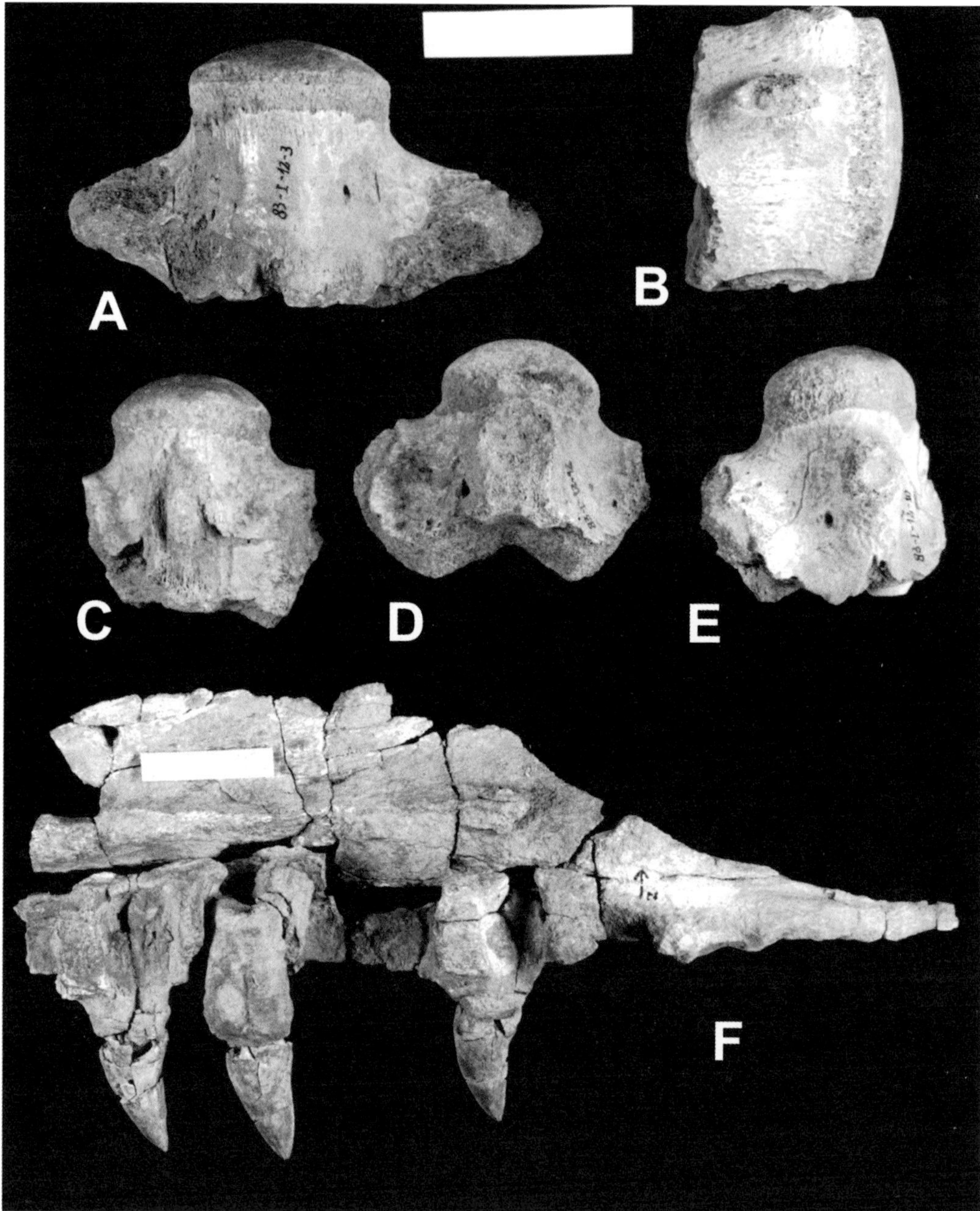

Fig. 5. (**A**) cf. *Moanasaurus* sp., trunk vertebra, ventral view (MLP 83-I-12-3); (**B**) Mosasaurinae, caudal vertebra, ventrolateral view (MLP 83-X-12-2); (**C**) cf. *Moanasaurus* sp., cervical vertebra, ventral view (MLP 83-I-12-11); (**D**) cf. *Moanasaurus* sp., cervical vertebra, ventral view (MLP 83-I-12-12); (**E**) cf. *Moanasaurus* sp., cervical vertebra, ventral view (MLP 83-I-12-10); (**F**) *Leiodon* sp., right maxillary fragment, medial view (MLP 98-I-10-1). Scales equal 5 cm.

(Fig. 3). Other specimens of *Mosasaurus* sp. indet. resemble in size those of *M. lemmonieri*, but are not well enough preserved for specific identification (Fig. 5B). These teeth (DJ.1053.14-A, DJ.1020.2-A and DJ.1020.2-B) are from the Late Maastrichtian López de Bertodano Formation of Seymour Island, and the latter two are from within 1 m of the Cretaceous–Tertiary boundary. Therefore, the stratigraphic distribution of *Mosasaurus* in Antarctica is from the Late Maastrichtian deposits of the López de Bertodano Formation of Vega and Seymour islands, with the highest stratigraphic occurrence just below the Cretaceous–Tertiary boundary (Figs 2 & 3). Both referred species, *Mosasaurus* sp. cf. *M. hoffmani* and *Mosasaurus* sp. aff. *M. lemmonieri*, are from the Late Maastrichtian.

Discussion

The stratigraphic distribution of mosasaurs in Antarctica mirrors that found elsewhere. For example, the tylosaurines existed from at least the Turonian (see Bell & VonLoh 1998) through to the Maastrichtian (Russell 1967). In the Northern Hemisphere, the group is represented by two genera: *Hainosaurus*, a genus identified from the type Maastrichtian of Europe (Dollo 1885) and from Campanian deposits in North America (Nicholls 1988); and *Tylosaurus*, known from the Coniacian into the Campanian (Russell 1967). The subfamily is possibly represented by *Tylosaurus* (Lingham-Soliar 1992) and by *Taniwhasaurus* (Hamurian = Maastrichtian, New Zealand: see Welles & Gregg 1971) in the Southern Hemisphere, with the latter genus recorded from the Late Campanian of Antarctica (Novas *et al.* 2002). The relationships among the southern genera remain unresolved, but, in any event, the tylosaurines from Antarctica occur during the same chronological interval as elsewhere. However, the endemism of the Antarctic tylosaurine genus, *Taniwhasaurus*, appears unique.

The Plioplatecarpinae range from the Turonian (see Bell & VonLoh 1998) or Coniacian (see Russell 1967) through to the Maastrichtian. *Plioplatecarpus* is known from the Early Campanian (South Dakota) through to the Maastrichtian (e.g. type Maastrichtian). The Antarctic range of the genus is currently only from the Late Maastrichtian up to within 1 m of the Cretaceous–Tertiary boundary. Earlier local occurrences might be expected with further Antarctic field investigations.

Mosasaurines are known from the ?Turonian (Martin & Stewart 1977; Bell & VonLoh 1998) through to the Maastrichtian, with *Mosasaurus* known from the Late Campanian through to the Maastrichtian. In Antarctica, the range of the genus is Late Maastrichtian, and, like that of *Plioplatecarpus*, ranges to within 1 m of the Cretaceous–Tertiary boundary. *Mosasarus.* sp. cf. *M. lemmonieri* occurs both in the Late Maastrichtian of Europe (Mulder *et al.* 2003) and of Antarctica (Martin *et al.* 2002). The other referred species, *Mosasaurus* sp. aff. *M. hoffmanni*, has the same range. *Leiodon* occurs from the Campanian through to the Maastrichtian in Europe and North America. Martin *et al.* (1999*a*, *b*, 2002) first recorded the genus from the Late Maastrichtian of Antarctica. The final mosasaurine described from the Late Maastrichtian of Antarctica is provisionally referred to cf. *Moanasaurus*, a genus found in the Maastrichtian of New Zealand (Wiffen 1980). This genus and the tylosaurine, *Taniwhasaurus*, are the only endemic taxa recorded from Gondwana. The remaining genera are known from the circum-Atlantic area, principally the type area of the Maastrichtian in The Netherlands–Belgium and from New Jersey in the United States. Overall, with the exception of two specimens, the remainder fit the previously known biostratigraphic pattern of mosasaurs. Additional specimens from lower in the section must be secured before the Late Campanian–Early Maastrichtian faunal turnover recorded elsewhere can be tested. Probably, the ranges of *Plioplatecarpus* and *Mosasaurus* will be found to extend downward when such collections are made, and recovery of additional mosasaur taxa such as *Prognathodon* and *Globidens* might be expected. The last Antarctic occurrences of *Plioplatecarpus* and *Mosasaurus* on Seymour Island that are within 1 m of the Cretaceous–Tertiary boundary represent some of the latest known mosasaur occurrences worldwide, confirming their survival up to the Cretaceous–Tertiary extinction event.

This research was funded principally through the United States National Science Foundation, Office of Polar Programs, (grants OPP #9815231 and OPP #0087972) without whose support this research would have been impossible. Sincere thanks go to Dr M. Reguero, Museuo de La Plata, Argentina, for the loan of important specimens and for providing stratigraphic occurrences. Dr J. A. Crame, British Antarctic Survey, likewise loaned important specimens, and our collaboration produced much of the data herein. I thank the Instituto Antártico Argentino through Dr S. Marenssi; Dr J. Case, St Mary's College, California; D. Chaney, US National Museum; and Dr A. Kihm, Minot State University, North Dakota, for collaboration on Vega Island. The contribution was enhanced by the reviews of Dr D. Pirrie, Camborne School of Mines, UK, and Dr B. Kear, South Australia Museum, Adelaide. In particular, I acknowledge field investigators from various countries who worked so diligently in Antarctica and kindly shared their knowledge.

References

Aubry, M.-P. 1995. From chronology to stratigraphy: interpreting the stratigraphic record. *In*: Berggren, W.A., Kent, D.V., Aubry, M.-P. & Hardenbol, J. (eds) *Geochronology, Time Scales and Global Stratigraphic Correlation*. Society of

Economic and Petroleum Geologists, Special Publications, **54**, 213–274.

BELL, G.L., JR & VONLOH, J.P. 1998. New records of Turonian Mosasauroids from the Western United States. *In*: MARTIN, J.E., HOGENSON, J.W. & BENTON, R.C. (eds) *Partners Preserving Our Past, Planning Our Future*. Museum of Geology, SD School of Mines & Technology, Dakoterra, **5**, 15–28.

CRAME, J.A., FRANCIS, J.E., CANTRILL, D.J. & PIRRIE, D. 2004. Maastrichtian stratigraphy of Antarctica. *Cretaceous Research*, **25**, 411–423.

CRAME, J.A., PIRRIE, D., RIDING, J.B. & THOMSON, M.R.A. 1991. Campanian–Maastrichtian (Cretaceous) Stratigraphy of the James Ross Island Area. *Journal of the Geological Society, London*, **148**, 1125–1140.

DOLLO, L. 1885. Le Hainosaure. *Revue des Questions Scientifiques*, **18**, 285–289.

GASPARINI, Z. & DEL VALLE, R. 1981. Mosasaurios: primer hallazgo en el continente Antártico. *Antártida*, **11**, 16–20.

LINGHAM-SOLIAR, T. 1992. The Tylosaurine Mosasaurs (Reptilia, Mosasauridae) from the Upper Cretaceous of Europe and Africa. *Bulletin de l'Institut Royal des Sciences Naturelles de Belgique, Sciences de la Terre*, **62**, 171–194.

MACELLARI, C.E. 1988. Stratigraphy, sedimentology and paleoecology of Upper Cretaceous/Paleocene shelf-deltaic sediments of Seymour Island. *In*: FELDMANN, R.M. & WOODBURNE, M.O. (eds) *Geology and Paleontology of Seymour Island, Antarctic Peninsula*. Geological Society of America, Memoirs, **169**, 25–53.

MARENSSI, S.A. & SANTILLANA, S.N. 1998. Revision of the Late Cretaceous stratigraphy of Cape Lamb, Vega Island, Antarctic Peninsula. *Actas X Congreso Latinoamérica de Geología Económica*, **1**, 91–94.

MARTIN, J.E. & FERNANDEZ, M. 2005. The synonymy of the Late Cretaceous mosasaur (Reptilia) genus *Lakumasaurus* from Antarctica with *Taniwhasaurus* from New Zealand and its bearing upon faunal similarity. *In*: PANKHURST, R.J. & VEIGA, G.D. (eds) *Gondwana 12: Geological and Biological Heritage of Gondwana*, Academia Nacional de Ciencias, Cordoba, Argentina, 244.

MARTIN, J.E. & CRAME, J.A. 2006. Palaeobiological significance of the high-latitude Late Cretaceous vertebrate fossils from the James Ross Basin, Antarctica. *In*: FRANCIS, J.E., PIRRIE, D. & CRAME, J.A. (eds) *Cretaceous–Tertiary High-latitude Palaeoenvironments: James Ross Basin, Antarctica*. Geological Society, London, Special Publications, **258**, 109–124.

MARTIN, J.E., BELL, G.L., JR & BERTOG, J.L. 1996. Biostratigraphic ranges and biozonation of mosasaurs (Reptilia) within the Late Cretaceous of the North American Epicontinental Seaway. *Geological Society of America, Abstracts with Programs, Rocky Mountain Section*, **28**, (4), 16.

MARTIN, J.E., BELL, G.L., JR *ET AL*. 2002. Mosasaurs (Reptilia) from the Late Cretaceous of the Antarctic Peninsula. *In*: GAMBLE, J.A., SKINNER, D.N.B. & HENRYS, S. (eds) *Antarctica at the Close of a Millennium, 8th International Symposium on Antarctic Earth Sciences. Bulletin of the Royal Society of New Zealand*, **35**, 293–299.

MARTIN, J.E., BELL, G.L., JR, FERNANDEZ, M.S., REGUERO, M., CASE, J.A. & WOODBURNE, M.O. 1999*a*. Mosasaurs from Antarctica and their bearing on global distributions. *European Union of Geosciences, Journal of Conference Abstracts*, **4**, 734.

MARTIN, J.E., BELL, G.L., JR, FERNANDEZ, M.S., REGUERO, M., CASE, J.A. & WOODBURNE, M.O. 1999*b*. Mosasaurs from the Late Cretaceous of Antarctica and their bearing on marine paleobiogeography. *In*: *8th International Symposium on Antarctic Earth Sciences, Wellington, New Zealand, Programme and Abstracts*, 201.

MARTIN, L.D. & STEWART, J.D. 1977. The Oldest (Turonian) Mosasaurs from Kansas. *Journal of Paleontology*, **51**, 973–975.

MULDER, E.W.A., JAGT, J.W.M., KUYPERS, M.M.M., PEETERS, H.H.G. & ROMPEN, P. 2003. Stratigraphic distribution of Late Cretaceous marine and terrestrial reptiles from the Maastrichtian type area. *In*: *On Latest Cretaceous Tetrapods From the Maastrichtian Type Area*. Publicaties Van Het Natuurhistorisch Genootschap In Limburg, **XLIV**, 157–163.

NICHOLLS, E.J. 1988. The first record of the Mosasaur *Hainosaurus* (Reptilia: Lacertilia) from North America. *Canadian Journal of Earth Sciences*, **25**, 1564–1570.

NOVAS, F.E., FERNANDEZ, M., GASPARINI, Z.B., LIRIO, J.M., NUNEZ, H.J. & PUERTA, P. 2002. *Lakumasaurus antarcticus*, n. gen. et sp., a new mosasaur (Reptilia, Squamata) from the Upper Cretaceous of Antarctica. *Ameghiniana*, **39**, 245–249.

OLIVERO, E.B., MARTINIONI, D.R. & MUSSEL, F.J. 1992. Upper Cretaceous sedimentology and biostratigraphy of western Cape Lamb (Vega Island, Antarctica). Implications on sedimentary cycles and evolution of the basin. *In*: RINALDI, C.A. (ed.) *Geologia de la Isla James Ross*. Instituto Antartico Argentino, Buenos Aires, 147–166.

OLIVERO, E.B., SCASSO, R.A. & RINALDI, C.A. 1986. Revision of the Marambio Group, James Ross Island, Antarctica. *Contribución Científica del Instituto Antártico Argentino*, **331**, 1–28.

PIRRIE, D., CRAME, J.A., LOMAS, S.A. & RIDING, J.B. 1997. Late Cretaceous Stratigraphy of the Admiralty Sound Region, James Ross Basin, Antarctica. *Cretaceous Research*, **18**, 109–137.

PIRRIE, D., CRAME, J.A. & RIDING, J.B. 1991. Late Cretaceous stratigraphy and sedimentology of Cape Lamb, Vega Island, Antarctica. *Cretaceous Research*, **12**, 227–258.

RUSSELL, D.A. 1967. Systematics and morphology of American mosasaurs. *Bulletin of the Yale Peabody Museum of Natural History*, **23**, 1–240.

WELLES, S.P. & GREGG, D.R. 1971. Late Cretaceous marine reptiles of New Zealand. *Records of the Canterbury Museum*, **9**, 1–111.

WIFFEN, J. 1980. *Moanasaurus*, a new genus of marine reptile (Family Mosasauridae) from the Upper Cretaceous of North Island, New Zealand. *New Zealand Journal of Geology and Geophysics*, **23**, 507–528.

WOODBURNE, M.O. 1996. Precision and resolution in Mammalian chronostratigraphy: Principles, practices, examples. *Journal of Vertebrate Paleontology*, **16**, 531–555.

Palaeobiological significance of high-latitude Late Cretaceous vertebrate fossils from the James Ross Basin, Antarctica

JAMES E. MARTIN[1] & J. ALISTAIR CRAME[2]

[1]*Museum of Geology, SD School of Mines and Technology, Rapid City, South Dakota 57701, USA (e-mail: James.Martin@sdsmt.edu)*

[2]*British Antarctic Survey, Natural Environment Research Council, High Cross, Madingley Road, Cambridge CB3 0ET, UK (e-mail: JACR@bas.ac.uk)*

Abstract: A diverse marine assemblage of vertebrate fossils has been collected in recent years under the auspices of the British Antarctic Survey from Seymour, James Ross and Vega islands east of the Antarctic Peninsula. The specimens were derived from the Late Campanian Santa Marta Formation, Early Maastrichtian Snow Hill Island Formation and the Early–Late Maastrichtian López de Bertodano Formation. Sharks, teleosts, plesiosaurs and mosasaurs are represented, but birds and sea turtles are absent from the BAS collections; neornithine birds have been previously reported from the Late Cretaceous deposits of Antarctica. Shark teeth are relatively abundant, but teleosts are seemingly under-represented. Plesiosaurs (Elasmosauridae) are more abundant and complete than mosasaurs, and juveniles of both marine reptile groups are relatively common. The marine lizards, mosasaurs, are taxonomically diverse as elsewhere in the world, but with relatively few individuals compared to the plesiosaurs, which are taxonomically limited. A converse relationship normally occurs at other lower latitude Late Cretaceous localities. Some of these abundances and appearances may be due to collection bias, particularly due to difficult collecting conditions and weathering, but certain distributions may be the result of high latitudes.

During a recent phase of stratigraphical investigations into the latest Cretaceous sedimentary rocks of the James Ross Island Group, NE Antarctic Peninsula (Fig. 1), vertebrate fossils were collected from a number of localities. These specimens and others from the James Ross Basin may be utilized to develop our understanding of the vertebrate palaeontological history of Antarctica. The study of these vertebrate fossils was undertaken to better understand the diversity of marine creatures in Antarctica during the Late Cretaceous, and their relationships with other Gondwanan and Nearctic palaeofaunas, to aid in the understanding of vertebrate biostratigraphy and to reveal the role of Antarctica in vertebrate dispersal. The assemblages of vertebrates from Antarctica are just becoming known from the Late Cretaceous and, although relatively few vertebrates occurred compared to invertebrate fossils, the collections described herein provide additional diversity to the known palaeofauna of Antarctica. Both endemic and immigrant taxa appear, the latter aid in comprehension of the biostratigraphy and distribution of Late Cretaceous marine taxa. Antarctica has produced numerous vertebrates that occur elsewhere and provides evidence for marine dispersal patterns during the Late Cretaceous. In particular, the collections may provide the basis for assessment of profound changes and endemism known to occur at high latitudes prior to the K/T (Cretaceous–Tertiary) mass extinction event (e.g. Zinsmeister 1982). The James Ross Basin produces one of the few known high-latitude assemblages. Moreover, the preservation of a continuous section through the K/T boundary provides the basis for understanding both extinction and radiation of groups such as sharks, teleosts and birds across the K/T boundary. Based on distribution and interpretations of vertebrate habits, palaeoclimate may be traced from the exceptionally warm Turonian–Coniacian thermal maximum to declining palaeotemperatures through the latest Campanian and Maastrichtian ages (Huber 1998; Crame *et al.* 2004).

The material described herein comes from a series of sites on Seymour, James Ross and Vega islands east of the Antarctic Peninsula (Fig. 1) collected under the auspices of the British Antarctic Survey (BAS) and reposed in Cambridge. The area has also been visited in recent years by Argentine, US, UK and other geological parties, all of whom made comparative collections. The specimens were derived predominantly from Maastrichtian deposits, although at a few localities the lowest strata

From: Francis, J. E., Pirrie, D. & Crame, J. A. (eds) 2006. *Cretaceous–Tertiary High-Latitude Palaeoenvironments, James Ross Basin, Antarctica.* Geological Society, London, Special Publications, **258**, 109–124. 0305–8719/06/$15

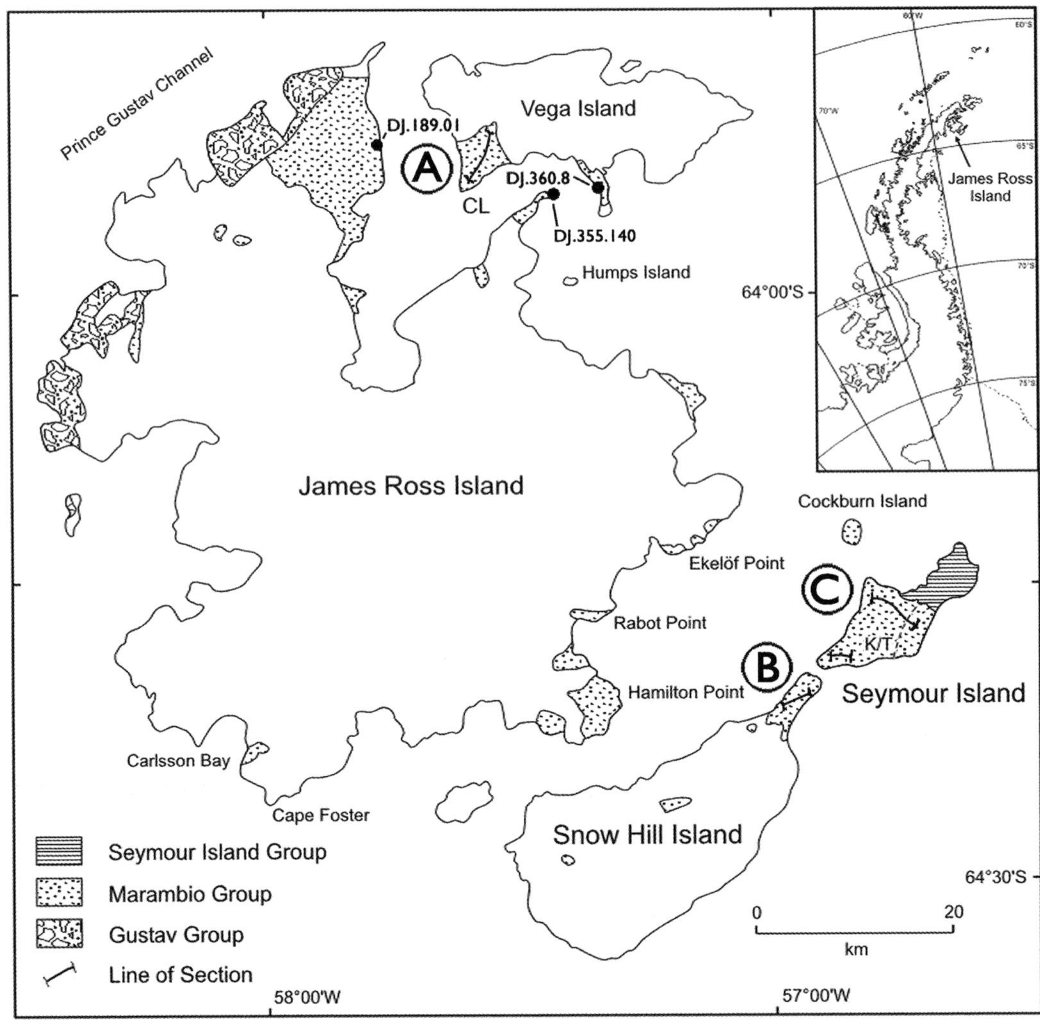

Fig. 1. Locality and sketch geological map of the James Ross Island Group. Based on Crame *et al.* (2004, fig. 1), with minor changes. A, principal line of section at Cape Lamb, Vega Island; B, principal line of section at Snow Hill Island; C, principal line of section at Seymour Island (i.e. the sections shown in Fig. 2). CL, Cape Lamb; K/T marks the position of the Cretaceous–Tertiary boundary on Seymour Island.

range into the latest Campanian. All sites were deposited during the Late Cretaceous under shallow-marine conditions and at a palaeolatitude of approximately 65°S (Lawver *et al.* 1992).

Stratigraphy

The sedimentary basin containing the James Ross Island Group (Fig. 1) includes an extremely thick sequence of Lower Cretaceous–Lower Cenozoic marine sedimentary rocks. These rocks accumulated initially in an extensional setting as the Weddell Sea (i.e. to the SE) opened, and subsequently in a back-arc position relative to the Antarctic Peninsula magmatic arc (Hathway 2000). The rocks are predominantly volcaniclastic and have been divided into a lower, coarser grained Gustav Group (Aptian–Coniacian) and an upper, finer grained Marambio Group (Coniacian–Danian) (e.g. Hathway 2000; Crame *et al.* 2004 and references therein). The regional dip throughout the greater part of the island group is approximately 10°SE, and the Seymour Island Group (Palaeocene–late Eocene) is only exposed on Seymour and Cockburn islands (Fig. 1).

Campanian–Maastrichtian strata of the Marambio Group are well exposed on northern James Ross Island and Cape Lamb, Vega Island. The units are repeated by a major ENE–WSW-trending reverse fault (or faults) on SE James Ross, Snow Hill and Seymour islands (Fig. 1). Indeed, the succession is considerably thicker here and includes the only known onshore exposure of the K/T boundary in Antarctica (Figs 1 & 2). As the Maastrichtian stratigraphy of Antarctica has recently been revised (Crame *et al.* 2004), only a brief synopsis of the three principal localities is given here.

Cape Lamb, Vega Island

The 480 m-thick stratigraphic section measured at Cape Lamb has been subdivided into three component lithostratigraphic units (Figs 1 & 2). The basal 52 m comprises massive silty mudstones with sparse interbedded thick sandstones assigned to the Herbert Sound Member of the Santa Marta Formation. The basal unit passes conformably upward into the principal unit at Cape Lamb, the 317 m-thick Cape Lamb Member of the Snow Hill Island Formation. Heavily bioturbated silty mudstones–silty sandstones predominate in this member, and abundant early diagenetic concretions have yielded a distinctive *Gunnarites antarcticus* ammonite assemblage. Strontium isotope dating on elements of this fauna indicates placement of the Campanian–Maastrichtian boundary (71 Ma) at a level very close to the base of the Cape Lamb Member (Fig. 2) (Crame *et al.* 1999, 2004).

The Cape Lamb Member is unconformably overlain by the 111 m-thick Sandwich Bluff Member of the López de Bertodano Formation (Fig. 2). This member is a lithologically distinctive unit composed of thin conglomerates, pebbly sandstones, sandstones and mudstones with a high volcaniclastic component and a very shallow-water, near-shore signature (Pirrie *et al.* 1991). Ammonites are rare within this unit, but a preliminary study of well-preserved palynofloras suggests a Late Maastrichtian age and correlation with a very high level in the Seymour Island section (Fig. 2) (Pirrie *et al.* 1991).

Snow Hill Island–Seymour Island

The base of the distinctive *Gunnarites antarcticus* assemblage has been used to effect a direct correlation between the lowermost levels of the Cape Lamb Member on Vega Island and approximately the base of the Karlsen Cliffs Member on Snow Hill Island (Figs 1 & 2). The latter unit, which is one of the two component members of the Snow Hill Island Formation in its type area, is a 170 m-thick succession of mudstones, sandy mudstones and heavily bioturbated sandstones with abundant early diagenetic concretions. The member is essentially Early Maastrichtian in age, although the lowermost levels may be latest Campanian (Crame *et al.* 2004).

The Karlsen Cliffs Member is unconformably overlain by the 175 m-thick Haslum Crag Member, which can be traced across the NE tip of Snow Hill Island onto the SW corner of Seymour Island (Figs 1 & 2). The Haslum Crag Member is a coarser grained unit, with medium- to coarse-grained, cross-bedded sandstones in the lower portion passing into intensely bioturbated siltstones and fine-grained sandstones. A prominent concretionary ammonite assemblage allows the correlation of the Haslum Crag Member with the uppermost levels of the Cape Lamb Member on Vega Island (Fig. 2) (Crame *et al.* 2004).

The Haslum Crag Member is in turn unconformably overlain by pale grey, recessive weathering mudstones comprising a basal unit of the López de Bertodano Formation on both Snow Hill and Seymour islands (Figs 1 & 2). This basal unit, which is 125 m thick and may be of member status (Crame *et al.* 2004), is succeeded by a 708 m-thick succession of remarkably uniform rusty-brown–tan and grey muddy siltstones that comprise the greater part of the López de Bertodano Formation as exposed on Seymour Island. A slight, but nonetheless significant, coarsening-upwards trend within the sequence occurs, and from approximately the 1000 m level (Fig. 2) dark-green weathering glauconitic sandstones become more prominent. The López de Bertodano Formation is intensely fossiliferous with marine invertebrate taxa such as ammonites, bivalves, gastropods, echinoids, decapod crustaceans and serpulids being particularly prominent. Vertebrate fossils and fossil wood occur throughout the Seymour Island section, but are particularly common in the uppermost 400 m (Fig. 2) (Macellari 1988; Zinsmeister & Macellari 1988; Zinsmeister *et al.* 1989). The K/T boundary occurs at the 1458 m level (Fig. 2), where it is marked by a prominent glauconitic horizon, a sudden loss of ammonites and a small iridium spike (Elliot *et al.* 1994; Zinsmeister 1998).

In total the Maastrichtian succession within the James Ross Basin is approximately 1150 m thick, making it one of the thickest onshore latest Cretaceous sequences anywhere in the Southern Hemisphere. A prominent belemnite

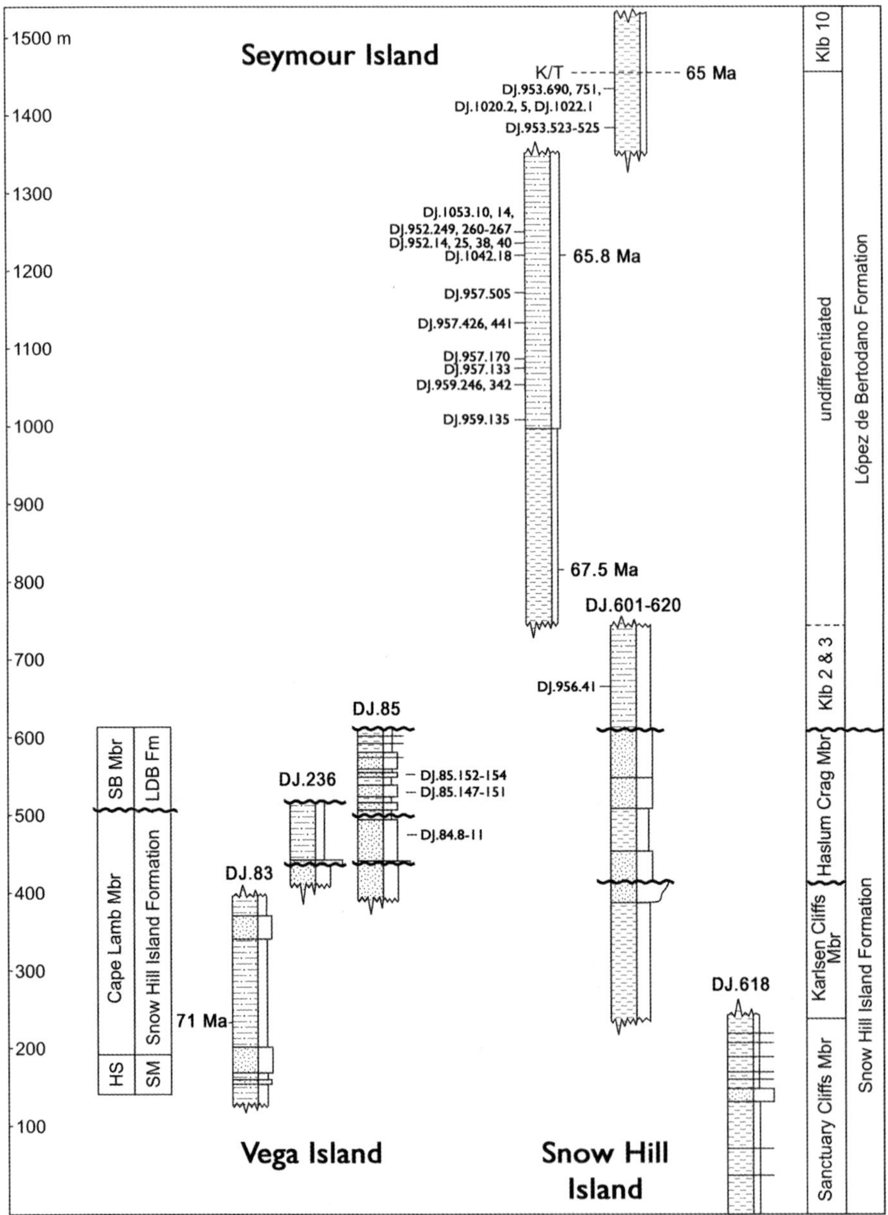

Fig. 2. Stratigraphic distribution of specimens within the Late Cretaceous sedimentary rocks of the James Ross Island Group. Based on Crame *et al.* (2004, fig. 2), with minor changes. HS, Herbert Sound Member; LDB, López de Bertodano Formation; SB Mbr., Sandwich Bluff Member; SM, Santa Marta Formation. Macellari's (1988) informal lithostratigraphic units Klb 2–3 and Klb 10 are shown. Prominent unconformities in the sections are marked by bold, wavy lines.

horizon in the lower levels of the López de Bertodano Formation on Seymour Island (820 m, Fig. 2) has been strontium-isotope dated at 67.5 Ma, and this age has provisionally been taken to mark the Early–Late Maastrichtian boundary in Antarctica (Crame *et al.* 2004). This boundary is obscured by the unconformity at the base of the Sandwich Bluff Member on Vega Island (Fig. 2), which is entirely Late Maastrichtian in age (Pirrie *et al.* 1991).

Previous investigations

Research concerning the Late Cretaceous deposits of the Antarctic Peninsula area has increased substantially in the last few decades. British, American and Argentinean campaigns have resulted in major increases in the knowledge of the lithostratigraphy, sedimentation and palaeontology of the Late Cretaceous interval. In comparison with many other taxonomic groups, Late Cretaceous vertebrate fossils from the Antarctic Peninsula have been much less intensively studied. Early work tended to concentrate on the conspicuous and relatively abundant reptilian taxa, with both the plesiosaurs and mosasaurs being identified (e.g. del Valle *et al.* 1977; Gasparini & del Valle 1981; Chatterjee & Zinsmeister 1982; Gasparini *et al.* 1984). More recent taxonomic revisions of this assemblage are contained in the studies by Chatterjee & Small (1989), Martin *et al.* (2002), Novas *et al.* (2002) and Gasparini *et al.* (2003). One significant trend to emerge from this research in Antarctica is that, in contrast to the lower latitudes of the Late Cretaceous Western Interior Seaway of North America or the type Maastrichtian area of Europe, plesiosaurs are more abundant than mosasaurs. Moreover, plesiosaurs are more commonly represented by partial skeletons, and mosasaurs are most often represented by tail sections (although some associated cranial material has been encountered). Interestingly, a relatively high number of juvenile mosasaurs and plesiosaurs are present. These observations may be related to relatively shallow water depths.

Fish and shark taxa have been noted (Cione & Medina 1987; Grande & Chatterjee 1987), and a recent review of these groups is provided by Kriwet *et al.* (2002, 2006). One biogeographically important trend is that teleosts are relatively under-represented compared to those in other Late Cretaceous marine deposits (e.g. Niobrara and lower Pierre formations of the North American Western Interior Seaway).

With the refinement of collecting techniques, a range of other vertebrate groups is becoming steadily more apparent in the latest Cretaceous of Antarctica. Theropod, ornithopod, ankylosaur and hadrosaurid dinosaurs have all been recorded from these marine strata (Olivero *et al.* 1991; Case *et al.* 2000), together with numerous neornithine birds (Noriega & Tambussi 1995; Case & Tambussi 1999; Cordes 2002). The latter are relatively abundant at higher levels in the Sandwich Bluff Member and represent a diversity of modern birds unknown elsewhere in the world in the Late Cretaceous, suggesting Antarctica may have been pivotal in avian dispersal and diversity.

Taphonomy

Vertebrate fossils from the Late Cretaceous deposits of Antarctica are normally fragmentary due to post-mortem disaggregation in the nearshore marine environment of deposition. The specimens themselves were originally well preserved, having been fossilized in a volcaniclastic sand environment in which early diagenetic carbonate cements predominated. Upon subaerial exposure and superficial weathering, the harsh Antarctic environment caused fracture and fragmentation of the brittle fossils. As a result, many specimens consist of a single element or tooth. Smaller shark and teleost fish teeth are relatively well preserved, but larger teeth of marine reptiles are normally fragmented.

Some specimens are represented by associated skeletal elements, particularly those of plesiosaurs and some fish, and some associated cranial elements occur. The collection is somewhat biased towards larger elements, and isolated elements are usually represented by resistant elements as exemplified by shark and bony fish teeth.

Systematic palaeontology

Table 1. *Vertebrate fossil assemblage in the BAS collections at Cambridge*

Chondrichthyes
- Chimaeridae
  - *Callorhinchus* sp.
- Chimaeridae
- Hexanchidae
  - *Notidanodon dentatus*
- Odontaspidae
  - *Odontaspis* sp.
  - cf. *Odontaspis* sp.

Osteichthyes
- Enchodontidae
  - cf. *Enchodus* sp.
- cf. Sphenocephalidae

Plesiosauria
- Elasmosauridae

Squamata
- Mosasauridae
  - Tylosaurinae
  - Plioplatecarpinae
    - *Plioplatecarpus* sp.
    - cf. *Plioplatecarpus* sp.
  - Mosasaurinae
    - *Leiodon* sp.
    - *Mosasaurus* sp. aff. *Mosasaurus hoffmanni*
    - *Mosasaurus* sp.

Chondrichthyes
Chimaeridae
Callorhinchus sp.

Referred specimen

DJ.1020.2-F, palatine tooth plate; within 1 m of the K/T boundary, Late Maastrichtian López de Bertodano Formation, Seymour Island.

Description

DJ.1020.2-F, a bifid palatine tooth plate, represents the first known occurrence of *Callorhinchus* from Antarctica (Fig. 3A) and occurs near the K/T boundary. Other records of the genus in Antarctica occur in the Eocene La Meseta Formation of Seymour Island (Kriwet & Gazdzicki 2003). Chimaerids are relatively common in the Late Cretaceous sedimentary rocks of the Antarctic Peninsula area and are also represented by *Ischyodus dolloi* and *Chimaera zangerli* (Stahl & Chatterjee 1999, 2002).

Chimaeridae

Referred specimen

DJ.1053.14-B, three broken tooth plates; from the Late Maastrichtian López de Bertodano Formation, Seymour Island.

Hexanchidae
Notidanodon dentatus

Referred specimens

DJ.1020.5, lower tooth, DJ.1022.1, lower tooth; within 1 m of K/T boundary, Late Maastrichtian López de Bertodano Formation, Seymour Island; DJ.85.147, tooth; Late Maastrichtian Sandwich Bluff Member, Vega Island.

Description

DJ.1022.1 (Fig. 3B) is a nearly complete lower tooth (4.2 cm) consisting of 10 preserved cusps. Only two small mesial cuspules are apicodistally directed and composed of laterally compressed, smooth cusps. However, a broken surface indicates another cuspule occurred mesially, resulting in a total of least 11 cusps. The principal cusp separates the mesial cuspules from seven distal cusps, but the principal cusp is only slightly larger than the adjacent distal cusps. Only the last few distal cusps decrease appreciably in size. As a result, the tooth is overall convex. DJ.1020.5 (3.7 cm) possesses three mesial cuspules, but the posterior portion is not as well preserved. Ten cusps are preserved, but broken cusps appear near the principal cusp. DJ.85.147 is a broken specimen preserved in a small concretion; six cusps are preserved. These teeth resemble those described by Cione & Medina (1987) derived from the López de Bertodano Formation of Seymour Island.

Odontaspidae
Odontaspis sp.

Referred specimen

DJ.1020.2-E3, tooth; from the Late Maastrichtian López de Bertodano Formation, Seymour Island.

Description

The smooth-sided, long, slender, smooth, slightly curved principal cusp (Fig. 3E) flanked by small lateral cusplets is characteristic of *Odontaspis*. The base is partially broken from this specimen.

cf. *Odontaspis* sp.

Referred specimens

DJ.1020.2-E1, broken tooth, DJ.1020.2-E2, broken tooth; from the Late Maastrichtian López de Bertodano Formation, Seymour Island.

Description

The base of DJ.1020.2-E1 (Fig. 3C) is highly abraded, and only a non-serrate, smooth, relatively short, curved, wide cusp remains. Any lateral cusplets were broken away. A lingual foramen is centrally positioned on the lingual tooth protuberance; poor preservation prevents definite assessment of a nutrient groove. DJ.1020.2-E2 (Fig. 3D) is poorly preserved,

Fig. 3. (**A**) *Callorhinchus*, upper tooth plate, DJ.1020.2-F; (**B**) *Notidanodon dentatus*, tooth, DJ.1022.1; (**C**) cf. *Odontaspis*, broken tooth, DJ.1020.2-E1; (**D**) cf. *Odontaspis*, broken tooth, DJ.1020.2-E2; (**E**) *Odontaspis*, broken tooth, DJ.1020.2-E3; (**F**) cf. *Enchodus*, six teeth, DJ.1020.2-D; (**G**) cf. Sphenocephalidae, posterior portion of skeleton; (**H**) Elasmosauridae, cervical vertebra, DJ.1042.18; (**I**) Elasmosauridae, propodial, DJ1042.18; (**J**) Elasmosauridae, vertebrae, DJ.953.690.

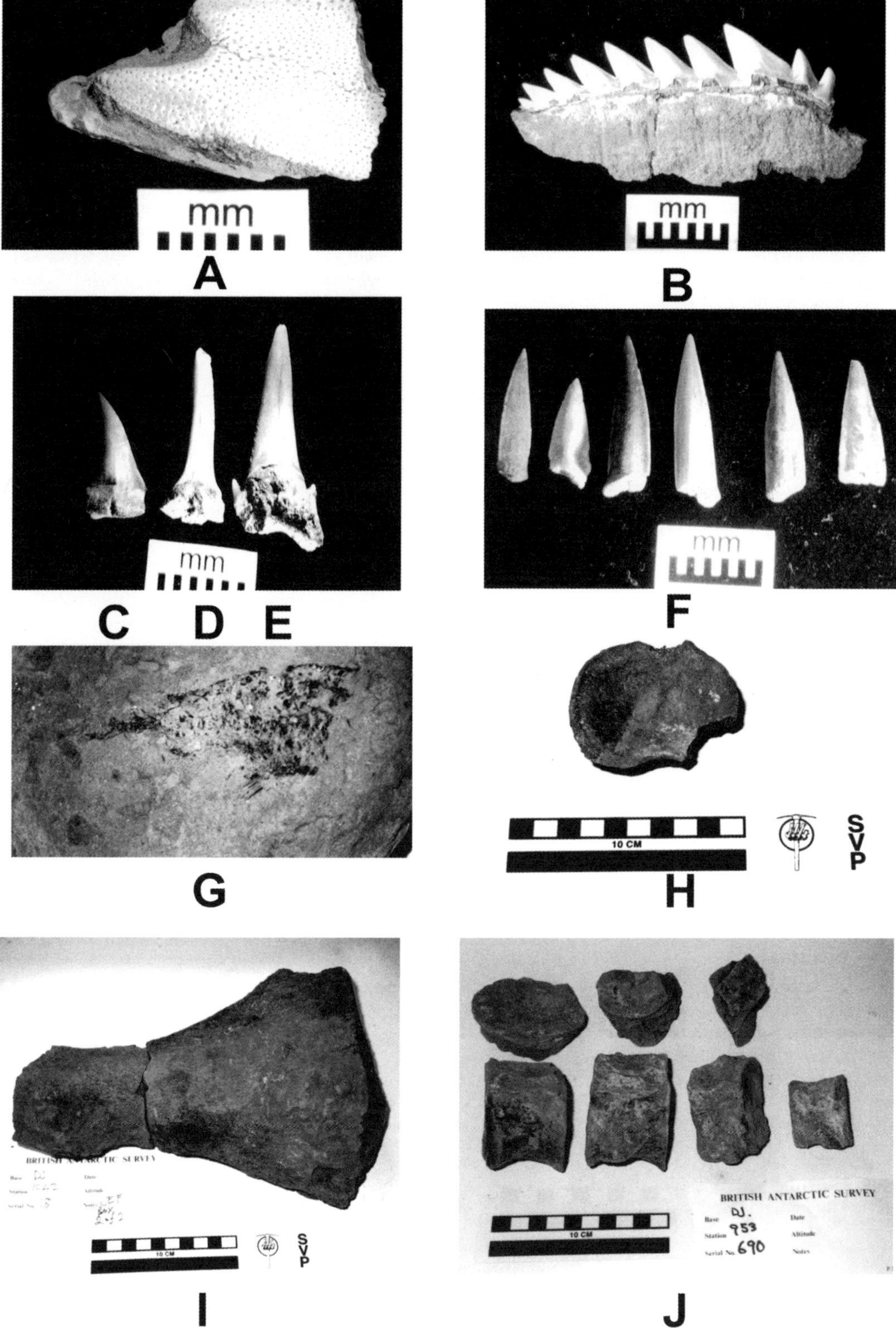
mm
A
mm
B
mm
C D E
mm
F
G
10 CM
SVP
H
10 CM
SVP
I
BRITISH ANTARCTIC SURVEY
Base DJ.
Station 953
Serial No. 690
Date
Altitude
Notes
10 CM
J

consisting only of the very long, non-serrate, smooth, slender principal cusp. The base is completely broken away, and whether lateral cusps existed cannot positively be ascertained. The long slender crown resembles that of *Odontapsis* or *Scapanorhynchus*. The cusp is relatively straight unlike the medially curved principal cusp of the latter genus. Therefore, a provisional assignment is made to *Odontaspis*, even though lateral cusplets are not preserved.

Squalomorphii, fam. indet.

Referred specimens

DJ.189.01, vertebra; from the Late Campanian Herbert Sound Member, Santa Marta Formation, Andreassen Point, James Ross Island. DJ.84.8, DJ.85.150, DJ.84.11, DJ.85.154, DJ.84.10, unassociated vertebrae; from the Late Maastrichtian Sandwich Bluff Member, López de Bertodano Formation, Vega Island. DJ.959.135, DJ.957.441, DJ.957.426, DJ.959.246, DJ.959.342, DJ.952.40, DJ.952.38, DJ.952.14, five vertebrae, DJ.957.170, three vertebrae; from the Late Maastrichtian López de Bertodano Formation, Seymour Island.

Description

All specimens are the typically round, doubly concave, and anteroposteriorly short ossified shark vertebrae. Lack of associated dentitions prevents lower taxonomic assignments.

Osteichthyes
Enchodontidae
cf. *Enchodus* sp.

Referred specimens

DJ.1020.2-D, six teeth; from the Late Maastrichtian López de Bertodano Formation, Seymour Island.

Description

These specimens (Fig. 3F) compare with the slender, pointed teeth of the teleost, *Enchodus*. The teeth are relatively large, somewhat asymmetrical, with both anterior and posterior cutting surfaces that extend the entire height of the tooth. This morphology can be attributed to the *E. petrosus* tooth type (Arambourg 1954; Goody 1976). However, compared to the Late Cretaceous taxa in North America, the distinctive minute serrations of *E. ferox* are not obvious, and the internal face is not striated as in *E. petrosus* (*E. lewesiensis* from the English Chalk does not exhibit striations). Because the teeth are isolated, and distinctive characters of vertebral count, number of tooth rows, ornamentation and proportions are unknown, the teeth are not assigned to a species. Nevertheless, they do corroborate the fact that the *Enchodus* was a pelagic fish distributed worldwide during the Late Cretaceous. Some of the teeth (Fig. 3F) also resemble those of ichthyodectids, but the small sample of isolated teeth and wide variation of enchodontid and ichthyodectid teeth observed in the Western Interior Seaway of North America prevents their assignment.

cf. Sphenocephalidae

Referred specimen

DJ.360.8, posterior half of fish skeleton; from the Early Maastrichtian Snow Hill Island Formation, Hill 177, False Island Point, Vega Island.

Description

DJ.360.8 (Fig. 3G) is a very small fish (maximum DV (dorsoventral) = 1.6 cm, DV through anal fin = 1.8 cm, preserved length = 3.9 cm) composed of approximately one-third of the posterior skeleton, including the anal fin, but the tail is poorly preserved. The scales possess a compressed concentric pattern, exhibiting circuli, but no annuli occur, suggesting provisional assignment to the family.

Plesiosauria
Elasmosauridae

Referred specimens

DJ.952.25, 67 associated postcranial elements, DJ.1042.18, partial postcranial skeleton, DJ.1053, partial propodial; from the Late Maastrichtian López de Bertodano Formation, Seymour Island.

Description

Most of the associated elements of DJ.952.25 are vertebral fragments with some rib fragments; three portions of propodial are included. One relatively wide cervical vertebra (AP (anteroposterior) = 6.9 cm, T (transverse) = 8.65 cm, DV = *c.* 6.5 cm)) is preserved well enough to exhibit a lateral ridge. Such a ridge has been observed to differentiate the Elasmosauridae (see Gasparini *et al.* 2003).

Although many elements of DJ.1042.18 are fragmentary, one cervical vertebra (AP = 5.9 cm, T = 6.85 cm, DV = 5.15 cm) exhibits the lateral ridge characteristic of the Elasmosauridae. The vertebra is composed of a dumb-bell shaped centrum (Fig. 3H), although the other cervicals are only incipiently so shaped. This feature is characteristic of *Aristonectes*, although not enough is preserved of this large individual for precise assignment. The smaller, well-preserved, oval anterior cervicals measure: AP = 5.6 cm, T = 4.28 cm, DV = 7.78 cm; AP = 6.0 cm, T = 7.1 cm, DV = *c.* 5.0 cm; whereas the well-preserved oval posterior cervicals measure: AP = 6.75 cm, T = 8.3 cm, DV = *c.* 7.0 cm; AP = 6.4 cm, T = 8.68 cm, DV = 6.8 cm. One round trunk vertebra measures AP = 6.68 cm, T = 8.7 cm, DV = 8.1 cm. The caudal vertebrae possess articulating haemal arches, which articulate between successive vertebrae. The largest caudal measures AP = 4.4 cm, T = 6.68 cm, DV = 6.1 cm. The propodial fragments (Fig. 3I) indicate a short, squat element, even though the propodial head is broken. The distal portion is wide (19.6 cm) and indicates that two elements articulated distally. The propodial is slightly arthritic and exhibits exotosis. The phalanges are relatively rectangular in proximal (particularly) and distal outline, rather than round.

The propodial, DJ.1053, is short and squat similar to those of other elasmosaurids. The distal width is 21–22 cm.

cf. Elasmosauridae

Referred specimens

DJ.953.690, six vertebrae and five associated vertebral fragments; from the Late Maastrichtian López de Bertodano Formation, Seymour Island.

Description

Most of the vertebrae (Fig. 3J) represent trunk and anterior caudals. None of the neural arches are fused onto the centra, indicating a relatively young age of the individual. Whereas the trunk vertebrae are large, round and spool-shaped, the caudals are compressed dorsoventrally. Most of the specimens are poorly preserved, but the centra of one trunk (AP = 4.45 cm, T = 6.6 cm, DV = 5.68 cm) and one anterior caudal (AP = 4.58 cm, T = 7.0 cm, DV = 5.35 cm) were preserved well enough to measure.

Plesiosauria, fam. indet.

Referred specimens

DJ.355.140, half of large water-worn vertebra, either anterior trunk or distal cervical vertebra; from the Early Maastrichtian Cape Lamb Member, Snow Hill Island Formation, Comb Ridge, the Naze, James Ross Island. DJ.953.523, ?neural spine, DJ.953.524, transverse process of vertebra, and DJ.953.525, vertebral centrum, associated; DJ.953.751, large proximal portion of a rib; DJ.952.267, partial propodial; from the Late Maastrichtian López de Bertodano Formation, Seymour Island. DJ.952.260, 2 articulated vertebrae in concretion; from 250 m below the K/T boundary, Late Maastrichtian López de Bertodano Formation, Seymour Island.

Description

Most of these specimens can only be ascribed to the Plesiosauria. DJ.953.751, DJ.952.267 and DJ.355.140 are very large, and may therefore represent the Elasmosauridae, but not enough morphology is preserved for precision in assignment. DJ.953.523–5 are associated, but very poorly preserved. The centrum, DJ.953.525, is approximately 5.7 cm anteroposteriorly, but is too poorly preserved for other measurements.

Squamata
Mosasauridae
Tylosaurinae

Referred specimens

DJ.956.41, two or three caudal vertebrae; from the basal portion of the Early Maastrichtian López de Bertodano Formation, Seymour Island.

Description

The vertebrae of DJ.956.41 are large and consist of articulated caudal vertebrae in a concretion. The proximal portion of one half of a complete chevron exhibits an articular surface, indicating a large tylosaurine mosasaur. Although diagnostic features for generic designation are not found among the caudal vertebrae, their large size and articulating haemel arches suggests assignment to *Taniwhasaurus* (= *Lakumasaurus*) (Martin & Fernandez 2005), *Hainosaurus* or, less likely, *Tylosaurus*. The former genus was described from Antarctica (Novas *et al.* 2002), so reference to that genus seems probable.

Plioplatecarpinae
Plioplatecarpus sp.

Referred specimens

DJ.952.249, recurved tooth; from the Late Maastrichtian López de Bertodano Formation, Seymour Island. DJ.1020.2-C, small curved tooth; from within 1 m of the K/T boundary, Late Maastrichtian López de Bertodano Formation, Seymour Island.

Description

DJ.952.249 (Fig. 4A) is a strongly recurved, laterally compressed small tooth (AP = 8.5 mm, DV = 10.5 mm), suggesting a tooth derived from the pterygoid. The tooth possesses the distinctly striated enamel indicative of *Plioplatecarpus* or *Taniwhasaurus*.

DJ.1020.2-C (Fig. 4B) is a small (AP = 4.2 mm, T = 7.2 mm, DV = 10.0 mm), squat, laterally compressed, recurved tooth, perhaps a pterygoid tooth. The enamel of the external side is faceted and striated; the enamel of the internal side is very striated. The striated enamel indicates *Plioplatecarpus* (e.g. Kuypers *et al.* 1998), and the very small size suggests an immature individual.

cf. *Plioplatecarpus* sp.

Referred specimen

DJ.1020.2-H, broken tooth; from within 1 m of the K/T boundary, Late Maastrichtian López de Bertodano Formation, Seymour Island.

Description

DJ.1020.2-H (Fig. 4C) is a small (AP = 5.5 mm, T = 4.0 mm, DV = 8.0 mm), recurved, broken tooth that exhibits striations. However, the tooth is abraded and the striations appear to have been in the enamel. Its association with DJ.1020.2-C would seem to indicate an assignment to *Plioplatecarpus*, but preservation prompts reference to the genus.

Discussion

All of the specimens ascribed to *Plioplatecarpus* could be assigned to *Taniwhasaurus* (Novas *et al.* 2002). Recent examination by the first author confirmed that the dentition of this tylosaurine mosasaur possesses distinct striations. Deeply etched striations had previously been considered characteristic of the Plioplatecarpinae. The BAS specimens are considered representative of the latter group due to the small size in conjunction with the striations. Of course, the possibility exists that they may represent juvenile specimens of *Taniwhasaurus*.

Mosasaurinae
Mosasaurus sp. aff. *Mosasaurus hoffmanni*

Referred specimen

DJ.1053.10, large, fragmentary skull; from the Late Maastrichtian López de Bertodano Formation, Seymour Island.

Description

The skull is preserved in a concretion, which subsequently was broken into numerous pieces, making identification difficult. No evidence of a quadrate was observed and the skull appears to have been broken away posterior to the frontals. When complete, the skull may have been nearly 1 m in length. Only broken teeth remain in the BAS collection, and most are relatively smooth, with slight facets. Some remnants are only the moulds of the interior of the teeth, and even some moulds are very well faceted. Interestingly, some years prior to the collection of the specimen by BAS, American parties under the direction of Michael Woodburne collected a few interlocked teeth and jaw fragments. The tip of a tooth in the BAS collection fits perfectly onto one of the interlocked teeth (Fig. 4I), proving that they are the same individual. Within the collection are teeth that are distinctly faceted (Fig. 4J) and appear to have offset carinae, rather than being anteroposteriorly oriented. Although broken, the teeth provide the most compelling evidence for the identity of the mosasaur. The teeth are high-crowned, conical (Fig. 4I), have relatively thin enamel, carinae that are not serrate, slight striations in the enamel internally and some teeth are faceted. Those that are faceted exhibit three external facets (Fig. 4J) and offset carinae. Although not well enough preserved for precise identification, the teeth greatly resemble those of *Mosasaurus*, and their great size and tooth morphology is suggestive of *Mosasaurus hoffmanni*, originally described from the type area of the Maastrichtian (Fig. 4J).

Mosasaurus sp.

Referred specimens

DJ.1053.14-A, large broken tooth; from the Late Maastrichtian López de Bertodano Formation, Seymour Island. DJ.1020.2-A, large broken

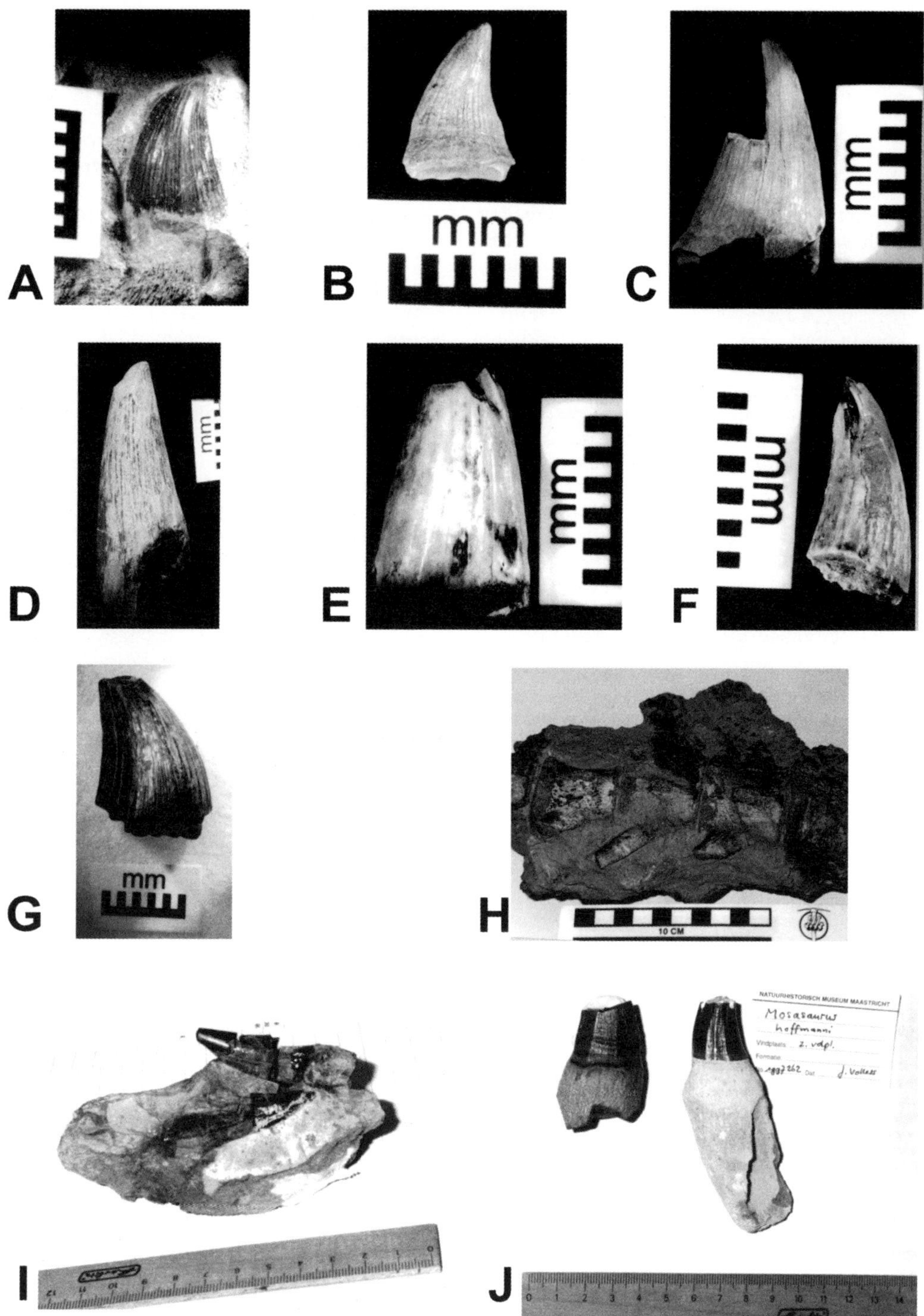

Fig. 4. (**A**) *Plioplatecarpus*, tooth, DJ.925.249; (**B**) *Plioplatecarpus*, tooth, DJ.1020.2-C; (**C**) cf. *Plioplatecarpus*, broken tooth, DJ.1020.2-H; (**D**) *Leiodon*, broken tooth, DJ.952.266; (**E**) *Mosasaurus*, broken tooth, DJ.1020.2-A; (**F**) *Mosasaurus*, broken tooth, DJ.1020.2-B; (**G**) *Mosasaurus*, broken tooth, DJ.1053.14-A; (**H**) Mosasaurinae, vertebrae, DJ.957.133; (**I**) *Mosasaurus* sp. aff. *M. hoffmanni*, teeth, DJ.1053-10, MOW 95-13; and (**J**) *Mosasaurus* sp. aff. *M. hoffmanni*, tooth on left compared to *M. hoffmanni* from the type Maastricht on the right.

tooth, DJ.1020.2-B, abraded tooth; from within 1 m of the K/T boundary, Late Maastrichtian López de Bertodano Formation, Seymour Island.

Description

DJ.1053.14-A (Fig. 4G) is a prismatic tooth (AP = *c.* 14.5 mm, T = *c.* 9.5 mm, DV = *c.* 24.0 mm) with both the base and tip broken away. Distinct lineation occurs within the dentine. The offset carinae and distinct tooth facets are features resembling those of *Mosasaurus.*

DJ.1020.2-A (AP = 17.0 mm, T = 16.0 mm) is broken in half (Fig. 4E), is faceted with no incised striations in the enamel, although lineation occurs within the enamel. The dentine exhibits no such lineation. The faceted tooth resembles those of the mosasaurines, and its long, conical outline suggests *Mosasaurus*.

DJ.1020.2-B (Fig. 4F) is smaller (AP = 7.0 mm, T = 6.5 mm, DV = *c.* 14.5 mm) than DJ.1020.2-A, and, although not much enamel remains, that which does has a similar enamel structure with lineated facets. DJ.1020.2-B possesses both anterior and posterior carinae, although the posterior is more prominent.

Leiodon sp.

Referred specimen

DJ.952.266, tooth; from the Late Maastrichtian López de Bertodano Formation, Seymour Island.

Description

The tooth (AP = 21 mm, T = 15.5 mm, DV = 43.0 mm) is not faceted, has smooth enamel and is laterally compressed (Fig. 4D). The carina is not serrate, and slight enamel striations appear internally. The lateral compression and smooth enamel are indicative of well-developed shear, a characteristic of *Leiodon* (e.g. Kuypers *et al.* 1998).

Mosasaurinae

Referred specimens

DJ.957.133, 18 partially articulated caudal vertebrae, four of which possess transverse processes; DJ.957.505, caudal vertebra; from the Late Maastrichtian López de Bertodano Formation, Seymour Island.

Description

The centra of the caudal vertebrae, DJ.957.133 (Fig. 4H) are ovate, not hexagonal, in outline. The haemal and neural spines are exceedingly long and relatively thin. For example, one vertebrae of T = 5.6 cm and AP = 4.25 cm has a neural spine of 6.5 cm in length. Although the haemal arches are broken, they are fused to the centra; one haemal has a preserved length of 7.1 cm, with an estimated length of approximately 10 cm. The anterior four caudal vertebrae are slightly larger with an AP length in the range of 4.5–4.7 cm.

The caudal vertebra, DJ.957.505, has an ovate outline and does not possess transverse processes. The ventral portion of the specimen is abraded, but remnant haemal knobs are suggestive of a mosasaurine. The specimen is of Moanasaurus from New Zealand, moderate size (centrum measurements: AP = 3.7 cm, T = 4.4 cm, DV = 5.2 cm).

All of these vertebrae appear to be those of adult individuals and are in the range of Moanasaurus from New Zealand, *Mosasaurus lemonnieri* from Europe, *M. mokoroa* from New Zealand and *M. conodon* or *M. missouriensis* from North America.

Discussion

The Late Cretaceous shark assemblage from Antarctica is comprised of chimaerids, hexanchids and odontaspids. This occurrence of the chimaerid, *Callorhinchus*, represents the first appearance of the genus in the Late Cretaceous of Antarctica and is represented by a diagnostic palatine tooth plate. Hexanchids are particularly known from the circum-Pacific area, as well as the Atlantic region, but are unknown from the North American Western Interior Seaway. In comparison, odontaspids were more widely distributed, including the North American seaway.

Interestingly, bony fish are poorly represented in the collections, suggesting that they may have been genuinely poorly represented in Antarctica during the Late Cretaceous. Only relatively few isolated teeth are assignable to the cosmopolitan teleost, *Enchodus*, and possibly to the ichthyodectids A single fish skeleton lacks the anterior half, and the tail is poorly preserved. Only on the basis of scale morphology can the small fish be tentatively assigned to the Sphenocephalidae, which also occur in the Late Cretaceous of Europe.

The plesiosaurs are not well preserved, and no diagnostic cranial elements were found in these collections. However, based on vertebral and propodial sizes and shapes, two plesiosaur taxa are probably present. One of these may be *Aristonectes*, a Late Cretaceous elasmosaurid recorded from Antarctica and Argentina that was adapted to strain the marine waters for small invertebrates (Gasparini *et al.* 2003). The other plesiosaur in the collections may be a more typical piscivorous elasmosaurid. No evidence exists of either short-necked forms such as the polycotylids, nor the very long-necked varieties of plesiosaurs such as *Styxosaurus* from North America.

Mosasaurs are the most taxonomically diverse group in the collections. Large vertebrae with articulating haemal arches suggest the presence of a tylosaurine mosasaur such as *Taniwhasaurus*, described from Antarctica and New Zealand, *Hainosaurus*, originally described from the type area of the Maastricht in the Netherlands–Belgium, or perhaps *Tylosaurus*, known principally from North America. These taxa are in need of revision, and, even then, these vertebrae will be difficult to assign. Nevertheless, they do indicate that the largest of mosasaurs inhabited the northern Antarctic Peninsula region during the Early Maastrichtian and corresponds to previous records of tylosaurines (Martin *et al.* 2002; Novas *et al.* 2002) from the Late Campanian portion of the Santa Marta Formation through the lower López de Bertodano (Early Maastrichtian). Another relatively closely related mosasaur taxon is *Plioplatecarpus*. This smaller piscivorous (Martin 1994) mosasaur is represented by small recurved, somewhat laterally compressed, distinctly striated teeth. This genus was first described from Antarctica by Martin *et al.* (2002) based on specimens from Seymour and Vega islands, their range was from the Early Maastrichtian Cape Lamb Member of the Snow Hill Island Formation to the Late Maastrichtian Sandwich Bluff Member of the López de Bertodano Formation. Some *Plioplatecarpus* specimens in the BAS collections were found within 1 m of the Cretaceous–Tertiary boundary (Fig. 2) and, if not juvenile *Taniwhasaurus* represent the highest known occurrence of the genus before their demise during the great extinction.

Mosasaurine mosasaurs were relatively diverse at the end of the Cretaceous, and this also appears to have been the case in Antarctica. *Leiodon* and at least two species of *Mosasaurus* can be differentiated in the BAS collections. *Leiodon* is known from Maastrichtian localities in the circum-Atlantic region and represents an extremely derived carnivorous reptile. Most previously reported specimens from Antarctica (Martin *et al.* 2002) occurred in the Late Maastrichtian Sandwich Bluff Member of the López de Bertodano Formation on Vega Island, but the specimen identified herein comes from Late Maastrichtian portion of the López de Bertodano Formation on Seymour Island.

Two sizes of *Mosasaurus* occur in the BAS collections: a medium-sized *Mosasaurus* (and/or *Moanasaurus*) and a very large species. The smaller *Mosasaurus* possesses teeth and vertebrae very similar to medium-sized species elsewhere, especially *M. lemonnieri.* This species was identified originally from the Maastricht area of Europe, and was provisionally identified from Antarctica, based on a quadrate fragment, cranial elements and vertebrae (Martin *et al.* 2002). The vertebrae are within the size range of *M. lemonnieri*, but the quadrate fragment is slightly larger than most known specimens. How this species relates to other medium-sized mosasaurs such as *M. conodon*, *M. missouriensis* or *M. mokoroa* remains to be determined. In any case, the occurrence of one *Mosasaurus* specimen within 1 m of the K–T boundary in Antarctica represents one of the highest known occurrences of the taxon globally.

The second, larger species of *Mosasaurus* was identified in two collections. Reports of a partial mosasaur jaw collected by the M. O. Woodburne parties had circulated for some years. Finally, a number of large interlocking teeth in matrix from both upper and lower jaws were found associated with two well-preserved isolated teeth, with the top of the crowns broken away. These teeth had characteristics of faceted enamel very similar to the large, Late Maastrichtian species, *M. hoffmanni*, which was the first mosasaur ever described. When these teeth were compared to those of a large skull in the BAS collections, the tip of a broken tooth was found to fit perfectly onto a tooth found by the American parties. This large skull appears to be similar to *M. hoffmanni.* Two possible morphs currently exist in collections from Europe; one with distinctly faceted teeth and with the posterior carina offset from the anteroposterior axis, and one with relatively smooth enamel, much like that of the tylosaurines. The Antarctic specimen exhibits characteristics of both, and is, therefore, provisionally referred to *M. hoffmanni.* This specimen was found in the Late Maastrichtian of Antarctica, which coincides with the age of the taxon in Europe.

Overall, the material within the Cambridge BAS collections provides new information concerning the palaeohistory of the Antarctic Peninsula region. Numerous records of Antarctic taxa provide biostratigraphic information, and the last appearances of some taxa occur just below the K/T boundary, particularly those of mosasaurs. The high-latitude occurrences of many taxa at a generic level found at lower latitudes elsewhere indicate a wide tolerance and cosmopolitan distribution of many marine taxa. However, some taxa such as the mosasaur,

Taniwhasaurus, indicate Gondwanan endemism. Sea turtles are a notable absence in Cretaceous assemblages from Antarctica, although they have been recorded from Eocene marine rocks that were deposited at approximately the same latitude (de la Fuente *et al.* 1995; Albright *et al.* 2003). Therefore, they could be expected from the Cretaceous rocks; however, their rarity has been noted at northern high latitudes (Nicholls & Russell 1990). Even if they are found in the Late Cretaceous rocks of Antarctica, their abundance will never rival that from deposits at lower latitudes. The BAS collections exhibit much greater superficial similarity with the European and eastern North American Late Cretaceous assemblages than with those from the Late Cretaceous North American Interior Seaway assemblages, suggesting greater isolation of the Western Interior Seaway.

Conclusions

- The BAS collections contain at least 14 taxa of sharks, bony fish, plesiosaurs and mosasaurs from the Maastrichtian of the James Ross Basin. Mosasaurs are the most taxonomically diverse family of vertebrates in the collections, whereas plesiosaurs are the most abundant specimens encountered. A similar increase in the number of plesiosaurs has been documented at higher northern latitudes (Nicholls & Russell 1990).
- The seeming rarity of teleost fish and sea turtles supports the view that they may have been relatively poorly represented in the Late Cretaceous high-latitude regions.
- Sharks are represented by the first Cretaceous record of *Callorhinchus* known from Antarctica. *Notiodanodon* and *Odontaspis* have been recorded from Antarctica previously, and are common elsewhere, although *Notiodanodon* does not occur in the Western Interior Seaway of North America.
- Plesiosaurs are common in Antarctica, ranging throughout the Late Campanian–Maastrichtian section. Two varieties of elasmosaurids occur, one adapted for straining small crustaceans and other small invertebrates, and the second more typically piscivorous. No short-necked plesiosaurs have yet been found in Antarctica.
- The taxonomically diverse mosasaurs include tylosaurines, plioplatecarpines and mosasaurines in the BAS collections. Both *Plioplatecarpus* (? *Taniwhasaurus*) and *Mosasaurus* are found within 1 m of the Cretaceous–Tertiary boundary, representing the highest occurrences of these taxa. A skull referred to *M. hoffmanni* suggests the first occurrence of the species in the Southern Hemisphere.
- Most specimens are relatively fragmentary due to intense surface weathering processes of poorly consolidated lithologies in the harsh Antarctic environment.
- Shallow-marine conditions may explain the abundance of plesiosaur specimens, as well as the great number of juvenile marine reptiles.

This research was partially funded by the United States National Science Foundation, Office of Polar Programs, grants OPP 9815231 and OPP 0087972. We thank Dr J. D. Stewart for identification of the cf. sphenocephalid and Dr J. Kriwet for identification of the chimaerids, Dr K. Shimada for discussions concerning sharks, and D. Parris and B. Grandstaff for opinions concerning teleosts. Drs J. A. Case and M. O. Woodburne lent important specimens for study, and Dr J. Jagt and A. Schulp made collections available at the Natuurhistorisch Museum Maastricht. Drs D. Brinkman, Z. Gasparini, J. Kriwet and D. Pirrie reviewed the manuscript and their efforts greatly enhanced the contribution. Dr D. Cantrill provided important locality information and consultation; we would like to thank him and a large number of other BAS geologists and colleagues for all their hard work in finding this material.

References

ALBRIGHT, B., WOODBURNE, M.O., CASE, J.A. & CHANEY, D.S. 2003. A leatherback sea turtle from the Eocene of Antarctica: implications for the antiquity of gigantothermy in Dermochelyidae. *Journal of Vertebrate Paleontology*, **23**, (3 Suppl.), 29A.

ARAMBOURG, C. 1954. Les poisons Cretaces du Jebel Tselfat (Maroc). *Notes et Mémoires. Services des Mines et de la Carte Géologique du Maroc*, **118**, 1–188.

CASE, J.A. & TAMBUSSI, C.P. 1999. Maestrichtian record of neornithine birds in Antarctica: comments on a Late Cretaceous radiation of modern birds. *Journal of Vertebrate Paleontology*, **19**, (3 suppl.), 37A.

CASE, J.A., MARTIN, J.E., CHANEY, D.S., REGUERO, M., MARENSSI, S.A., SANTILLANA, S.M. & WOODBURNE, M.O. 2000. The first duck-billed dinosaur (Family Hadrosauridae) from Antarctica. *Journal of Vertebrate Paleontology*, **20**, 612–614.

CIONE, A.L. & MEDINA, F. 1987. A record of *Notidanodon pectinatus* (Chondrichthyes, Hexanchiformes) in the Upper Cretaceous of the Antarctic Peninsula. *Mesozoic Research*, **1**, 79–88.

CHATTERJEE, S. & SMALL, B.J. 1989. New plesiosaurs from the Upper Cretaceous of Antarctica. *In*: CRAME, J.A. (ed.) *Origin and Evolution of the Antarctic Biota*. Geological Society, London, Special Publications, **47**, 197–215.

CHATTERJEE, S. & ZINSMEISTER, W.J. 1982. Late Cretaceous marine vertebrates from Seymour Island, Antarctic Peninsula. *Antarctic Journal of the United States*, **27**, (5), 66.

CORDES, A.H. 2002. A new charadriiform avian specimen from the early Maastrichtian of Cape Lamb, Vega Island, Antarctic Peninsula. *Journal of Vertebrate Paleontology*, **22**, (3 Suppl.), 46A.

CRAME, J.A., FRANCIS, J.E., CANTRILL, D.J. & PIRRIE, D. 2004. Maastrichtian stratigraphy of Antarctica. *Cretaceous Research*, **25**, 411–423.

CRAME, J.A., MCARTHUR, J.M., PIRRIE, D. & RIDING, J.B. 1999. Strontium isotope correlation of the basal Maastrichtian Stage in Antarctica to the European and US biostratigraphic schemes. *Journal of the Geological Society, London*, **156**, 957–964.

DE LA FUENTE, M., SANTILLANA, S.R. & MARENSI, S. 1995. An Eocene leatherback turtle (Cryptodira: Dermochelydae) from Seymour Island, Antarctica. *Studia Geologica Salmanticensia*, **31**, 17–30.

DEL VALLE, R., MEDINA, F. & DE BRANDONI, Z. 1977. Nova preliminar sobre el hallazgo de reptiles fósiles marinos del suborden Plesiosauria en las islas James Ross y Vega, Antártida. *Contribución Insituto Antártico Argentino*, **212**, 1–13.

ELLIOT, D.H., ASKIN, R.A., KYTE, F.T. & ZINSMEISTER, W.J. 1994. Iridium and dinocysts at the Cretaceous-Tertiary boundary on Seymour Island, Antarctica: Implications for the K/T event. *Geology*, **22**, 675–678.

GASPARINI, Z. & DEL VALLE, R. 1981. Mosasaurios: primer hallazgo en el continente Antártico. *Antártida*, **11**, 16–20.

GASPARINI, Z., BARDET, N., MARTIN, J.E. & FERNANDEZ, M. 2003. The elasmosaurid plesiosaur *Aristonectes* Cabrera from the latest Cretaceous of South America and Antarctica. *Journal of Vertebrate Paleontology*, **23**, (1), 105–116.

GASPARINI, Z., DEL VALLE, R. & GONI, R. 1984. An *Elasmosaurus* (Reptilia, Plesiosauria) of the Upper Cretaceous in the Antarctic. *Contribucion Instituto Antartico Argentino*, **305**, 1–24.

GOODY, P.C. 1976. *Enchodus* (Teleostei: Enchodontidae) from the Upper Cretaceous Pierre Shale of Wyoming and South Dakota with an evaluation of the North American enchodontid species. *Palaeontographica*, **152**, 91–112.

GRANDE, L. & CHATTERJEE, S. 1987. New Cretaceous fish fossils from Seymour Island, Antarctic Peninsula. *Palaeontology*, **30**, 829–837.

HATHWAY, B. 2000. Continental rift to back-arc basin: Jurassic–Cretaceous stratigraphical and structural evolution of the Larsen Basin, Antarctic Peninsula. *Journal of the Geological Society, London*, **157**, 417–432.

HUBER, B.T. 1998. Tropical paradise at the Cretaceous poles? *Science*, **282**, 2199–2200.

KRIWET, J. & GAZDZICKI, A. 2003. New Eocene Antarctic chimaeroid fish (Holocephali, Chimaeriformes). *Polish Polar Research*, **24**, 29–51.

KRIWET, J., LIRIO, J.M. NUÑEZ, H.J., PUCEAT, E. & LECUYER, C. 2006. Late Cretaceous Antarctic fish diversity. *In*: FRANCIS, J.E., PIRRIE, D. & CRAME, J.A. (eds) *Cretaceous–Tertiary High-latitude Palaeoenvironments, James Ross Basin, Antarctica.* Geological Society, London, Special Publications, **258**, 83–100.

KRIWET, J., NUÑEZ, H.J. & LIRIO, J.M. 2002. Late Cretaceous fish faunas from the Antarctic Peninsula. *Journal of Vertebrate Paleontology*, **23**, (3 Suppl.), 76A.

KUYPERS, M.M.M. & JAGT, J.W.M. *ET AL.* 1998. Laatkretaceisce mosasauriers uit Luik-Limburg: nieuwe vondsten leiden tot nieuwe inzichten. *Publications van het Natuurhistorisch Genootschap in Limburg*, **41**, 4–47.

LAWVER, L.A. GAHAGAN, L.M. & COFFIN, M.F. 1992. The development of paleoseaways around Antarctica. *In*: KENNETT, J.P. & WARNKE, D.A. (eds) *The Antarctic Paleoenvironment: A Perspective on Global Change*. American Geophysical Union, Antarctic Research Series, **56**, 7–30.

MACELLARI, C.E. 1988. Stratigraphy, sedimentology and paleoecology of Upper Cretaceous/Paleocene shelf-deltaic sediments of Seymour Island (Antarctica Peninsula). *In*: FELDMAN, R.M. & WOODBURNE, M.O. (eds) *Geology and Palaeontology of Seymour Island, Antarctic Peninsula.* Geological Society of America, Memoirs, **169**, 25–53.

MARTIN, J.E. 1994. Gastric residues in marine reptiles from the Late Cretaceous Pierre Shale in South Dakota: their bearing on extinction. *Journal of Vertebrate Paleontology*, **14**, (3 Suppl.), 36A.

MARTIN, J.E. & FERNANDEZ, M. 2005. The synonymy of the Late Cretaceous mosasaur (Reptilia) genus *Lakumasaurus* from Antarctica with *Taniwhasaurus* from New Zealand and its bearing upon faunal similarity. *In*: PANKHURST, R.J. & VEIGA, G.D. (eds) *Gondwana 12: Geological and Biological Heritage of Gondwana*, Academia Nacional de Ciencias, Cordoba, Argentina, 244.

MARTIN, J.E., BELL, G.L., JR. *ET AL.* 2002. Mosasaurs (Reptilia) from the Late Cretaceous of the Antarctic Peninsula. *In*: GAMBLE, J.A., SKINNER, D.N.B. & HENRYS, S. (eds) *Antarctica at the Close of a Millennium, 8th International Symposium on Antarctic Earth Sciences. Bulletin of the Royal Society New Zealand*, **35**, 293–299.

NICHOLLS, E.L. & RUSSELL, A.P. 1990. Paleobiogeography of the Cretaceous Western Interior Seaway of North America: the vertebrate evidence. *Palaeogeography, Palaeoclimatology, Palaeoecology*, **79**, 19–169.

NORIEGA, J.I. & TAMBUSSI, C.P. 1995. A Late Cretaceous Presbyornithidae (Aves: Anseriformes) from Vega Island, Antarctic Peninsula: Paleobiogeographic implications. *Ameghiniana*, **32**, 57–61.

NOVAS, F.E., FERNANDEZ, M., GASPARINI, Z.B., LIRIO, J.M., NUÑEZ, H.J. & PUERTA, P. 2002. *Lakumasaurus antarcticus*, n. gen. et sp., a new mosasaur (Reptilia, Squamata) from the Upper Cretaceous of Antarctica. *Ameghiniana*, **39**, 245–249.

OLIVERO, E.B., GASPARINI, Z., RINALDI, C.A. & SCASSO, R.A. 1991. First record of dinosaurs in Antarctica (Upper Cretaceous, James Ross Island): paleogeographical implication. *In*:

THOMSON, M.R.A., CRAME, J.A. & THOMSON, J.W. (eds) *Geological Evolution of Antarctica*, Cambridge University Press, Cambridge, 617–622.

PIRRIE, D., CRAME, J.A. & RIDING, J.B. 1991. Late Cretaceous stratigraphy and sedimentology of Cape Lamb, Vega Island, Antarctica. *Cretaceous Research*, **12**, 227–258.

STAHL, B. & CHATTERJEE, S. 1999. A Late Cretaceous chimaerid (Chondrichthyes, Holocephali) from Seymour Island, Antarctica. *Paleontology*, **42,** 979–989.

STAHL, B. & CHATTERJEE, S. 2002. A Late Cretaceous callorhynchid (Chondrichthyes, Holocephali) from Seymour Island, Antarctica. *Journal of Vertebrate Paleontology*, **22**, 848–850.

ZINSMEISTER, W.J. 1982. Late Cretaceous–Early Tertiary molluscan biogeography of the southern circum-Pacific. *Journal of Paleontology*, **56**, 84–102.

ZINSMEISTER, W.J. 1998. Discovery of a fish mortality horizon at the K–T boundary on Seymour Island: re-evaluation of events at the end of the Cretaceous. *Journal of Paleontology*, **72**, 556–571.

ZINSMEISTER, W.J. & MACELLARI, C.E. 1988. Bivalvia (Mollusca) from Seymour Island, Antarctic Peninsula. *In*: FELDMANN, R.M. & WOODBURNE, M.O. (eds) *Geology and Palaeontology of Seymour Island, Antarctic Peninsula.* Geological Society of America, Memoirs, **169**, 253–284.

ZINSMEISTER, W.J., FELDMANN, R.M., WOODBURNE, M.O. & ELLIOT, D.H. 1989. Latest Cretaceous/earliest Tertiary transition on Seymour Island, Antarctica. *Journal of Paleontology*, **63**, 731–738.

Eustatically controlled sedimentation recorded by Eocene strata of the James Ross Basin, Antarctica

SERGIO A. MARENSSI

Instituto Antártico Argentino, Universidad de Buenos Aires and CONICET, Cerrito 1248, Buenos Aires (1010), Argentina (e-mail: smarenssi@dna.gov.ar)

Abstract: The Eocene La Meseta Formation is an unconformity-bounded unit that records the geological evolution of the James Ross Basin, NE Antarctic Peninsula, during a period of decreasing tectonism and a lull in volcanic activity. This unit represents a composite incised valley, filled with deltaic, estuarine and shallow-marine deposits showing a landwards facies shift that indicates deposition during an overall sea-level rise. The six unconformity-based internal units (Valle de las Focas, Acantilados, Campamento, *Cucullaea* I, *Cucullaea* II and Submeseta allomembers) are interpreted to represent minor-scale regressive–transgressive events. Geological, palaeontological and new strontium isotopic ages allow the correlation of base-level changes with second- and third-order eustatic sea-level fluctuations. The base of the La Meseta Formation is correlated with a global 56 Ma lowstand in sea level followed by a main episode of flooding between 54.3 and 52.4 Ma. The base of the *Cucullaea* I Allomember is correlated with the well-known late Ypresian (49 Ma) lowstand, and the base of the Submeseta Allomember with the 36 Ma lowstand. Correlation of Eocene sea-level fluctuations in the northern Antarctic Peninsula with the global sea-level curve strengthens the concept of global syncroneity of the eustatic sea-level curve.

Sea-level fluctuations have a significant influence on stratal architecture of continental margins and interior basins. Eustacy also controls hydrographic–climatic patterns and, indirectly, biotic patterns as well. Differentiating between the effects of eustacy and tectonics in the sedimentary record has been a major goal of sedimentologists and stratigraphers. Although the chronology of fluctuating global sea levels is regarded as a key factor in the development of sequence stratigraphy, there is no consensus about the global synchroneity of these changes. Furthermore, differentiating eustatic from tectonic events is contentious, even in passive margin settings.

The aim of this paper is to present a case study from the Eocene of the James Ross Basin, northern Antarctic Peninsula (Fig. 1). The database for this paper is derived exclusively on outcrop studies on Seymour (Marambio) and Cockburn islands, which are the only places where the topmost sedimentary fill of the basin crops out. The chronology and timescales are those from Berggren *et al.* (1995), while the sea-level chart is from Haq *et al.* (1987).

Geological setting

The Larsen Basin (Fig. 1) is located on the continental shelf off the coast of the northern Antarctic Peninsula (Macdonald *et al.* 1988). The better-known James Ross Basin (del Valle *et al.* 1992) is the northern sub-basin of the Larsen Basin (Fig. 1). A 6–7 km-thick sedimentary succession was deposited between Jurassic and Eocene times in continental–marine environments related to the evolution of the Larsen Basin from a continental-rift to a back-arc setting (Hathway 2000).

The sedimentary fill of the James Ross Basin has been divided into a series of megasequences separated by boundaries that record main periods of change in basin configuration. The youngest of these cycles consists of a ?Hauterivian–Eocene aggradational–progradational deep- to shallow-marine clastic wedge, deposited along a faulted western margin of the Antarctic Peninsula during a phase of arc uplift and extension (Hathway 2000).

Pirrie *et al.* (1991) suggested that marginal and intrabasinal tectonism in the James Ross Basin was on the decline after Coniacian times. Sedimentation was subsequently controlled by base-level changes, with a magnitude of 50–100 m (and probably less than 50 m), but the biostratigraphic resolution at that time prevented them from constraining whether these changes were tectonically or eustatically driven. The former can tentatively be recognized during the latest stages of basin development as tectonic activity declined.

Coniacian–Eocene sediments in the James Ross Basin reflect mainly shallow-marine and coastal environments (Macellari 1988a;

From: Francis, J. E., Pirrie, D. & Crame, J. A. (eds) 2006. *Cretaceous–Tertiary High-Latitude Palaeoenvironments, James Ross Basin, Antarctica.* Geological Society, London, Special Publications, **258**, 125–133. 0305–8719/06/$15

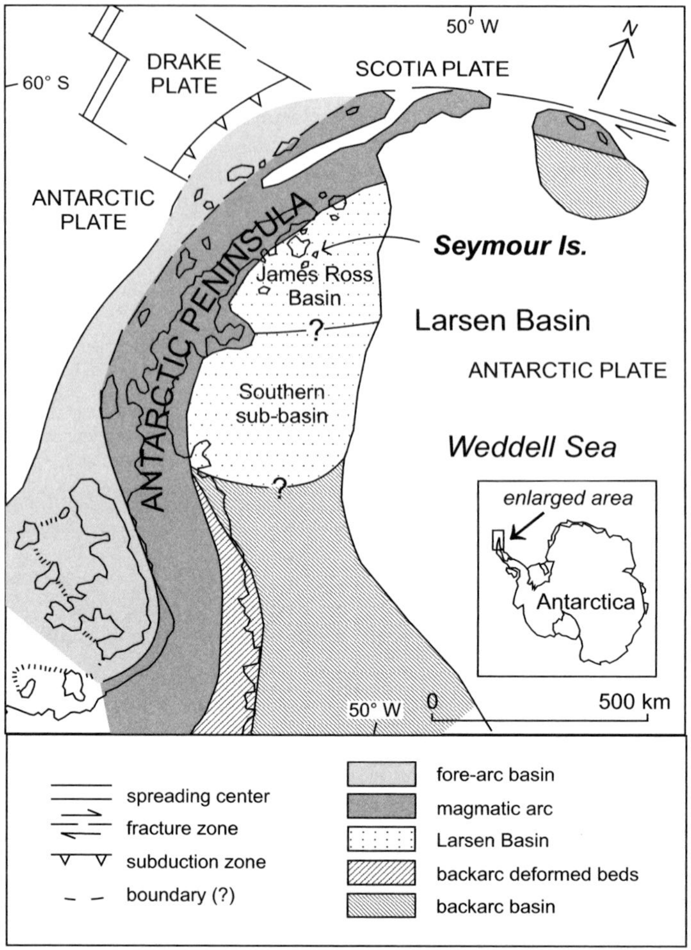

Fig. 1. Sketch map of the Antarctic Peninsula region showing the major tectonic settings.

Marenssi *et al.* 1998*a*; Pirrie *et al.* 1991). Basin uplift or decreased basin subsidence outpaced by sedimentation led to the development of a broad shallow shelf (Pirrie *et al.* 1991) that was sporadically emergent during the Palaeogene (Sadler 1988; Marenssi *et al.* 1998*b*). As a consequence of the development of this broad, stable shelf, decreased synsedimentary tectonic deformation and decreased coeval arc volcanism, sedimentation was most probably controlled by eustatic sea-level changes. There is now considerable speculation that elsewhere in the Late Cretaceous major sea-level fluctuations changes might in fact be glacioeustatic (Crame *et al.* 2004). The repeated pattern of regressions and transgressions within the Late Cretaceous of the James Ross Basin, broadly comparable to depositional cycles within the Austral Basin in southern South America (Macellari 1988*b*; Macellari *et al.* 1989; Pirrie *et al.* 1991), may support this hypothesis. In addition, Abreu & Anderson (1998) presented evidence of eustatic episodes related to the evolution of an Antarctic ice sheet at least since the base of the Lutetian stage (lower–middle Eocene boundary). During the Eocene there was no synsedimentary tectonic activity in the basin and the only recorded deformation can be related to slope failure in delta-front depositional settings within the La Meseta Formation (Elliot & Trautman 1982; Doktor *et al.* 1988; Marenssi *et al.* 1998*b*).

Sedimentology and stratigraphy of the La Meseta Formation

The La Meseta Formation (Elliot & Trautman 1982; Marenssi *et al.* 1998*a*), represents a composite incised-valley system cut into an emergent marine shelf with more than one episode of incision (Marenssi 1995; Marenssi *et al.* 1998*b*). The sedimentary succession cropping out on Seymour Island (Fig. 2) has been subdivided into six erosionally based allomembers (Marenssi & Santillana 1994; Marenssi 1995; Marenssi *et al.* 1998*a*, *b*) (Fig. 3) that can be grouped into three sedimentary cycles previously reported by Elliot & Trautman (1982) and Porebski (1995).

The base of this unit on Seymour Island defines a steep-sided valley 7 km wide with more than 70 m of relief (Sadler 1988; Marenssi 1995; Marenssi *et al.* 1998*b*). Sadler (1988), and later Marenssi (1995) and Marenssi *et al.* (1998*b*) attributed the origin of the basal unconformity to fluvial erosion. This is consistent with the onlapping stratal geometry of the valley fill (Sadler 1988; Marenssi 1995; Marenssi *et al.* 1998*a*, *b*). Further, Marenssi (1995) and Marenssi *et al.* (1998 *a*, *b*) indicated that the base of each of the six allomembers represent a fluvial erosional surface reshaped during the following transgression.

On the other hand, Porebski (1995, 2000), based on his interpretation of the sedimentary environments as being 'invariably estuarine/shallow marine' (Porebski 1995, p. 84) and a 'tectonic' origin of synsedimentary internal deformational features (cf. Pezzetti & Krissek 1987; Sadler 1988; Doktor *et al.* 1988; Pirrie *et al.* 1991; Marenssi 1995; Marenssi *et al.* 1998*b*), favoured a deformational–erosional origin for the basal surface. However, Porebski (1995, p. 86) suggested that although 'the relative sea level rises bound to the individual valley-fill increments would have resulted from tectonic and tectonically-enhanced eustatic causes, . . . the sequence boundaries would probably reflect a predominantly eustatic signal'. The La Meseta Formation is indeed a composite valley fill containing not only estuarine, but also deltaic and shallow-marine sediments (Stilwell & Zinsmeister 1992; Marenssi 1995; Marenssi *et al.* 1998*b*, 2002) (Fig. 4). The deformation features are restricted to delta-front environments and to cut bank collapses of estuarine channels of

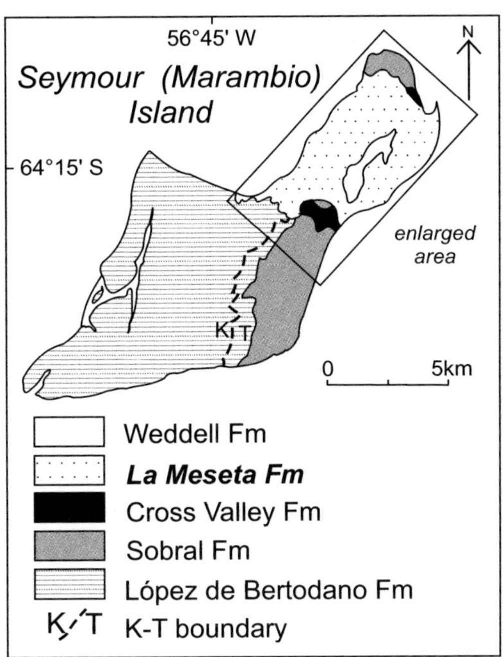

Fig. 2. Geological map of Seymour (Marambio) Island. After Marenssi *et al.* (2002).

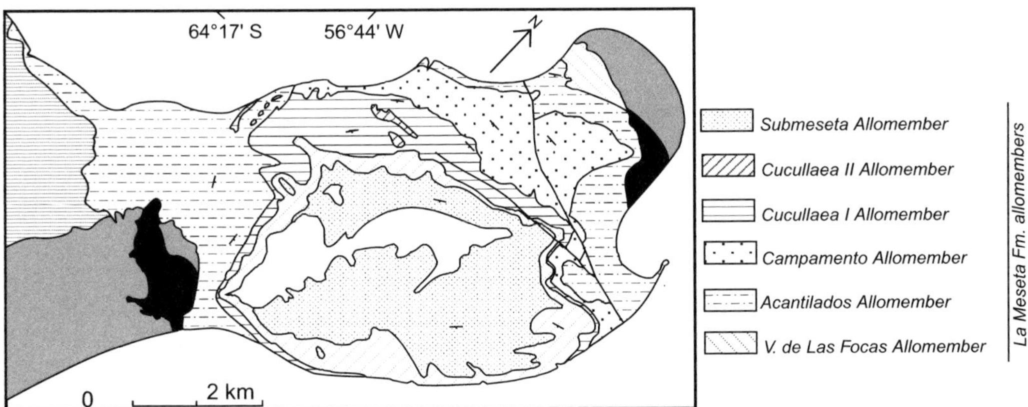

Fig. 3. Geological map of the northern third of Seymour (Marambio) Island showing the distribution of the internal units of the La Meseta Formation. After Marenssi *et al.* (2002).

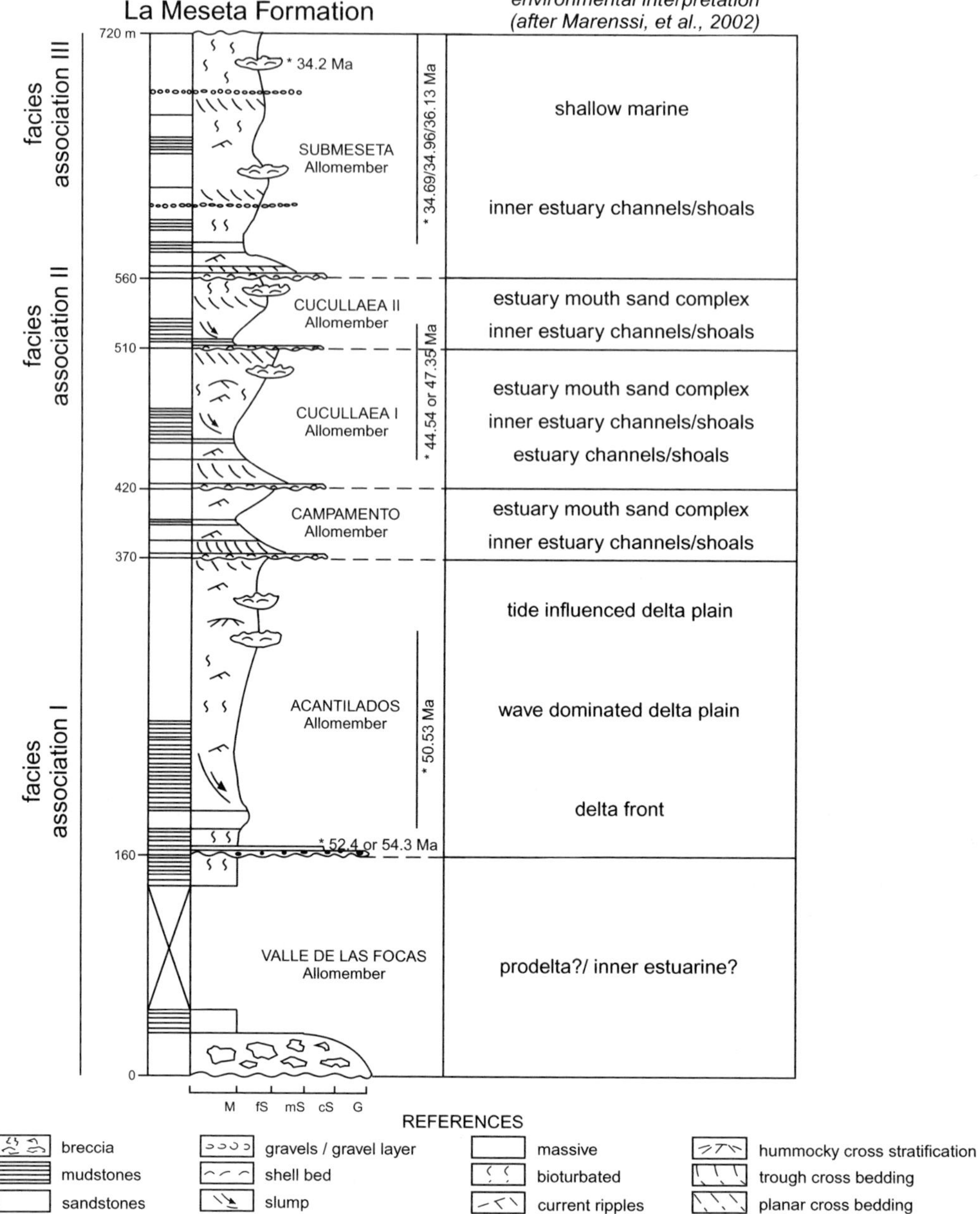

Fig. 4. Stratigraphic column for the La Meseta Formation on Seymour Island (modified from Marenssi *et al.* 2002).

fluvial valleys (Pezzetti & Krissek 1987; Stilwell & Zinsmeister 1992; Marenssi 1995; Marenssi *et al.* 1998*b*).

The basal sedimentary increment ('Sequence' of Porebski 1995) is at least 250 m thick, representing a thickening- and coarsening-upwards succession that includes the allomembers Valle de las Focas, Acantilados and Campamento (Marenssi *et al.* 1998*a*). The base is not exposed anywhere, so the axial facies of all but the last allomember are not known. The Acantilados Allomember represents sedimentation in a prodelta of a bayhead delta or a low-energy mid-estuary setting within the incised valley

(Marenssi *et al.* 1998*b*). The erosional boundary between this and the succeeding unit is overlain by transgressive, glauconitic cross-bedded clean sandstones containing trace and body fossils that suggest fully marine conditions (Casadío *et al.* 2000, 2001). Accordingly, the Acantilados and Campamento allomembers represent the progradation of a deltaic system within the drowned valley after the flooding event (Marenssi *et al.* 1998*b*; also see Elliot & Trautman 1982; Coccozza & Clarke 1992).

The second cycle is about 180 m thick and comprises allomembers *Cucullaea* I and *Cucullaea* II. Within these allomembers Porebski (1995) recognized a total of eight transgressive–regressive (T–R) cycles. The basal unconformity bounding the *Cucullaea* I Allomember represents a major event within the Eocene. It shows a prominent basal incision with up to 50–60 m of relief (Sadler 1988) floored by a thick coquina with cobbles, phosphatic clasts and a few reworked Cretaceous invertebrates (Telm 4 of Sadler 1988), bearing all of the hallmarks of a major erosional hiatus followed by slow sediment accumulation. Stilwell & Zinsmeister (1992) stated that this unit may represent an expression of a transgression following a lowstand in sea level. In this cycle, each allomember records sedimentation from inner estuarine settings to estuary mouth–shoreface environments, including barrier island–lagoon facies (Stilwell & Zinsmeister 1992; Marenssi 1995; Marenssi *et al.* 1998*b*).

The youngest cycle is about 140 m thick and is represented by the Submeseta Allomember. The basal incision is up to 60 m deep. This erosional surface is covered by 20–40 m-thick inclined heterolithic facies (IHS in the sense of Thomas *et al.* 1986) composed of estuarine, very fine sandstones and mudstones. The inclined heterolithic packet passes up to a stacked set of 25–40 m-thick sandy, aggradational to possibly retrogradational, shoreface parasequences (Porebski 1995; Marenssi *et al.* 1998*b*).

Overall, the La Meseta Formation represents a good example of a punctuated transgression (Cattaneo & Steel 2003), where the long-term landwards shift of facies is punctuated by short regressive periods.

Age of the La Meseta Formation

The age of the La Meseta Formation has received much attention by numerous authors (Stilwell & Zinsmeister 1992). Based on the molluscan faunas, Zinsmeister & Camacho (1980) proposed a Late Eocene–early Oligocene age for this unit. Stilwell & Zinsmeister (1992) also sustained this age. Based on this assumption the basal erosion surface was related to the late Ypresian (49 Ma) sea-level lowstand by Sadler (1988). However, Cocozza & Clarke (1992) and later Askin (1997) indicated that the lower third of this unit (Valle de las Focas, Acantilados and the lower part of Campamento allomembers) is late Early Eocene in age.

In close agreement with the former statement, a new absolute age derived from $^{87}Sr/^{86}Sr$ ratios in *Ostrea antarctica* shells from the transgressive sands at the base of the Acantilados Allomember is presented. The Radiogenic Isotopes Laboratory in the Department of the Geological Sciences of the Ohio State University performed the analyses. $^{87}Sr/^{86}Sr$ determinations were made using a dynamic multicollector of all Sr isotopes on a Finnigan MAT 261A thermal ionization mass spectrometer. Measured values of $^{87}Sr/^{86}Sr$ were normalized assuming normal Sr with $^{87}Sr/^{86}Sr$ = 0.119400. The reference value of $^{87}Sr/^{86}Sr$ for the SRM987 is 0.710242 ± 0.000010 (1 σ external reproductivity). The $^{87}Sr/^{86}Sr$ ratio was converted into a numerical age using the SIS (Strontium Isotope Stratigraphy), version 3: 10/99 of the look-up table of McArthur *et al.* (2001). The reference value $^{87}Sr/^{86}Sr$ used (= 0.710242) was corrected to make the data concordant with SRM987 of 0.710248 used in the construction of this look-up table. Average $^{87}Sr/^{86}Sr$ values of 0.707709 ± 0.000003 indicate an age of between 52.40 and 54.33 Ma (Early Eocene) for the analysed samples. Therefore, the erosional event at the base of La Meseta Formation has to be older than the late Ypresian lowstand.

Based on the study of palynofloras, Askin (1997) considered the middle part of the La Meseta Formation (*Cucullaea* I and *Cuccullaea* II allomembers) to be Middle Eocene. This age corresponds well with the age derived from fossil mammals (Bond *et al.* 1993) and $^{87}Sr/^{86}Sr$ derived ages of 44.54 or 47.35 Ma reported by Dutton *et al.* (2002) for Telm 5 (*Cucullaea* I or *Cucullaea* II allomembers).

The upper third of the unit (Submeseta Allomember) is regarded as Late Eocene–earliest Oligocene by Askin (1997). However, Dingle *et al.* (1998) reported a $^{87}Sr/^{86}Sr$ derived age of 34.2 Ma (late Late Eocene) for the topmost few metres of the unit, bracketing it into the Eocene. In addition, Dutton *et al.* (2002) presented $^{87}Sr/^{86}Sr$ derived ages of 36.13, 34.96 and 34.69 Ma for Telm 7 (Submeseta Allomember).

Global sea-level chart and the age of the La Meseta unconformities

The La Meseta Formation unconformably rests on top of all the other sedimentary units of the island. Sedimentary rocks as old as 50.53 Ma (Dutton *et al.* 2002) or even 52.4–54.3 Ma (base of Acantilados Allomember) lap onto Maastrichtian and Palaeocene rocks (Sadler 1988; Marenssi *et al.* 1998*a*). However, the microflora of the Valle de las Focas Allomember lacks any diagnostic Palaeocene elements (Wrenn & Hart 1988; Askin 1997) constraining the age into the early Early Eocene.

The youngest underlying unit, the Cross Valley Formation (Elliot & Trautman 1982), has been regarded as late Late Palaeocene (Wrenn & Hart 1988). Therefore, the erosional episode at the base of the La Meseta Formation may be correlated with the short-term lowstands between 55 and 56 Ma (56 Ma sequence boundary of Haq *et al.* 1987), very close to the Palaeocene–Eocene boundary (Fig. 5).

The next most prominent erosional surface at the base of the *Cucullaea* I Allomember has to be younger than the late Early Eocene sedimentary rocks of the underlying allomembers and older than the Middle Eocene mammal-bearing sedimentary rocks of *Cucullaea* I Allomember. Therefore, this erosional surface is now correlated with the late Ypresian lowstand (49.5 Ma sequence boundary of Haq *et al.* 1987). Given a Late Eocene age for the Submeseta Allomember, its basal unconformity may correspond to the 36 Ma lowstand in sea level (Fig. 5).

Estimates of amplitudes of the sea-level fluctuations during the Eocene can be derived from the Haq *et al.* (1987) chart. They range between 70 and 20 m, and are therefore consistent with the erosional depth and sedimentary thickness of the valley confined units deposited within the La Meseta Formation (Marenssi *et al.* 1998*b*).

If these main erosional surfaces are, in fact, correlative with lowstands in sea level, then longer fluctuations should also be recognized in the La Meseta Formation. The long-term eustatic curve of Haq *et al.* (1987) can be compared with the general evolution of the La Meseta Formation to test this hypothesis.

The Haq *et al.* (1987) sea-level curve (Fig. 5) indicates a relatively rapid sea-level rise from 56 to 52.5 Ma, followed by a very slow fall until 39.5 Ma. This was followed by a moderate rise that lasted until 35 Ma. The rapid rise is consistent with the interpreted flooding of the incised valley, reaching a maximum at about the age suggested by the transgressive sandstone of the Valle de las Focas–Acantilados boundary (54.3–52.4 Ma reported herein). Deposition first took place in a low-energy middle-estuarine setting represented by the Valle de las Focas mudstones. Shortly before or during the maximum flooding event, the progradation of a bayhead delta (Acantilados and Campamento allomembers) filled up the valley. The erosion surface at the base of the Campamento Allomember may be interpreted as a tidal ravinement surface, and could have been formed during the minor 51.5 Ma lowstand. During Middle Eocene time, the most 'regressive' facies were deposited (cf. Stilwell & Zinsmeister 1992). Abundance of land-derived fossils like tree trunks, leaves, a flower and mammal teeth occur within the *Cucullaea* I Allomember. A combination of open-marine, protected and estuarine environments provided suitable habitats and preservation potential, recorded by high fossil diversity and abundance within this unit. The eight transgressive–regressive cycles described by Porebski (1995) are confined to the *Cucullaea* I and *Cucullaea* II allomembers, and they may be correlated with the same number of minor sequences or lowstands in the Haq *et al.* (1987) curve (Fig. 5). The base of the Submeseta Allomember is then correlated with the main lowstand at 36 Ma. Finally, as the sea level rose again, estuarine heterolithic sediments were deposited first, and followed by a set of aggradational–retrogradational shoreface parasequences that developed at the top of the unit during the Late Eocene.

Conclusions

The NW margin of the Antarctic Peninsula acted as the extensional margin of the Weddell Sea, and, therefore, the James Ross Basin may have evolved from a continental-rift to a back-arc setting during the Jurassic–?Late Cretaceous (Hathway 2000) and finally into a continental shelf in the Eocene (latest Cretaceous–Palaeogene?). A wide stable shelf developed during the Eocene to the east of the northern Antarctic Peninsula. A period of tectonic quiescence and a lull in volcanism allowed sedimentation to be controlled mostly by sea-level changes. Correlation of these fluctuations with global sea-level charts indicate that they were eustatic in origin. Sedimentological, palaeontological and geochemical data strongly support this correlation. Eocene sedimentation on the James Ross Basin was mainly controlled by eustatic sea-level changes.

Eocene rocks of the James Ross Basin represent sedimentation in coastal and

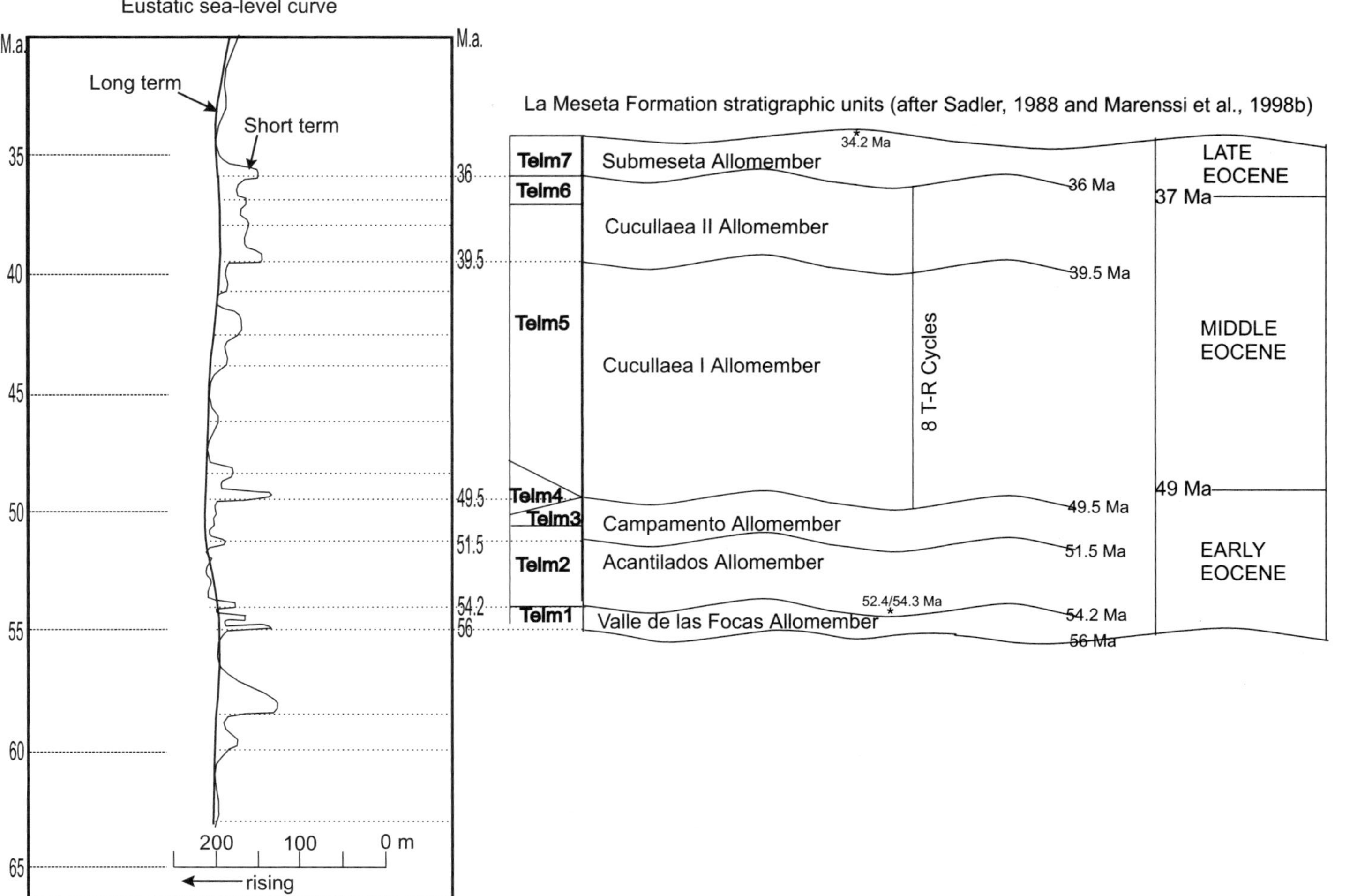

Fig. 5. Comparison between the global seal-level curve (left) from Haq *et al.* (1987) and the La Meseta Formation unconformity-bounded internal units (right) showing proposed ages for the major erosive events at the base of each allomember and the eustatic control on sedimentation.

shallow-marine environments. The La Meseta Formation fills an incised valley cut into an emergent shelf during a lowstand of the sea level. The composite fill represents a good example of a main transgression punctuated by short regressive periods.

New biostratigraphic and strontium isotopic data allow correlation of the erosive events of the La Meseta Formation, with lowstands shown in the Haq *et al.* (1987) sea-level curve (third order). The base of the La Meseta Formation is now placed at 56 Ma, and a main episode of flooding took place between 54.3 and 52.4 Ma. The base of the *Cucullaea* I Allomember is correlated with the well-known late Ypresian (49 Ma) lowstand and the base of the Submeseta Allomember with the 36 Ma lowstand.

The palaeoenvironmental evolution of the La Meseta Formation also agrees well with the long-term (second-order) sea-level curve of Haq *et al.* (1987). A moderately quick sea-level rise, between 56 and 52.5 Ma, is coincident with valley inundation and the development of estuarine conditions within that valley. After maximum water depths were reached at about 52.5 Ma, a thick bayhead delta sequence filled the valley. High rates of sedimentation coupled with a slow sea-level fall between 52.5 and 39.5 Ma led to the development of the most regressive facies of the section represented by barrier island–lagoon and/or estuarine settings. A subsequent slow sea-level rise from 39.5 up to 35 Ma produced aggradational–retrogradational shoreface parasequences at the top of the section.

I wish to acknowledge the logistic support of the Dirección Nacional del Antártico and Fuerza Aérea Argentina during field work in Antarctica. The advice and guidance of Dr E. Olivero during the last part of my PhD studies is warmly acknowledged. Prof. J. McArthur kindly provided an upgrade of the 'look-up table' version 3: 10/99, 2001. Useful reviews from J. B. Anderson, D. Pirrie and S. Porebski helped to greatly improve the manuscript. The Instituto Antártico Argentino supported this project along with grants from National Geographic Society (6615-99 and 7125-01 to the author).

References

ABREU, V.S. & ANDERSON, J.B. 1998. Glacial eustasy during the Cenozoic: Sequence stratigraphic implications. *AAPG Bulletin*, **82**, 1385–1400.

ASKIN, R.A. 1997. Eocene–?Earliest Oligocene terrestrial palynology of Seymour Island, Antarctica. *In*: RICCI, C.A. (ed.) *The Antarctic Region: Geological Evolution and Processes. Proceedings of the VII International Symposium on Antarctic Earth Sciences, Siena, 1995*. Terra Antartica Publication, Siena, 993–996.

BERGGREN, W.A., KENT, D.V., SWISHER, C.C., III & AUBRY, M.P. 1995. A revised Cenozoic geochronology and chronostratigraphy. *In*: BERGGREN, W.A., KENT, D.V., AUBRY M.P. & HARDENBOL, J. (eds) *Geochronology, Time Scales and Global Stratigraphic Correlation*. Society of Economic Paleontologists and Mineralogists, Special Publications, **54**, 129–212.

BOND, M., REGUERO, M.A. & VIZCAÍNO, S.F. 1993. Mamíferos continentales de la Formación La Meseta (Terciario, Antártida). *In*: *Biocronología. XIII Congresso Brasileiro de Paleontologia y I Simpósio Paleontológico do Cone Sul, Porto Alegre, Brasil, Actas*, 93. Universidade do Vale do Rio dos Sinos, São Leopoldo.

CASADÍO, S., MARENSSI, S.A. & SANTILLANA, S.N. 2000. Trazas bioerosivas endolíticas debidas a briozoos perforantes (Ctenostomata) en el Eoceno de Antártida. *In*: *II Congreso Latinoamericano de Sedimentología y VIII Reunión Argentina de Sedimentología, Mar del Plata, Marzo 2000. Resúmenes*, Asociación Argentina de Sedimentología, La Plata, 59–60.

CASADÍO, S., MARENSSI, S.A. & SANTILLANA, S.N. 2001. Bioerosive endolithic traces attributed to boring bryozoans (Ctenostomata) in the Eocene of Antarctica. *Ameghiniana*, **38**, 321–329.

CATTANEO, A. & STEEL, R. 2003. Transgressive deposits: a review of their variability. *Earth Science Reviews*, **62**, 187–228.

COCCOZZA, C. & CLARKE, C. 1992. Eocene microplankton from La Meseta Formation. *Antarctic Science*, **4**, 355–362.

CRAME, J.A., FRANCIS, J.E., CANTRILL, D.J. & PIRRIE, D. 2004. Maastrichtian stratigraphy of Antarctica. *Cretaceous Research*, **25**, 411–423.

DEL VALLE, R.A., ELLIOT, D.H. & MACDONALD, D.I.M. 1992. Sedimentary basins on the east flank of the Antarctic Peninsula: proposed nomenclature. *Antarctic Science*, **4**, 477–478.

DINGLE, R., MARENSSI, S.A. & LAVELLE, M. 1998. High latitude Eocene climatic deterioration: Evidence from the northern Antarctic Peninsula. *Journal of South American Earth Sciences*, **11**, 571–579.

DOKTOR, M., GAZDZICKI, A., MARENSSI, S.A., POREBSKI, S.J., SANTILLANA, S.N. & VRBA, A.V. 1988. Argentine–Polish geological investigations on Seymour (Marambio) Island, Antarctica, 1988. *Polish Polar Research*, **9**, 521–541.

DUTTON, A.L., LOHMANN, K. & ZINSMEISTER, W.J. 2002. Stable isotope and minor element proxies for Eocene climate of Seymour Island, Antarctica. *Paleoceanography*, **17**, (2), 1–13.

ELLIOT, D.H. & TRAUTMAN, T.A. 1982. Lower Tertiary strata on Seymour Island, Antarctic Peninsula. *In*: CRADDOCK, C. (ed.) *Antarctic Geoscience*. University of Wisconsin Press, Madison, WI, 287–297.

HAQ, B.U., HARDENBOL, J. & VAIL, P.R. 1987. Chronology of fluctuating sea levels since the Triassic. *Science*, **235**, 1156–1166.

HATHWAY, B. 2000. Continental rift to back-arc basin: Jurassic–Cretaceous stratigraphical and structural evolution of the Larsen Basin, Antarctic Peninsula. *Journal of the Geological Society, London*, **157**, 417–432.

MACDONALD, D.I.M., BAKER, P.F. ET AL. 1988. A preliminary assessment of the hydrocarbon potential of the Larsen Basin, Antarctica. *Marine and Petroleoum Geology*, **5**, 34–53.

MACELLARI, C.E. 1988*a*. Stratigraphy, sedimentology and paleoecology of Upper Cretaceous Paleocene shelf-deltaic sediments of Seymour Island (Antarctic Peninsula). *In*: FELDMANN, R.M. & WOODBURNE, M.O. (eds) *Geology and Paleontology of Seymour Island, Antarctic Peninsula*. Geological Society of America, Memoirs, **169**, 25–53.

MACELLARI, C.E. 1988*b*. Cretaceous paleogeography and depositional cycles of western South America. *Journal of the South American Earth Sciences*, **1**, 373–418.

MACELLARI, C.E., BARRIO, C.A. & MANASSERO, M.J. 1989. Upper Cretaceous to Paleocene depositional sequences and sandstone petrography of southwestern Patagonia (Argentina and Chile). *Journal of South American Earth Sciences*, **2**, 223–239.

MARENSSI, S.A. 1995. *Sedimentología y paleoambientes de sedimentación de la Formación La Meseta, isla Marambio, Antártida*. PhD Thesis, Universidad de Buenos Aires.

MARENSSI, S.A. & SANTILLANA, S.N. 1994. Unconformity-bounded units within the La Meseta Formation, Seymour Island, Antarctica: a preliminary approach. *In*: *XXI Polar Symposium, Warszawa, Poland, Abstracts*. Polish Academy of Sciences, Warsaw, 33–37.

MARENSSI, S.A., NET, L.I. & SANTILLANA, S.N. 2002. Provenance, depositional and paleogeographic controls on sandstone composition in an incised valley system: the Eocene La Meseta Formation, Seymour Island, Antarctica. *Sedimentary Geology*, **150**, 301–321.

MARENSSI, S.A., SANTILLANA, S.N. & RINALDI, C.A. 1998*a*. Stratigraphy of La Meseta Formation (Eocene), Marambio Island, Antarctica. *In*: CASADÍO, S. (ed.) *Paleógeno de América del Sur y de la Península Antártica*. Revista de la Asociación Paleontológica Argentina, Publicación Especial, **5**, 137–146.

MARENSSI, S.A., SANTILLANA, S.N. & RINALDI, C.A. 1998*b*. *Paleoambientes sedimentarios de la Aloformación La Meseta (Eoceno), isla Marambio (Seymour), Antártida*. Instituto Antártico Argentino, Contribución, **464**.

MCARTHUR, J.M., HOWARTH, R.J. & BAILEY, T.R. 2001. Strontium Isotope Stratigraphy: LOWESS Version 3: Best fit to marine Sr-isotope curve for 0–509 Ma and accompanying Look-Up Table for deriving numerical age. *Journal of Geology*, **109**, 155–170. (Look-up Table version 3:10/99.)

PEZZETTI, T.F. & KRISSEK, L.A. 1987. Sedimentology and provenance of the Tertiary La Meseta Formation, Seymour Island, Antarctica. *Geological Society of America, Abstracts with Programs*, **18**, 319.

PIRRIE, D., WHITHAM, A.G. & INESON, J.R. 1991. The role of tectonics and eustacy in the evolution of a marginal basin: Cretaceous–Tertiary Larsen Basin, Antarctica. *In:* MACDONALD, D.I.M. (ed.) *Sedimentation, Tectonics and Eustacy: Sea Level Changes at Active Margins*. International Association of Sedimentologists, Special Publications, **12**, 293–305.

POREBSKI, S.J. 1995. Facies architecture in a tectonically controlled incised-valley estuary: La Meseta Formation (Eocene) of Seymour Island, Antarctic Peninsula. *Studia Geologica Polonica*, **107**, 7–97.

POREBSKI, S.J. 2000. Shelf-valley compound fill produced by fault subsidence and eustatic sea-level changes, Eocene La Meseta Formation, Seymour Island, Antarctica. *Geology*, **28**, 147–150.

SADLER, P. 1988. Geometry and stratification of uppermost Cretaceous and Paleogene units on Seymour Island, northern Antarctic Peninsula. *In:* FELDMANN, R.M. & WOODBURNE, M.O. (eds) *Geology and Paleontology of Seymour Island, Antarctic Peninsula*. Geological Society of America, Memoirs, **169**, 303–320.

STILWELL, J.D. & ZINSMEISTER, W.J. 1992. *Molluscan Systematics and Biostratigraphy. Lower Tertiary La Meseta Formation, Seymour Island, Antarctic Peninsula*. American Geophysical Union, Antarctic Research Series, **55**.

THOMAS, R.G., SMITH, D.G., WOOD, J.M., VISSER, J., CALVERLEY-RANGE, E.A. & KOSTER, E.H. 1986. Inclined heterolithic stratification – terminology, description, interpretation and significance. *Sedimentary Geology*, **53**, 123–179.

WRENN, J.H. & HART, G.F. 1988. Paleogene dinoflagellate cyst biostratigraphy of Seymour Island, Antarctica. *In*: FELDMANN, R.M. & WOODBURNE, M.O. (eds) *Geology and Paleontology of Seymour Island, Antarctic Peninsula*. Geological Society of America, Memoirs, **169**, 303–320.

ZINSMEISTER, W.J. & CAMACHO, H.H. 1980. Late Eocene Struthiolariidae (Mollusca: Gastropoda) from Seymour Island, Antarctic Peninsula and their significance to the biogeography of early Tertiary shallow-water faunas of the Southern Hemisphere. *Journal of Paleontology*, **54**, 1–14.

First gondwanatherian mammal from Antarctica

FRANCISCO J. GOIN[1], MARCELO A. REGUERO[1], ROSENDO PASCUAL[1], WIGHART VON KOENIGSWALD[2], MICHAEL O. WOODBURNE[3], JUDD A. CASE[4], SERGIO A. MARENSSI[5], CAROLINA VIEYTES[1] & SERGIO F. VIZCAÍNO[1]

[1]*División Paleontología Vertebrados, Museo de La Plata, Paseo del Bosque s/n, 1900 La Plata, Argentina (e-mail: fgoin@museo.fcnym.unlp.edu.ar)*

[2]*Institut für Paläontologie, Universität Bonn, Nussallee 8, D-53115 Bonn, Germany*

[3]*Department of Earth Sciences, University of California, Riverside, CA 92521, USA*

[4]*Department of Biology, St Mary's College, Moraga, CA 94575, USA*

[5]*Instituto Antártico Argentino, Cerrito 1248, 1010 Buenos Aires, Argentina*

Abstract: Gondwanatherians are an enigmatic group of extinct non-therian mammals apparently restricted to some of the western Gondwanan continents (Late Cretaceous–early Palaeocene of South America, and Late Cretaceous of Madagascar and India). They developed rodent-like incisors and the earliest known hypsodont cheek-teeth among mammals. Recently, a small rodent-like dentary fragment was recovered from middle Eocene beds on the Antarctic Peninsula, preserving part of the incisor; both the incisor enamel structure and the mandibular morphology suggest close affinities with *Sudamerica ameghinoi* from the early Palaeocene of Patagonia, up to now the youngest known Gondwanatheria. Thus, the new specimen becomes the youngest occurrence of a gondwanathere, adding significant direct and indirect evidence on: (1) the already documented cosmopolitanism of gondwanatheres among Gondwanan mammals; and (2) the crucial biogeographical role of Antarctica during the Cretaceous–Tertiary mammalian transition.

Our knowledge of gondwanatherian mammals is relatively recent. They were first reported in 1984 and were alternatively regarded as edentates (Scillato Yané & Pascual 1985; Mones 1987), paratherians (Scillato Yané & Pascual 1985; Bonaparte 1986), multituberculates (Bonaparte *et al.* 1989; Krause 1990, 1993; Krause & Bonaparte 1990, 1993; Krause *et al.* 1992; Kielan-Jaworowska & Bonaparte 1996) and, recently, as dubious Allotheria (Krause *et al.* 1997); that is, as probably related to the Multituberculata as a sister-group. A study of the most complete gondwanatherian specimen known up to now (a fragmentary dentary with part of the incisor, two cheek-teeth *in situ* and two more alveoli) led Pascual *et al.* (1999) to regard them as Mammalia *incertae sedis*. Their distribution seems to have been restricted to southern continents, as all known taxa come from Upper Cretaceous levels in Patagonia, Madagascar and India, and from Lower Palaeocene levels in Patagonia. With the exception of the more generalized *Ferugliotherium windhauseni* (Family Ferugliotheriidae: Krause *et al.* 1992; Kielan-Jaworowska & Bonaparte 1996), all other gondwanatherians are grouped in the family Sudamericidae, characterized by the possession of high-crowned cheek-teeth, covered with a cementum layer. Among the latter, *Gondwanatherium* and *Sudamerica* at least have the transverse ridges transformed in well-separated lophs/lophids. *Sudamerica* at least has four lower molariforms and the most rodent-like pattern seen among gondwanatherians (Pascual *et al.* 1999). Lower molars of *Ferugliotherium*, *Gondwanatherium* and *Sudamerica* show identical patterns of wear and transverse ridges (Krause & Bonaparte 1993). All three of them are also known to have enlarged, rodent-like incisors, which in *Sudamerica* are followed posteriorly by a large diastema. The remaining gondwanatherians are: (1) *Lavanify miolaka* from Madagascar (Krause *et al.* 1997); (2) an unnamed taxon from India (Krause *et al.* 1997); and (3) an also unnamed ?sudamericid from Tanzania, Africa (Krause *et al.* 2003), all of them known by a few isolated specimens. *Lavanify* and the Indian gondwanatherian have their enamel structure quite derived with respect to the South American taxa (Krause *et al.* 1997; Koenigswald *et al.* 1999).

Here we report the youngest and southernmost discovery of a gondwanatherian mammal,

From: FRANCIS, J. E., PIRRIE, D. & CRAME, J. A. (eds) 2006. *Cretaceous–Tertiary High-Latitude Palaeoenvironments, James Ross Basin, Antarctica.* Geological Society, London, Special Publications, **258**, 135–144. 0305–8719/06/$15

coming from Middle Eocene levels in Antarctica. One of the specimens consists of a fragmentary dentary bearing part of the rodent-like incisor; an isolated, fragmentary upper incisor is tentatively referred to the same taxon. As shown below, both their general anatomy and the enamel microstructure of the incisors suggest that the Antarctic specimens are closely related to the early Palaeocene sudamericid *Sudamerica ameghinoi*.

All specimens were collected by picking screened sediment concentrate. A preliminary comment on the presence of gondwanatherian mammals in Antarctica was given recently by Reguero *et al.* (2002). This discovery adds a new Gondwanan continent to the distribution of gondwanatherians, extending in time and space the already known extensive distribution of these mammals. In addition, it sheds new light on the biogeographical role of Antarctica in the evolution of Gondwanan mammals.

Stratigraphical setting and terrestrial palaeoenvironment inferred for the *Cucullaea* I allomember

The abbreviations used are: MLP, Departamento Paleontología Vertebrados, Museo de La Plata, Argentina; IAA, fossil Antarctic localities discovered by MLP and Instituto Antártico Argentino researchers. All measurements are in mm.

Known by the Argentinean paleontologists as the 'Ungulate site' (Marenssi *et al.* 1994), the locality IAA (Instituto Antártico Argentino) 1/90 is the richest Antarctic mammal-bearing locality known up to now (Bond *et al.* 1990; Goin *et al.* 1994; Case *et al.* 1996; Vizcaíno *et al.* 1998*a*, *b*) (Fig. 1), and has also produced the new specimens described here. Locality IAA 1/90 is within the *Cucullaea* I Member of the La Meseta Formation.

The late Early–latest Late Eocene La Meseta Formation (Elliot & Trautman 1982; Marenssi *et al.* 1998) crops out in the northern third of Seymour (Marambio) Island, some 100 km off the northern Antarctic Peninsula. This 710 m-thick clastic unit records sedimentation in a deltaic and estuarine setting within an incised valley capped by shallow-marine deposits. The mammal-bearing levels are composed of thick shelly conglomerates, well-sorted sands and interlaminated sand/mud channel fills with thin shelly conglomeratic intervals. The bioclastic fraction comprises mainly gastropods (naticids), but marine and land vertebrate remains, plant fragments and other marine invertebrates also occur. Marenssi *et al.* (1994) considered these beds as reworked, moderate- to high-energy facies of a subtidal shallow-marine environment.

The *Cucullaea* I Allomember, formerly Telm 4 plus 5 of Sadler (1988), is most probably middle Eocene in age (Reguero *et al.* 2002) and it is the richest terrestrial mammal-bearing unit in Antarctica (Reguero *et al.* 1998). This allomember has also provided the largest collection of leaves from the Eocene of Antarctica (locality C/88, Gandolfo *et al.* 1998) and most of the wood remains reported up to now from the La Meseta Formation (Torres *et al.* 1994; Brea 1998). However, all the terrestrial remains were transported and deposited in a shallow-marine setting, being concentrated by means of sedimentological processes and mixed with a normal marine macrofauna (Marenssi *et al.* 1998).

Physiognomical analysis of the leaves collected from this interval indicates a temperate–cool temperate, and seasonally moist climate with mean annual temperatures of between 11 and 13° C (locality C/88, Gandolfo *et al.* 1998). Clay mineral and geochemical data show that the northern Antarctic Peninsula (Seymour Island) experienced a climatic deterioration from very warm non-seasonal wet conditions in the Early Eocene to a cold, frost-prone and dry regime at the end of the Late Eocene (Dingle *et al.* 1998). By Middle Eocene times, the Seymour Island climate was experiencing a rapid cooling from very warm to cold through a strongly seasonal period (climatic episodes E2 and E3, Dingle *et al.* 1998).

Fossil wood recovered from the *Cucullaea* I Allomember indicates the presence of a nearby forest mainly composed of Araucariaceae, Podocarpaceae, Cupressaceae and Nothofagaceae trees (Torres *et al.* 1994; Brea 1998), while the leaves attest to the presence of Dilleniaceae, Myricaceae, Myrtaceae and Lauraceae plants. They most closely resemble species of extant plants growing in the Valdivian (Chile and Argentina) and New Zealand forests.

Systematic palaeontology

Class MAMMALIA Linnaeus 1758
Order GONDWANATHERIA Mones 1987
Family SUDAMERICIDAE Scillato-Yané & Pascual 1984
Genus and species indet., cf. *Sudamerica ameghinoi* Scillato-Yané & Pascual 1984

Referred specimen

MLP 95-I-10-5 (Figs 2 A–C & 3A, B), anterior part of a left dentary with the rodent-like incisor partially preserved.

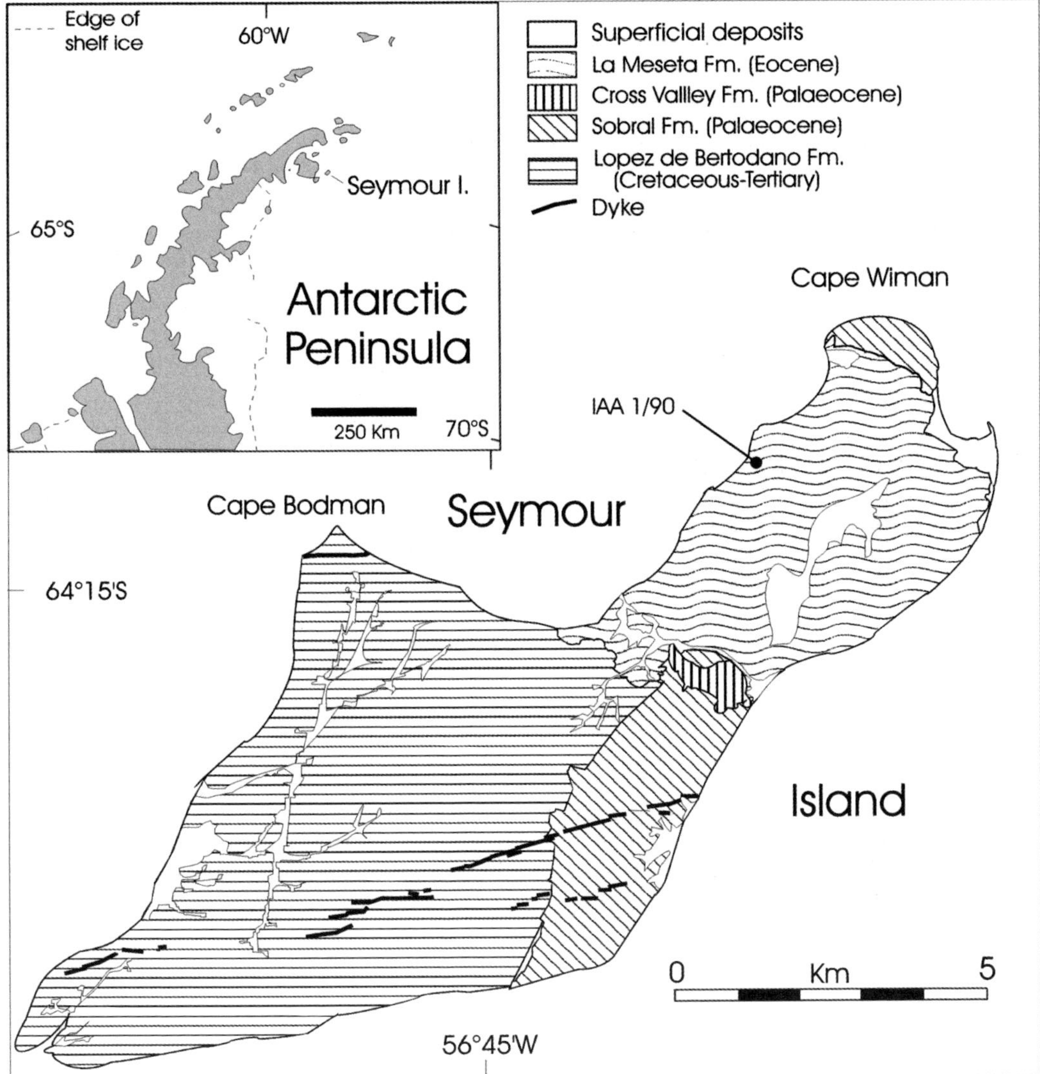

Fig. 1. Map of Seymour (Marambio) Island (Antarctic Peninsula) showing locality IAA 1/90.

Tentatively referred specimen

MLP 96-I-5-47, an isolated ?left upper incisor (Fig. 3A, B) broken at its proximal end. Both specimens were collected by picking screened sediment concentrate matrix worked at locality IAA 1/90.

Measurements

MLP 95-I-10-5: dentary width at the mental foramen 3.49 mm; dentary height at the mental foramen 6.05 mm; height of the lower incisor 3.65 mm; width of the lower incisor 1.53 mm. MLP 96-I-5-47 (isolated upper incisor fragment): mesiodistal width 1.61; buccolingual width 2.29; length 5.37. This last measurement does not reflect the tooth's real length, as it is broken at its proximal end.

Locality

IAA 1/90 (Fig. 1), Seymour (Marambio) Island, Antarctic Peninsula. 64°14′ 04.672″ S and 56°39′ 56.378″W. Elevation is 57.19 m above sea level (Lusky *et al.* 1994). Levels at this locality correspond to the La Meseta Formation, more precisely to the lowest third of Unit II (Elliot &

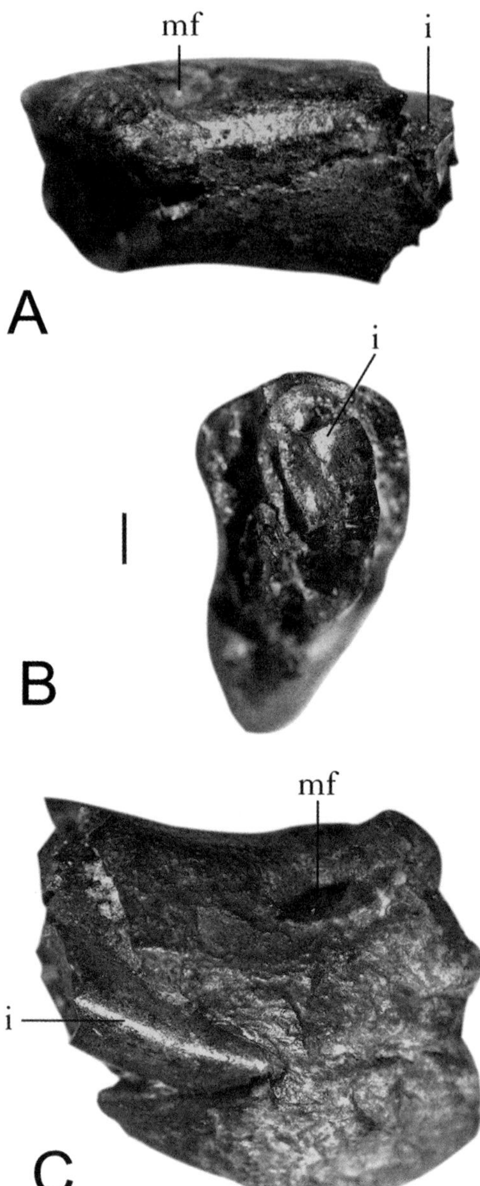

Fig. 2. Sudamericidae genus et species indet., cf. *Sudamerica ameghinoi* Scillato Yané & Pascual 1984. Specimen MLP 95-I-10-5, an anterior portion of a left dentary showing the enlarged, rodent-like incisor partially preserved. (**A**) dorsal, (**B**) anterior and (**C**) lateral views. References: i, incisor; mf, mental foramen. The scale bar is 1 mm.

Trautman 1982), or Telm 5 (Sadler 1988) or Allomember *Cucullaea* I (Marenssi & Santillana 1994).

Description

The dentary fragment (MLP 95-I-10-5; Fig. 2A–C) is wider dorsally than ventrally; the symphysis is flat and vertical, while the lateral face is slightly convex. At the lateral face the dentary is broken in such a way that the intra-alveolar portion of the incisor can be seen. In section, this tooth resembles the dentary: medially flat, laterally convex, sharply angled ventromedially and rounded dorsally. A thin enamel layer at the ventral surface covers it. The preserved dorsal surface of the dentary bears no alveoli, indicating that there was a diastema between the incisor and the molariform teeth. The mental foramen is suboval in shape and is placed very high on the lateral face of the dentary, close to the diastema. Immediately behind the mental foramen the horizontal ramus rises sharply. The isolated upper incisor (MLP 96-I-5-47; Fig. 3A, B) is suboval in section, and has the enamel layer restricted to the buccal and distal faces. There is a small, horizontal wear facet at the apex. Even though smaller than the lower incisor, it matches well the size expected for an occlusal antagonist of the lower one.

Enamel microstructure

Small enamel fragments belonging to specimens MLP 95-I-10-5 and MLP 96-I-5-47 have been subjected to microstructural investigations carried out at the Institut für Paläontologie, Universität Bonn. For the enamel description we use the terms and abbreviations listed in a recent glossary by Koenigswald & Sander (1997). The enamel of the lower incisor is very thin (about 50 μm). From transverse and longitudinal sections the schmelzmuster was reconstructed (Fig. 4A, B). The enamel structure consists of two well-defined layers of tangential enamel, with prisms rising or descending only very little. Although the prism direction changes, there is no prism decussation, typical for rodent incisors. The tangential enamel is divided into a mesial and a lateral field by a neutral area with straight prism orientation, close to the inflection of the band towards the mesial side. On the lateral side of the neutral area the prisms are laterally oriented in the inner layer, and mesially in the outer layer. In the mesial part, however, the prisms of the inner layer are directed mesially, and laterally in the outer layer. This pattern is seen when the entire length of the section is studied. Between the inner and the outer layers, prisms change direction in a simultaneous prism deviation. The

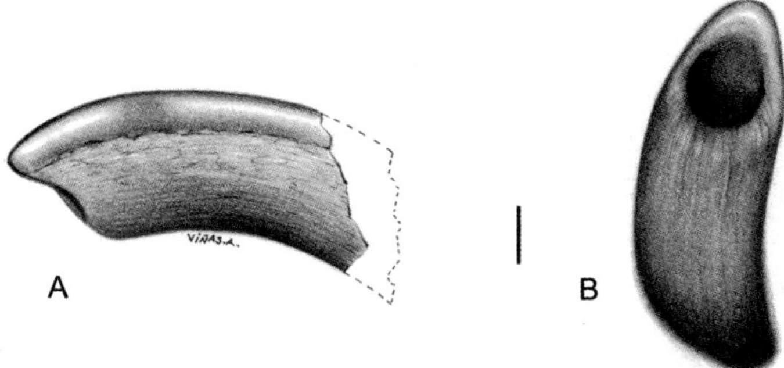

Fig. 3. MLP 96-I-5-47, an isolated ?left upper incisor tentatively referred to the same Sudamericidae genus and species indet cf. *Sudamerica ameghinoi*. (**A**) Lateral and (**B**) occlusal view. The scale bar is 1 mm.

prisms are surrounded by a thick interprismatic matrix (IPM), which runs straight from the enamel–dentine junction (EDJ) to the outer enamel surface (OES) without inclination. In the outer layer several prisms disappear so that the IPM becomes dominant, but does not form a continuous prismless outer enamel (PLEX). Prisms are isodiametrical and relatively small (of around 5 µm each), and are surrounded by incomplete prism sheaths. Close to the EDJ several tubules could be seen in the enamel. In the isolated upper incisor (Fig. 3C) the enamel is even thinner (20 µm). The schmelzmuster consists only of one layer of radial enamel. The prisms are straight, but from the sections available we cannot decide how much they are rising. As in the lower incisor, prisms are surrounded by a very thick IPM set at a slight angle in relation to the prisms. The prism cross-section is open and many prisms show a distinct seam. In an almost tangential section prisms seem to be, in some areas, vertically aligned. The IPM between prisms is separated by a discontinuity normally not seen in interrow sheets.

Comments

Although superficially similar, the dentary does not belong to a polydolopine marsupial, the most abundant mammal in the La Meseta Formation: (1) it is very wide dorsally and narrow ventrally, while in polydolopines the cross-section is fusiform, reaching its maximum width at the middle portion of the ramus; (2) in marsupials, the mental foramen is never placed so dorsally on the labial face of the mandible; (3) relative to the symphyseal plane, the plane of the mandible in polydolopines is obliquely oriented, not vertical as seen in anterior view; (4) the lower incisor is proportionally too large compared to any known 'pseudodiprotodont' marsupial; and (5) from the preserved material it is clear that the alveolar plane is placed much higher than the diastema, a feature that does not occur in polydolopimorphians. However, all these features agree well with known remains of the gondwanatherian mammal *Sudamerica ameghinoi* (Pascual *et al.* 1999), known only from early Palaeocene levels at Punta Peligro, in Central Patagonia. Additionally, the cross-section of this lower incisor is similar to that of gondwanatherians, and particularly to that of *Sudamerica ameghinoi*: medially flat, laterally convex, angled ventromedially and rounded dorsally (Koenigswald *et al.* 1999). It also agrees well in the presence of a thin, labial enamel layer. Regarding its size, the dentary fragment from the La Meseta Formation is much smaller than that of the only known dentary of *S. ameghinoi* (Pascual *et al.* 1999). Finally, the attribution of the upper incisor is based on its similarity to assigned isolated upper incisors of other gondwanatherians, such as *Feruglio-therium windhauseni* (Krause *et al.* 1992), although no upper incisor is known from *Sudamerica ameghinoi* itself.

Regarding the enamel structure of the Antarctic specimens, the upper and the lower incisors show a different schmelzmuster. Such a difference is also known in several rodents (Koenigswald 1997), where the lower incisor always shows the more derived state. In our material the lower incisor, with two layers of tangential enamel, is definitely more derived than the upper one. Two-layered schmelzmusters are frequent among enlarged incisors of various mammals, such as primates, rodents, insectivores, etc. (Koenigswald 1996). The absence of Hunter–Schreger bands in the schmelzmuster of the Antarctic specimen, however, rules out its affinities with rodents and most other eutherians. In ptilodontoid multituberculates (Sahni 1979)

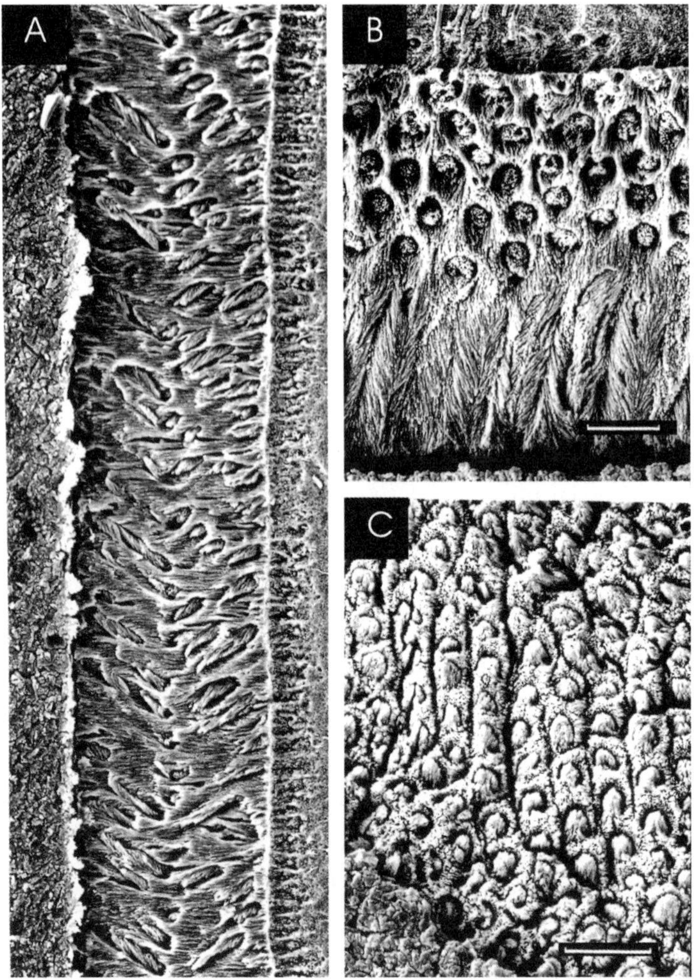

Fig. 4. Enamel of the lower incisor of the Antarctic gondwanathere (specimen MLP 95-I-10-5) in (**A**) longitudinal and (**B**) transverse sections. The longitudinal section shows the two layers and the thick interprismatic matrix (IPM) surrounding the prisms (P). The prism orientation in the two layers of tangential enamel can be seen from the transverse section. In the outer layer the IPM dominates due to the disappearance of several prisms. (**C**) Semi-tangential section of the upper incisor (specimen MLP 96-I-5-47), showing the incomplete prism sheath surrounding the prisms, and the alignment of prisms with the central discontinuity. The scale bar is 10 μm.

and various South American marsupials such as *Groeberia* and *Argyrolagus*, and most Australian kangaroos, a combination of radial enamel and tangential enamel was found (Koenigswald & Pascual 1990; Koenigswald 1994). This pattern differs distinctly from the schmelzmuster found in the Antarctic specimens, in which the prisms do not show any inclination. Conversely, the enamel structure of the Antarctic specimens shows clear correspondences with *Sudamerica ameghinoi* (Koenigswald *et al.* 1999; Pascual *et al.* 1999). Unfortunately, we cannot compare the schmelzmuster of the Antarctic specimen with gondwanatherians from Madagascar and India as incisors are only known for *Gondwanatherium* and *Sudamerica*. However, judging from their cheek-tooth enamel, both the Madagascan and the Indian forms are substantially more derived than all other gondwanatherians in the possession of continuous inter-row sheets of interprismatic matrix. In turn, the lower incisor enamel of the Antarctic specimen shares with that of *Sudamerica* almost all characteristics seen in the schmelzmuster: the low inclination of the prisms, the position of the neutral area, the two layers of tangential enamel, the specific prism orientation in each layer, the straight orientation of the crystallites of the IPM and the predominance of the IPM in the outer layer, due to the disappearance of prisms. The similarity of this unique pattern in the two taxa indicates a close relationship. Minor differences, such as the more pronounced angle between the inner and outer tangential layers and the better definition of the prisms, indicate that the schmelzmuster of the Antarctic specimen is somewhat more derived than that of *Sudamerica ameghinoi*. The incisor of the Late Cretaceous *Gondwanatherium patagonicum* shows a quite different schmelzmuster, indicating a more distant relationship compared with *Sudamerica* (Koenigswald *et al.* 1999).

In short, all macro- and microscopic evidence at hand suggests that the Antarctic specimens pertain to a Gondwanatherian mammal, and that they are more closely related to the Early Palaeocene *Sudamerica ameghinoi* than to any other member of this group. While assigning these specimens to the Sudamericidae, owing to their fragmentary nature we do not enter into a formal taxonomy at the generic or specific level.

Discussion

The discovery of Antarctic (this paper), as well as Indian and Madagascan (Krause *et al.* 1997), sudamericid gondwanatherians confirms the broad distribution of this group of mammals in Gondwanan continents during Mesozoic and the earliest Cenozoic times, and gives new insights into our limited knowledge of the history of Gondwanan mammals (Pascual 1997*a*, *b*, 1998). It also confirms previous hypotheses on the crucial role of Antarctica in the biogeographical evolution of southern hemisphere mammals (Keast 1972; Elliot & Trautman 1982; Krause *et al.* 1997). The Patagonian Early Palaeocene ornithorhynchid monotreme is yet one more example (Pascual *et al.* 1992*a*, *b*). Monotremes and marsupials had to have been active parts of inter-Gondwanan faunal interchanges involving Antarctica as a continental 'crossroad' or 'stepping-stone'. We do not know whether gondwanatheres played a similar role as so far they have not been recorded either in the Antarctic or Australian Cretaceous. Nevertheless, Krause *et al.* (1997) suggested that Antarctica might have served as an important Cretaceous biogeographical link between South America and Indo-Madagascar. In recognition of the derived enamel structure of the Antarctic sudamericid and the endemism of the contemporaneous ungulate mammals with respect to South American relatives (Bond *et al.* 1995), the most parsimonious hypothesis is that the La Meseta Formation mammal fauna was relictual in the Antarctic Peninsula relative to a biota shared between the Antarctic Peninsula and South America in the Early Palaeocene. This vicariant hypothesis appears to reflect the subsequent separation of the Antarctic and South American continents. The remarkably good fossil record of late Palaeocene (Itaboraian Age) South American mammals, both in Patagonia and Brazil (see, for example, Pascual *et al.* 1996), adds support to this hypothesis: no gondwanatheres have been recorded in Itaboraian beds, or in the subsequent Tertiary land mammal-bearing beds.

Up to now, the Middle Eocene mammals recovered from the Antarctic La Meseta Formation included frugivorous polydolopimorphian and insectivorous didelphimorphian marsupials (Goin *et al.* 1994, 1999), tardigrade xenarthrans (which are likely to have been folivorous) and ungulate browsers, such as a trigonostylopid astrapothere (Vizcaíno *et al.* 1998*b*). As indicated elsewhere, hypsodonty cannot be related only to grazing as it also occurs in burrowing and semi-aquatic mammals such as the ctenomyids (Pascual *et al.* 1965) and the castorid rodents, respectively.

It is evident that the record of a sudamericid gondwanatherian in the Eocene Antarctic beds does not contribute to our understanding of the presence of this group in Madagascar and India during the Late Cretaceous. First, the Antarctic record is geologically too young; second, the La Meseta Formation taxon seems more closely allied to the Early Palaeocene *Sudamerica ameghinoi* than to any other gondwanatherian. The hypothesis of Krause *et al.* (1997) that Antarctica may have served as an important Cretaceous biogeographical link between South America and Indo-Madagascar requires the record of a gondwanatherian in Cretaceous Antarctic beds. The apparent absence of gondwanatheres during this time in Antarctica, and in Late Cretaceous and younger beds in Australia, is probably due to lack of discoveries. Such remains should be diligently sought.

We thank the personnel and authorities of the Instituto Antártico Argentino, and especially to S. Santillana for their logistic support during field work at locality IAA 1/90 in the Antarctic Peninsula; Mr J. J. Moly, for his fieldwork in Antarctica and picking efforts at the Museo de la Plata that led to the discovery of the specimens studied here; A. Sahni, from Panjab University, for loaning to us the indeterminate Indian gondwanatherian mentioned in the text for examination at the Institut für Paläontologie in Bonn; M. Tomeo, A. Viñas, G. Oleschinski and D. Kranz for their contributions to the artwork; D. Kalthof, who helped F. J. Goin use the SEM facilities at the Institut für Paläontologie in Bonn; the 'Alexander von Humboldt Stiftung', who generously provided F. J. Goin with a fellowship in order to study in Bonn during 1998, and with the stereomicroscope equipment and software used in the making of Figure 2; the 'Deutsche Forschungs Gemeinschaft', who supported the background of the work of W. v. Koenigswald in the form of several grants to study the structure of mammalian enamel; and J. J. Hooker and D. W. Krause for their very useful critical comments on the original manuscript. Financial support to R. Pascual and F. J. Goin was derived from the Consejo Nacional de Investigaciones Científicas y Técnicas, Argentina (PMT-PICT 0227) and National Geographic Society, USA (grant 5905-97).

References

BONAPARTE, J.F. 1986. A new and unusual late Cretaceous mammal from Patagonia. *Journal of Vertebrate Paleontology*, **6**, 264–270.

BONAPARTE, J.F., KRAUSE, D.W. & KIELAN-JWOROWSKA, Z. 1989. *Ferugliotherium windhauseni* Bonaparte, the first known multituberculate from Gondwanaland. *Journal of Vertebrate Paleontology*, **10**, 14A.

BOND, M., CARLINI, A.A., GOIN, F.J., LEGARRETA, L., ORTIZ-JAUREGUIZAR, E., PASCUAL, R. & ULIANA, M.A. 1995. Episodes in South American land mammal evolution and sedimentation: testing their apparent concurrence in a Palaeocene succession from central Patagonia. *In*: *IV Congreso Argentino de Paleontología y Bioestratigrafía, Trelew, Actas*, Asociación Paleontológica Argentina, Trelew, 47–58.

BOND, M., PASCUAL, R., REGUERO, M.A., SANTILLANA, S.H. & MARENSSI, S.A. 1990. Los primeros ungulados extinguidos sudamericanos de la Antártida. *Ameghiniana*, **26**, 240.

BREA, M. 1998. Análisis de los anillos de crecimiento en leños fósiles de coníferas de la Formación La Meseta, Isla Seymour (Marambio), Antártida. *In*: CASADÍO, S. (ed.) *Paleógeno de América del Sur y de la Península Antártica*. Asociación Paleontológica Argentina, Publicación Especial, **5**, 163–175.

CASE, J.A., GOIN, F.J. & WOODBURNE, M.O. 1996. Eocene Antarctic ameridelphian marsupials with tribosphenic molars: implications upon marsupial biogeography. *Journal of Vertebrate Paleontology*, **16**, (3 Suppl.), 26A.

DINGLE, R., MARENSSI, S. & LAVELLE, M. 1998. High latitude Eocene climate deterioration: evidence from the northern Antarctic Peninsula. *Journal of South American Earth Sciences*, **11**, 571–579.

ELLIOT, D.H. & TRAUTMAN, T. 1982. Lower Tertiary strata on Seymour Island. *In*: CRADDOCK, C. (ed.) *Antarctic Geoscience*. University of Wisconsin Press, Madison, WI, 287–297.

GANDOLFO, M.A, MARENSSI, S.A. & SANTILLANA, S.N. 1998. Flora y paleoclima de la Formación La Meseta (Eoceno medio), isla Marambio (Seymour), Antártida. *In*: CASADÍO, S. (ed.) *Paleógeno de América del Sur y de la Península Antártica*. Asociación Paleontológica Argentina, Publicación Especial, **5**, 155–162.

GOIN, F.J., CASE, J.A., WOODBURNE, M.O., VIZCAÍNO, S.F. & REGUERO, M.A. 1999. New Discoveries of 'Opossum-like' Marsupials from Antarctica (Seymour Island, Middle Eocene). *Journal of Mammalian Evolution*, **6**, 335–365.

GOIN, F.J., REGUERO, M.A. & VIZCAÍNO, S.F. 1994. Novedosos hallazgos de 'comadrejas' (Marsupialia) del Eoceno Medio de Antártida. *In*: *III Jornadas de Comunicaciones sobre Investigaciones Antárticas*. Dirección Nacional del Antártico–Instituto Antártico Argentino, Buenos Aires, Abstracts, 59–61.

KEAST, A. 1972. Introduction: The Southern Continents as Backgrounds for Mammalian Evolution. *In*: KEAST, A., ERK, F.C. & GLASS, B. (eds) *Evolution, Mammals, and Southern Continents*. State University of New York Press, Albany, NY, 19–22.

KIELAN-JAWOROWSKA, Z. & BONAPARTE, J.F. 1996. Partial dentary of a multituberculate mammal from the late Cretaceous of Argentina and its taxonomic implications. *Revista del Museo Argentino de Ciencias Naturales 'Bernardino Rivadavia' Nueva Serie*, **145**, 1–9.

KOENISGWALD, W.v. 1994. Differenzierungen im Zahnschmelz der Marsupialia im Vergleich zu den Verhõltnissen bei den Placentalia (Mammalia). *Berliner Geowissenschaftliche Abhandlungen (Bernard Krebs Festschrift)*, **E13**, 45–81.

KOENIGSWALD, W.v. 1996. Die Zahl der Schmelzschichten in den Inzisiven bei den Lagomorpha und ihre systematische Bedeutung. *Bonner Zoologische Beiträge (Festschrift J. Niethammer)*, **46**, 33–57.

KOENIGSWALD, W.v. 1997. The variability of the enamel at the dentition level. *In*: KOENIGSWALD, W.v. & SANDER, P.M. (eds) *Tooth Enamel Microstructure*. Balkema, Rotterdam, 193–201.

KOENIGSWALD, W.v. & PASCUAL, R. 1990. The Schmelzmuster of the Paleogene South American rodentlike marsupials *Groeberia* and *Patagonia* compared to rodents and other Marsupialia. *Paläontologische Zeitschrift*, **64**, 345–358.

KOENIGSWALD, W.v. & SANDER, M.P. 1997. Glossary. *In*: KOENIGSWALD, W.v. & SANDER, P.M. (eds) *Tooth Enamel Microstructure*. Balkema, Rotterdam, 267–280.

KOENIGSWALD, W.v., GOIN, F.J. & PASCUAL, R. 1999. Hypsodonty and enamel microstructure in the Palaeocene gondwanatherian mammal *Sudamerica ameghinoi*. *Acta Palaeontologica Polonica,* **44**, 263–300.

KRAUSE, D.W. 1990. The Gondwanatheria, a new suborder of Multituberculata from South America. *Journal of Vertebrate Paleontology*, **10**, (3 Suppl.), 1S.

KRAUSE, D.W. 1993. *Vucetichia* (Gondwanatheria) is a junior synonym of *Ferugliotherium* (Multituberculata). *Journal of Paleontology*, **67**, 321–324.

KRAUSE, D.W. & BONAPARTE, J.F. 1990. The Gondwanatheria, a new suborder of Multituberculata from South America. *Journal of Vertebrate Paleontology*, **10**, 13A.

KRAUSE, D.W. & BONAPARTE, J.F. 1993. Superfamily Gondwanatherioidae: A previously unrecognized radiation of multituberculate mammals in South America. *Proceedings of the National Academy of Science*, **90**, 9379–9383.

KRAUSE, D.W., KIELAN-JAWOROWSKA, Z. & BONAPARTE, J.F. 1992. *Ferugliotherium* Bonaparte, the first known multituberculate from South America. *Journal of Vertebrate Paleontology*, **12**, 351–376.

KRAUSE, D.W., GOTTFRIED, M.D., O'CONNOR, P.M. & ROBERTS, E.M. 2003. A Cretaceous mammal from Tanzania. *Acta Palaeontologica Polonica*, **48**, 321–330.

KRAUSE, D.W., PRASAD, G.V.R., KOENIGSWALD, W.V., SAHNI, A. & GRINE, F.E. 1997. Cosmopolitanism among Gondwana Late Cretaceous mammals. *Nature*, **390**, 504–507.

LINNAEUS, C. 1758. *Systema Naturae*, X edn. (*Systema naturae per regna tria naturae, secundum classes, ordines, genera, species, cum characteribus, differentiis, synonymis, locis. Tomus I. Editio decima, reformata.*) Holmiae (Laurentii Salvii), [1–4], 1–824.

LUSKY, J.C., REGUERO, M.A. & VIZCAÍNO, S.F. 1994. Geographical position applying Global Position System (GPS) in the Eocene land-vertebrate bearing localities from Seymour (Marambio) Island, Antarctic Peninsula. *In*: *III Jornadas de Comunicaciones sobre Investigaciones Antárticas*. Dirección Nacional del Antártico–Instituto Antártico Argentino, Buenos Aires, Abstracts, 53–54.

MARENSSI, S.A. & SANTILLANA, S.N. 1994. Unconformity bounded units within La Meseta Formation, Seymour Island, Antarctica: a preliminary approach. *In*: ZALEWSKI, M. (ed.) *XXI Polar Symposium*, Institute of Geophysics, Warsaw, 33–37.

MARENSSI, S.A., REGUERO, M.A., SANTILLANA, S.N. & VIZCAÍNO, S.F. 1994. Eocene land mammals from Seymour Island, Antarctica: palaeobiological implications. *Antarctic Science*, **6**, 3–15.

MARENSSI, S.A., SANTILLANA, S.N. & RINALDI, C.A. 1998. Stratigraphy of the La Meseta Formation (Eocene), Marambio (Seymour) Island, Antarctica. *In*: CASADÍO, S. (ed.) *Paleógeno de América del Sur y de la Península Antártica.* Asociación Paleontológica Argentina, Publicación Especial, **5**, 137–146.

MONES, A. 1987. Gondwanatheria, un Nuevo Orden de Mamíferos Sudamericanos (Mammalia: Edentata: ?Xenarthra). *Comunicaciones Paleontológicas del Museo de Historia Natural de Montevideo*, **1**, 237–240.

PASCUAL, R. 1997*a*. Fossil land mammals and the geobiotic history of Southern South America. *In*: *II Southern Connection Congress. Noticiero de Biología*, **5**, 59.

PASCUAL, R. 1997*b*. The Gondwanan history of mammals, the other history. *In*: *7th International Theriological Congress, Universidad Nacional Autónoma de México, Acapulco, México*, Abstracts, 278–279.

PASCUAL, R. 1998. The history of South American Land Mammals: the seminal Cretaceous–Paleocene transition. *In*: CASADÍO, S. (ed.) *Paleógeno de América del Sur y de la Península Antártica.* Asociación Paleontológica Argentina, Publicación Especial, **5**, 9–18.

PASCUAL, R., ARCHER, M., ORTIZ-JAUREGUIZAR, E., PRADO, J.L., GODTHELP, H. & HAND, S.H. 1992*a*. First discovery of monotremes in South America. *Nature*, **356**, 704–705.

PASCUAL, R., ARCHER, M., ORTIZ-JAUREGUIZAR, E., PRADO, J.L., GODTHELP, H. & HAND S.H. 1992*b*. The first non-Australian monotreme: an early Palaeocene South American Platypus (Monotremata, Ornitorhynchidae). *In*: AUGEE, M.L. (ed.) *Platypus and Echidnas*. Royal Zoological Society of New South Wales, Sydney, 1–14.

PASCUAL, R., GOIN, F.J., KRAUSE, D.W., ORTIZ-JAUREGUIZAR, E. & CARLINI, A.A. 1999. The first gnathic remains of *Sudamerica*: implications for Gondwanathere relationships. *Journal of Vertebrate Paleontology*, **19**, 373–382.

PASCUAL, R., ORTEGA HINOJOSA, E.J. & PISANO, J. 1965. Un nuevo Octodontidae (Rodentia, Caviomorpha) de la Formación Epecuén, Plioceno medio de Hidalgo (Provincia de La Pampa). *Ameghiniana*, **4**, 19–30.

PASCUAL, R., ORTIZ-JAUREGUIZAR, E. & PRADO, J.L. 1996. Land mammals: paradigm for Cenozoic South American geobiotic evolution. *Münchner Geowissenschaftlich Abhandlungen (A)*, **30**, 265–319.

REGUERO, M.A., MARENSSI, S.A. & SANTILLANA, S.N. 2002. Antarctic Peninsula and Patagonia Paleogene terrestrial environments: biotic and biogeographic relationships. *Palaeogeography, Palaeoclimatology, Palaeoecology*, **2776**, 1–22.

REGUERO, M.A., VIZCAÍNO, S.F., GOIN, F.J., MARENSSI, S.A. & SANTILLANA, S.N. 1998. Eocene high-latitude terrestrial vertebrates from Antarctica as biogeographic evidence. *In*: CASADIO, S. (ed.) *Paleógeno de América del Sur y de la Península Antártica.* Asociación Paleontológica Argentina, Publicación Especial, **5**, 185–198.

SADLER, P.M. 1988. Geometry and stratification of uppermost Cretaceous and Paleogene units on Seymour Island, northern Antarctic Peninsula. *In*: FELDMANN, M. & WOODBURNE, M.O. (eds) *Geology and Paleontology of Seymour Island, Antarctic Peninsula.* Geological Society of America, Memoirs, **169**, 303–320.

SAHNI, A. 1979. Enamel ultrastructure of certain North American Cretaceous Mammals. *Palaeontographica Abteilung A*, **166**, 37–49.

SCILLATO YANÉ, G.J. & PASCUAL, R. 1984. *Un peculiar Paratheria, Edentata (Mammalia) del Paleoceno de Patagonia (Argentina).* Primeras Jornadas Argentinas de Paleontología de Vertebrados, Resúmenes, **15**.

SCILLATO YANÉ, G.J. & PASCUAL, R. 1985. Un peculiar Xenarthra del Paleoceno medio de Patagonia (Argentina). Su importancia en la sistemática de los Paratheria. *Ameghiniana*, **21**, 173–176.

TORRES, T., MARENSSI, S.A. & SANTILLANA, S.N. 1994. Maderas fósiles de la isla Seymour, Formación La

Meseta, Antártica. *Serie Científica del INACH, Santiago de Chile*, **44**, 17–38.

VIZCAÍNO, S.F., PASCUAL, R., REGUERO, M.A. & GOIN, F.J. 1998*a*. Antarctica as a background for mammalian evolution. *In*: CASADÍO, S. (ed.) *Paleógeno de América del Sur y de la Península Antártica*. Asociación Paleontológica Argentina, Publicación Especial, **5**, 199–209.

VIZCAÍNO, S.F., REGUERO, M.A., GOIN, F.J., TAMBUSSI, C.P. & NORIEGA, J.I. 1998*b*. An approach to the structure of the Eocene terrestrial vertebrate community from Antarctic Peninsula. *In*: CASADÍO, S. (ed.) *Paleógeno de América del Sur y de la Península Antártica*. Asociación Paleontológica Argentina, Publicación Especial, **5**, 177–183.

Late Eocene penguins from West Antarctica: systematics and biostratigraphy

C. P. TAMBUSSI[1], C. I. ACOSTA HOSPITALECHE[1], M. A. REGUERO[1] & S. A. MARENSSI[2]

[1]*División Paleontología Vertebrados, Museo de La Plata, Paseo del Bosque s/nro, 1900 La Plata, Argentina (e-mail: tambussi@museo.fcnym.unlp.edu.ar)*

[2]*Instituto Antártico Argentino, Cerrito 1248, 1010 Buenos Aires, Argentina*

Abstract: Penguins are by far the most dominant group of marine vertebrates in the Eocene La Meseta Formation (Seymour Island, Antarctica). We analysed the penguin fauna recovered there from both a systematic and a biostratigraphic point of view. We have added two new species (*Tonniornis mesetaensis* and *T. minimum*) and have defined a biostratigraphic unit, the *Anthropornis nordenskjoeldi* Biozone. This interval of strata, easily distinguishable by the numerous occurrence of penguin bones and the phosphatic brachiopod *Lingula*, is located nearly 30–35 m below the top of the 145 m-thick Submeseta Allomember. The highest morphological and taxonomic penguin diversity living sympatrically (organisms that live simultaneously in the same place), including giant and tiny species, is documented in this interval. Fossil penguins bones studied in this paper, recovered from rocks interpreted as shallow-marine deposits, accumulated between 34.2 and 36.13 Ma (late Late Eocene).

Seymour (Marambio) Island is a small, ice-free island located close to the tip of the Antarctic Peninsula. Highly fossiliferous sedimentary rocks from a crucial period in Earth history – the Eocene – are exposed in the La Meseta Formation in the northern part of the island. Consequently, the La Meseta Formation provides a unique opportunity to learn about composition, dynamics and faunal turnover in Eocene ecosystems.

Penguins are by far the most dominant group of marine vertebrates within this unit and we are particularly interested in these wonderful diving birds. The purpose of this work is to analyse the Seymour Island penguin faunal diversity, as well as to provide new information regarding the biostratigraphy of this group.

The first fossil penguin bones from Antarctica were collected by the Swedish South Polar Expedition in 1901–1903. Through the study of these specimens, Wiman (1905) identified five species, totally new to science, and provided the first stratigraphic locations of the deposits. One of the most characteristic features of the Eocene Antarctic penguin fauna is the presence of giant forms, the largest known in the world, in horizons that are dated as latest Eocene. In particular, one of these horizons yields the largest and sturdiest penguin known, *Anthropornis nordenskjoeldi*, associated with other small to medium-sized penguins (Myrcha *et al.* 2002).

Within the past decades, numerous, mainly systematic, works have been devoted to the fossil penguins of the Antarctic Peninsula (Simpson 1946, 1971; Marples 1953; Case 1992; Myrcha *et al.* 2002). For almost a century, the number of species and their stratigraphic provenance have been a matter of debate. Currently, the systematics of fossil spheniscids is based on isolated bones due to the fragmentary nature of the findings and more commonly on tarsometatarsi (Jadwiszczak 2001, 2003) and humeri (Simpson 1946). Moreover, most of the species are only known by one of these elements. Myrcha *et al.* (2002), for example, studied exclusively the tarsometatarsi of the penguins of the La Meseta Formation and identified four new species.

Although the tarsometatarsus is certainly one of the most valuable bones within a penguin skeleton and useful in systematics at the interspecific level, it also shows a high degree of intraspecific and taxonomically non-significant variation. In this sense, to consider the taxonomic diversity of the group based just on isolated tarsometatarsi could be a questionable exercise. Yet, in the absence of complete skeletons or associated bones, the study of isolated tarsometatarsi has increased the knowledge of the Antarctic spheniscids. In addition, we think that the study of humeri is potentially as suitable as that based on tarsometatarsi. To the degree

From: Francis, J. E., Pirrie, D. & Crame, J. A. (eds) 2006. *Cretaceous–Tertiary High-Latitude Palaeoenvironments, James Ross Basin, Antarctica.* Geological Society, London, Special Publications, **258**, 145–161. 0305–8719/06/$15

that we have been able to analyse, the humerus exibits a relatively smaller intraspecific variation and thus seems to be a more appropriate systematic tool. In addition, humeri are known for most of the species, except those based entirely on an isolated tarsometatarsus (*Palaeeudyptes klekowskii*, *Delphinornis arctowskii*, *D. gracilis*, *Mesetaornis polaris* and *Marambiornis exilis*).

After many palaeontological investigations on Seymour Island, the penguin-bearing localities have increased significantly. The bulk of the penguin-bearing localities are situated within the upper part of the Submeseta Allomember in rocks belonging to the Facies Association III of Marenssi *et al.* (1998*a*). These localities have produced a large number of fossil penguin remains that document the primary diversity known at present. More than 2000 penguin elements are housed at the Museo de La Plata (MLP), and from that collection, the humeri are the basis for the present work.

Geographical and geological setting

The Eocene La Meseta Formation (Rinaldi *et al.* 1978; Elliot & Trautman 1982; Marenssi *et al.* 1998*b*) crops out in Seymour and Cockburn islands, close to the northern tip of the Antarctic Peninsula (Fig. 1a). This unit is the topmost exposed part of the sedimentary fill of the Late Jurassic–Tertiary James Ross Basin (del Valle *et al.* 1992) and was interpreted as the filling of an incised-valley system (Marenssi *et al.* 1998*a*).

The La Meseta Formation is composed of sandstones and mudstones with interbedded shell-rich conglomerates, organized into six erosionally based internal units (Fig. 2a), named from base to top: Valle de Las Focas, Acantilados, Campamento, *Cucullaea* I, *Cucullaea* II and Submeseta allomembers. These units were deposited during the Eocene in deltaic, estuarine and shallow-marine settings, mostly within a NW–SE-trending valley (Marenssi *et al.* 1998*a*, *b*).

Based on the description of 12 facies (Marenssi *et al.* 1998*a*), three main facies associations are distinguished in the La Meseta Formation. The basal facies association (Facies Association I) extends from the Valle de Las Focas, through to the Acantilados and the Campamento allomembers. It is fine-grained, mainly composed of mudstones and very fine sandstones, with prominent breccias and faults as products of synsedimentary deformation. Facies Association I represents valley-confined deposition in progradational–aggradational tide-dominated and wave-influenced delta-front–delta-plain environments at the beginning of the infill of the incised valley. Accommodation space decreased and energy increased as the delta built up to sea level; the incision of third-order surfaces in the upper part of the Acantilados Allomember indicates the change from a wave-reworked delta front to a tide-dominated delta-plain environment.

The intermediate facies association (Facies Association II) includes the *Cucullaea* I, *Cucullaea* II and the lower part of the Submeseta allomembers, ranging from conglomeratic beds to mudstones with a diverse and abundant macrofauna that correspond to a valley-confined estuary-mouth to inner-estuary complex. Tidal channels and mixed flats, tidal inlets and deltas, and washover and beach environments represent the interfingering of high- and low-energy environments.

Finally, the uppermost facies association (Facies Association III) is characterized by a more uniform sandy lithology that represents non-confined tide- and storm-influenced near-shore environments (Fig. 3). Tidal sandwaves and intercalated flat scoured surfaces with pebble lags characterize normal and event deposition, respectively, and the increased abundance of glauconite on the top of the Submeseta Allomember suggests an overall sea-level rise.

Facies Association III transitionally rests on top of muddy estuarine deposits of Facies Association II (Marenssi *et al.* 1998*a*) and is composed mainly of fine- to medium-grained sandstones, either well sorted and cross-bedded with mud drapes as muddy and massive. They intercalate with thin shell beds, gravel beds and clay levels. Beds are mostly tabular and thickly bedded, although some channelled deposits (mainly shell beds) also occur.

Most penguin bones recovered from Facies Association III are disarticulated, some are broken but most of them are complete, although with various degrees of abrasion (Fig. 4). They are found on the surface ('float') of loose, massive and usually bioturbated fine to medium sandstones, and only a few bones are found in conglomerates. Only a single partially articulated penguin skeleton was found in this level. Most if not all of them were transported, at least for a short time, before burial and therefore the accumulations represent parauthocthonous assemblages. However, we agree with Stilwell & Zinsmeister (1992) in that time-averaging is at a minimum and fossils are approximately contemporaneous with the entombing sediments. Phosphatic inarticulate brachiopods (*Lingula* sp.) attached to some of the bones also support this idea.

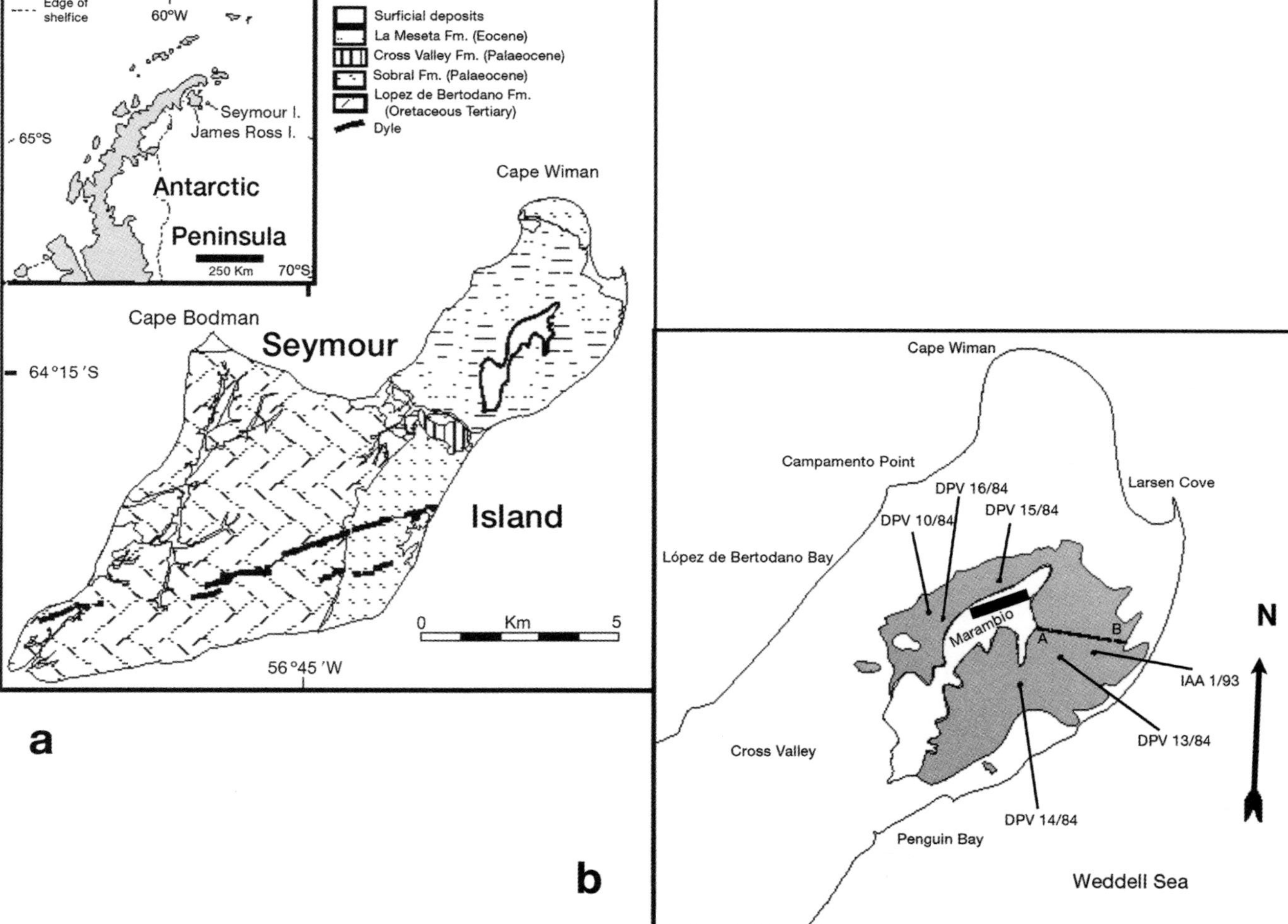

Fig. 1. (**a**) Map showing the location of Seymour Island, Antarctic Peninsula. (**b**) Sketch map of the northern part of Seymour Island showing the distribution of the Submeseta Allomember and the fossil penguin-bearing localities cited in the text. A–B measured section (stratotype) of the *Anthropornis nordenskjoeldi* Biozone.

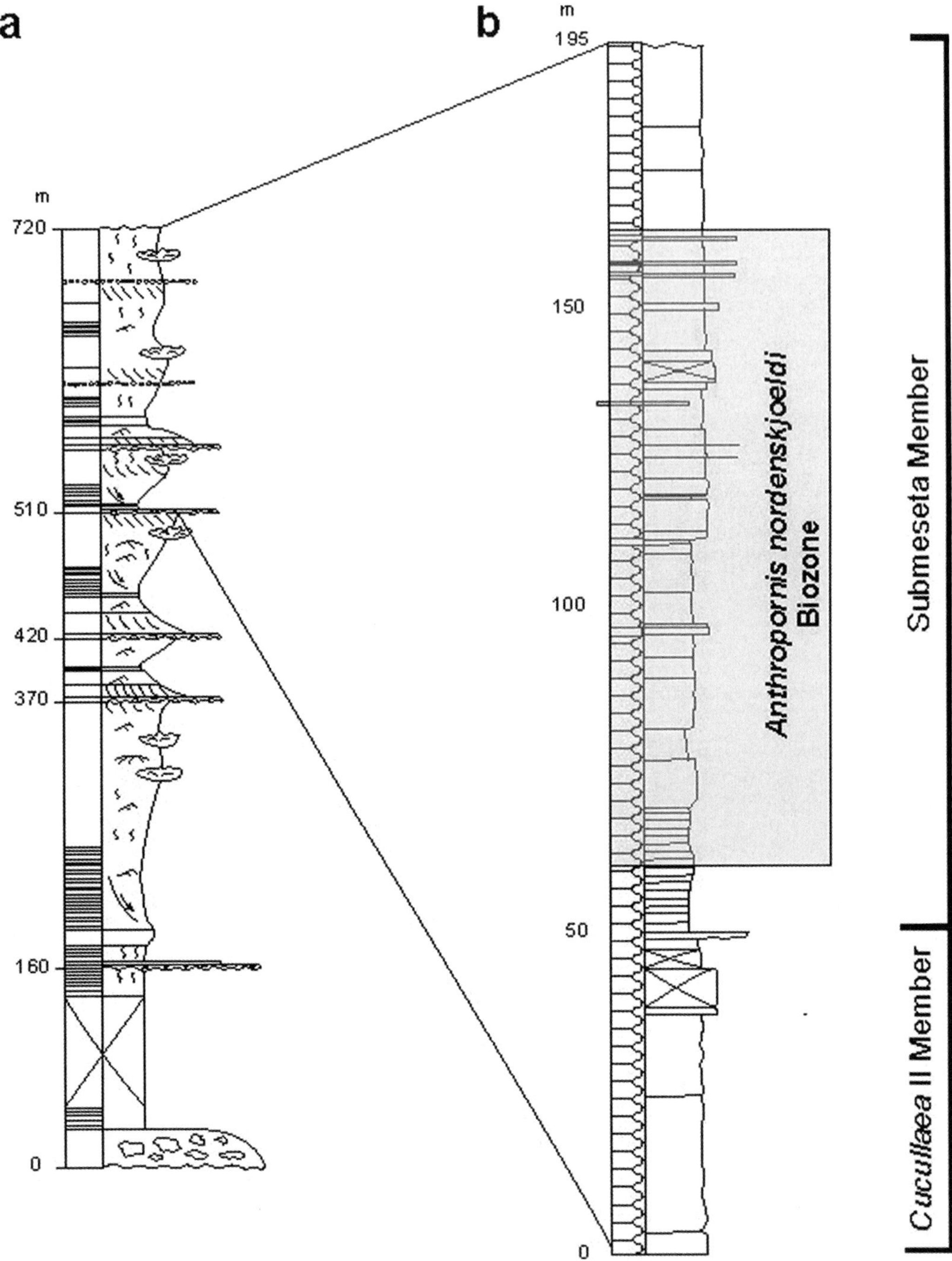

Fig. 2. (**a**) Stratigraphic section of the La Meseta Formation, Seymour Island, Antarctic Peninsula (modified from Reguero *et al.* 2002). (**b**) Measured section from the eastern flank of the plateau (stratotype of the *Anthropornis nordenskjoeldi* Biozone) showing stratigraphic levels of the *Cucullaea* II and Submeseta members (modified from Marenssi *et al.* 1998*a*). Shaded rectangular area delimits the *Anthropornis nordenskjoeldi* Biozone.

Fig. 3. General view of the uppermost part of Facies Association III (Submeseta Allomember) and *Anthropornis nordenskjoeldi* Biozone. Person (arrowed) for scale. Note the occurrence of well-cemented sandstones (detail in Fig. 2) and loose sands.

The stacking of the above-described facies associations suggests a major transgressive cycle, whereas the internal, second-order surfaces that generally floored each allomember reflect minor base-level falls and/or decreases in accommodation space.

Sedimentary structures and sediment textures allow a comparison with sedimentary models proposed either by Anderton (1976) in the Jura Quartzite shelf or by Bellosi (1987) in the Patagonian beds of the San Jorge Basin, Argentina. Both models can be summarized as interactive tide–storm systems. In these models, as well as in the La Meseta Formation, large- and medium-scale sandy bed-forms formed extensive fields of sandwaves and dunes laterally, associated with lower energy areas where muddy sand sheets and isolated patches of small-scale rippled sandstone developed. Eventually, channelled areas representing either rip (storm-related) or ebb (tide-generated) currents concentrated more vigorous flows.

During fair weather, tidal sandwaves slowly migrated in areas of higher current strength, while fields of sandy small ripples and muddy sand blankets accumulated in areas of lower energy. Cross-bedding, reactivation surfaces and mud drapes are all indicative of tidal action. Between these large bedforms and/or in the lower energy areas a prolific benthic fauna, mainly represented by infaunal and epifaunal filter-feeding and infaunal deposit-feeding organisms, proliferated. During high-energy periods, probably related to storms, depositional zones migrated ‘downcurrent’ (i.e. towards offshore), and proximal areas were eroded or at least winnowed of fine-grained particles leaving a coarse-grained lag (gravel beds) or scours subsequently filled with reworked shells and gravels (channel lags). Immediately after major storms a thin mud/sand blanket settled out from suspension.

Thus, Facies Association III represents sedimentation on a sandy tidal shelf influenced by storms. The transitional relationship with underlying estuarine deposits indicate a causal link and suggests an overall sea-level rise with backsteeping of sandy marine facies onto muddy estuarine ones.

The Submeseta Allomember has been regarded as Late Eocene since Dingle *et al.* (1998) reported a $^{87}Sr/^{86}Sr$ derived age of 34.2 Ma (late Late Eocene) for the topmost metres of the La Meseta Formation. In addition,

Fig. 4. Penguin humerus weathered in poorly consolidated pebbly sands showing a fresh breakage.

Dutton *et al.* (2002) also presented ages of 36.13, 34.96 and 34.69 Ma for different levels within Telm 7 (Submeseta Allomember).

In summary, fossil penguin bones studied in this paper were recovered from rocks of Facies Association III accumulated between 34.2 and 36.13 Ma (late Late Eocene). All specimens are stored in the Museo de La Plata, La Plata, Argentina.

Systematic palaeontology

Osteological terminology follows Baumel & Witmer (1993) and, when necessary, Simpson (1946) and O' Hara (1989). We use the classification proposed by Simpson (1946, 1971). Figure 5 shows the main terms used in the description of the bones. Measurements were taken with Vernier calipers with 0.01 mm increments and are included in Table 1.

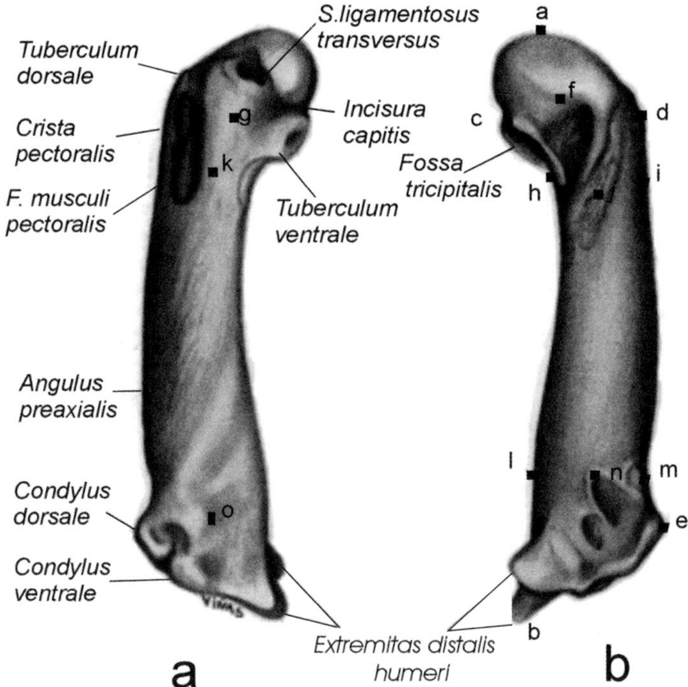

Fig. 5. Terms and measurement points of the spheniscid humerus used in the text and in Table 1.

Sphenisciformes Sharpe, 1891
Spheniscidae Bonaparte, 1831
Subfamily Anthropornithinae Simpson, 1946

Anthropornis nordenskjoeldi Wiman, 1905

Materials (humerus)

MLP 93-X-1-4 (proximal epiphysis), MLP 82-IV-23-4 (proximal epiphysis), MLP 83-I-1-190 (proximal epiphysis) and MLP 88-I-1-463 (proximal epiphysis).

Occurrence

Submeseta Allomember.

Description

Large and strong humerus (for measurements see Table 1). Non-flattened caput humeri. Small and undivided *fossa tricipitalis*. Conspicuous insertion of the *musculi brachialis internus*. Width and shallow *incisura capitis*. Eliptic and shallow *facies musculi pectoralis*. Undivided *sulcus ligamentosus transversus*. *Sulcus humerotricipitalis* narrower than the *sulcus escapulotricipitis*.

Anthropornis grandis (Wiman, 1905)

Materials (humerus)

MLP CX-60-25 (proximal epiphysis), MLP 83-V-30-5 (diaphysis) and MLP 93-X-1-104 (complete humerus).

Occurrence

Submeseta Allomember.

Description

The humerus is clearly smaller (Table 1) than that of *A. nordenskjoeldi*, whereas the humerus morphology is very similar between both species.

Anthropornis sp.

Materials (humerus)

MLP 83-V-20-25 (proximal and distal epiphysis), MLP 83-V-20-28 (proximal epiphysis), MLP 93-X-1-105 (proximal epiphysis), MLP 83-V-20-402 (fragmentary diaphysis), MLP 93-X-1-4 (distal epiphysis), MLP 83-V-30-4

Table 1. *Measurements of the humerus of penguin species of the La Meseta Formation, Seymour Island, Antarctic Peninsula*

Taxa[†]	Measurements[*]											
	a–b (cm)	c–d (cm)	b–e (cm)	f–g (cm)	h–i (cm)	j–k (cm)	l–m (cm)	n–o (cm)	Angle 1	Angle 2	Angle 3	Angle 4
Anthropornis nordenskjoeldi	18.28 (1)			3.61 (3)	2.83 (1)	2.05 (3)	3.82	2.24	8°	47°	12°	
Anthropornis grandis	19.42	5.4	4.21 (2)		3.18 (3)	2.02 (2)	3.11 (2)	1.96 (2)	6°	42°	7.6° (2)	25°
Delphinornis larseni	9.32	2.37 (3)	3.54 (4)	1.39 (3)	1.32 (3)	0.76 (5)	1.98 (6)	0.89 (6)	16° (2)	32° (2)	7.2° (6)	34.3° (6)
Tonniornis minimum	9.36		2.55		1.64	0.9	1.73 (2)	0.86 (2)			10.2° (2)	29.75°
Tonniornis mesetaensis	10.5		2.73	1.97	1.86	0.96	1.8	0.85	22.0°		9.0°	37.0°
Palaeeudyptes klekowskii	14.36 (3)	4.8	4.03 (5)	2.66 (7)	2.56 (9)	1.36 (11)	2.41 (11)	1.27 (11)	16.3° (3)	39°	8.8° (5)	29.1° (5)
Palaeeudyptes gunnari		4.52 (4)		2.86 (8)	2.70 (7)	1.49 (7)	2.43	1.18	15.66 (3)		7°	
Archaeospheniscus lopdelli			3.44 (2)	2.28	2.25 (2)	1.2 (3)	2.21 (3)	1.12 (3)			8.2° (2)	20°

Measurements of the humerus of penguin species recorded in the La Meseta Formation, Seymour Island, Antarctic Peninsula (The measurement lines are depicted in Fig. 5).

[*] a–b: total length; c–d: proximal width; b–e: condilar width; f–g: anteroposterior width of the head; h–i: lateral width taken under the fossa tricipitalis; j–k: anteroposterior width taken under the fossa tricipitalis; l–m: distal width; n–o: anteroposterior distal width; angle 1: degree of torsion of the head; angle 2: curvature degree of the head; angle 3: preaxial angle; angle 4: shaft trochlear angle.

[†] Measured materials. *Anthropornis nordenskjoeldi* MLP 82-IV-23-4, MLP 88-I-1-463, MLP 83-I-1-190; *Anthropornis grandis* MLP 83-V-30-5, MLP CX-60-25, MLP 93-X-1-104; *Delphinornis larseni* MLP 84-II-1-1, MLP 94-III-15-177, MLP 84-II-1-16, MLP 93-X-1-147, MLP 93-X-1-146, MLP 93-X-1-21, MLP 93-X-1-144, MLP 93-X-1-32, MLP 84-II-1-169, MLP 91-II-4-263; *Tonniornis minimum* MLP 93-I-6-3, MLP 93-X-1-22; *Tonniornis mesetaensis* MLP 93-X-1-145; *Palaeeudyptes klekowskii* MLP 82-IV-23-1, MLP CX-60-223, MLP CX-60-232, MLP 95-I-10-149, MLP 94-III-15-175, MLP 95-I-10-217, MLP 93-X-1-172, MLP 83-V-30-3, MLP CX-60-201, MLP 84-II-1-12[a], MLP 87-II-1-44, MLP 94-III-15-17, MLP 93-X-1-174, MLP 84-II-1-2, MLP 83-V-30-14, MLP 83-V-30-7, MLP 83-V-20-30, MLP 82-IV-23-2; *Palaeeudyptes gunnari* MLP 82-IV-23-59, MLP 95-I-10-226, MLP 82-IV-23-60, MLP 82-IV-23-64, MLP 84-II-1-41, MLP 83-V-20-51, MLP 83-V-20-403, MLP 84-II-1-115, MLP 86-V-30-15, MLP 86-V-30-16, MLP 88-I-1-464, MLP 88-I-1-469, MLP 91-II-4-262, MLP 93-X-1-3, MLP 93-X-1-30; *Archaeospheniscus lopdelli* MLP 84-II-1-111, MLP 93-X-1-27, MLP 93-X-1-97, MLP 95-I-10-227, MLP 95-I-10-231, MLP 95-I-10-233, MLP 95-I-10-236.

(proximal epiphysis) and MLP 87-II-1-42 (proximal epiphysis).

Occurrence

Submeseta Allomember, but MLP 87-II-1-42 was found in *Cucullaea* I Allomember.

Delphinornis larseni Wiman, 1905

Materials (humerus)

MLP 93-X-1-147 (near complete, lacks the distal end), MLP 93-X-1-146 (complete), MLP 84-II-1-169 (diaphysis and fragmentary proximal epiphysis), MLP 93-X-1-21 (diaphysis), MLP 84-II-1-16 (diaphysis and fragmentary proximal epiphysis), MLP 93-X-1-32 (diaphysis and proximal epiphysis), MLP 93-X-1-144 (diaphysis and distal epiphysis), MLP 94-III-15-177 (near complete, lacks the proximal end) and MLP 91-II-4-263 (proximal epiphysis).

Occurrence

Submeseta Allomember, with the exception of MLP 94-III-15-177 and MLP 91-II-4-263 that come from the *Cucullaea* I Allomember.

Description

The humerus is small (Table 1) and slender, with a straight shaft and an *angulus preaxialis* absent. The shaft trochlear angle is small, a characteristic shared with *Palaeeudyptes antarcticus*. The *caput humeri* is flattened and the *incisura capitis* is short and deep. The *fossa tricipitalis* is big and with an oval external edge. The *musculi brachialis internus* does not have well-marked insertions, while the *facies musculi pectoralis* is deep. The *facies musculi supracoracoideus* is oriented parallel to the main axis of the diaphysis. The *sulcus ligamentosus transversus* is divided in two unequal portions by an osseous partition.

Comments

Although Marples (1952) considered that Anthropornithinae and the Palaeeudyptinae belong to the same subfamily (Palaeeudyptinae), we follow Simpson (1946) who assigned *Delphinornis* into the Anthropornithinae.

The humerus of *Delphinornis larsenii* is less robust and smaller (Table 1) than that of *Platydyptes*, *Pachydyptes* and *Archaeosphenicus*. Also, *Delphinornis* presents a narrower shaft than that of the genus *Platydyptes*, *Pachydyptes* and *Archaeospheniscus*. Following Kandefer (1994), both *Mesetaornis* and *Marambiornis* possess a diaphysis narrower distally, whereas, conversely, the materials studied here show a diaphysis that is narrower proximally. Consequently, the materials studied here are assigned to *Delphinornis*. The three species of *Delphinornis* can be ordered according to their decreasing size as follows: *D. larsenni*, *D. gracilis* and *D. arctowskii*, and this characteristic – discrete sized group – facilitates their distinction. Kandefer (1994) figured and measured materials that were assigned to the two last species, but the bones studied here are bigger and they can be assigned to *D. larsenni*.

Delphinornis cf. *arctowskii* Myrcha, Jadwiszczak, Tambussi, Noriega, Gazdzicki, Tatur & del Valle, 2002

Material

MLP 93-X-1-70, near complete humerus.

Occurrence

Submeseta Allomember.

Description

The morphology more closely resembles that of *Delphinornis arctowski*, but the specimen is considerably smaller (Table 1). In the absence of articulated skeletons and/or morphometric studies that allow for the confirmation of the correlations between non-homologous bones, we prefer to be conservative in the systematic placement of MLP 93-X-1-70 as *Delphinornis* cf. *arctowski*.

Subfamily Palaeeudyptinae Simpson, 1946
Palaeeudyptes antarcticus Huxley, 1859

Material (humerus)

MLP 84-II-1-1 (humerus without the proximal epiphysis).

Occurrence

Submeseta Allomember.

Description

The humerus corresponds to a medium-sized penguin (Table 1). The diaphysis is sigmoid and narrower distally. The *angulus preaxialis* is well marked. The *caput humeri* is flattened. The *fossa tricipitalis* is proportionally large in relation to the size of the proximal end of the humerus, with a rounded edge and undivided. The *musculi brachialis internus* is very evident. The *incisura capitis* and the *facies musculi pectoralis* are deep. As in *Archaeospheniscus* (see below) the *sulcus ligamentosus transversus* is undivided. The *extremita distalis caudalis* is narrow and it does not extended beyond the diaphysis. The *sulcus humerotricipitalis* and the *sulcus scapulotricipitis* are subequal in size.

Comments

P. antarcticus was previously found in sediments of the early Oligocene of New Zealand (Simpson 1971). The material described herein constitutes the first occurrence of this species in Antarctica.

Palaeeudyptes gunnari (Wiman, 1905)

Materials (humerus)

MLP 82-IV-23-64 (diaphysis and proximal epiphysis), MLP 93-X-1-31 (complete humerus), MLP 82-IV-23-60 (proximal epiphysis), MLP 88-I-1-464 (proximal epiphysis), MLP 86-V-30-15 (proximal epiphysis), MLP 84-II-1-115 (proximal epiphysis), MLP 84-II-1-6 (proximal epiphysis), MLP 84-II-1-66 (proximal epiphysis), MLP 83-V-20-403 (proximal epiphysis), MLP 86-V-30-16 (proximal epiphysis), MLP 82-IV-23-59 (proximal epiphysis), MLP 84-II-1-41 (proximal epiphysis), MLP 83-V-20-51 (proximal epiphysis), MLP 95-I-10-226 (proximal epiphysis), MLP 93-X-1-30 (proximal epiphysis), MLP 91-II-4-262 (proximal epiphysis) and MLP 88-I-1-469 (proximal epiphysis).

Occurrence

Except MLP 91-II-4-262 and MLP 88-I-1-469, which come from *Cucullaea* I Allomember, all specimens come from the Submeseta Allomember.

Description

Marples (1953) published a detail description of very well preserved humerus of *Palaeeudyptes gunnari* (except for the *fossa tricipitalis* zone). For this reason, we only give a brief description of this area. There is a large and undivided *fossa tricipitalis*, with a rounded edge. In both *P. antarcticus* and *P. gunnari* the *incisura capitis* and the *facies musculi pectoralis* are deep. As in *P. antarcticus* and *Archaeospheniscus* (see below), the *sulcus ligamentosus transversus* is undivided.

Comments

The humerus corresponds to a medium-sized penguin, smaller (Table 1) than *P. antarcticus* (Simpson 1971) and *P. klekowskii.* As in other cases, we respected the designations made by Kandefer (1994) to distinguish *P. klekowskii*, one of the species that has been identified by an isolated tarsometatarsus.

Palaeeudyptes klekowskii Myrcha, Tatur and del Valle, 1990

Materials (humerus)

MLP CX-60-201 (complete humerus), MLP 93-X-1-172 (complete humerus), MLP 93-X-1-3 (incomplete humerus), MLP CX-60-223 (complete humerus), MLP 82-IV-23-2 (diaphysis and proximal epiphysis), MLP 84-II-1-11 (diaphysis and proximal epiphysis), MLP 95-I-10-149 (diaphysis and proximal epiphysis), MLP 83-V-30-7 (diaphysis), MLP 83-V-30-3 (diaphysis and proximal epiphysis), MLP 82-IV-23-3 (proximal epiphysis), MLP 83-V-30-14 (proximal epiphysis), MLP 82-IV-23-1 (diaphysis and proximal epiphysis), MLP 83-V-20-30 (proximal epiphysis), MLP 84-II-1-2 (diaphysis and distal epiphysis), MLP CX-60-232 (diaphysis), MLP 84-II-1-12a (distal epiphysis), MLP 91-II-4-227 (distal epiphysis), MLP 93-X-1-174 (distal epiphysis), MLP 94-III-15-175 (complete humerus), MLP 95-I-10-217 (distal epiphysis) and MLP 87-II-1-44 (distal epiphysis).

Occurrence

All specimens from Submeseta Allomember, except the last three were found in the lower levels of *Cucullaea* I Allomember.

Description

It is the largest species of the genus. A very detailed description of the species was given by Myrcha *et al.* (1990, 2002).

Archaeospheniscus lopdelli Marples, 1952

Materials (humerus)

MLP 94-III-15-17 (complete humerus), MLP 93-X-1-123 (proximal epiphysis), MLP 93-X-1-27 (proximal epiphysis), MLP 95-I-10-231 (diaphysis and distal epiphysis), MLP 95-I-10-236 (proximal epiphysis), MLP 84-II-1-110 (diaphysis and distal epiphysis), MLP 95-I-10-227 (diaphysis and proximal epiphysis), MLP 84-II-1-111 (diaphysis and proximal epiphysis), MLP 93-X-1-97 (diaphysis and distal epiphysis) and MLP 95-I-10-233 (diaphysis and distal epiphysis).

Occurrence

Submeseta Allomember.

Description

The diaphysis is sigmoid and curved with subequal distal and proximal widths. The shaft–trochlear angle is approximately 14° (angle between a tangent touching the dorsal and ventral condyles and a line joining the midpoints of proximal and distal ends of the shaft). The *caput humeri* is flattened and the *fossa tricipitalis* is undivided and relatively large. The *incisura capitis* is deep. The insertion of the *pectoralis secundus* is oblique. An incipient division of the *sulcus ligamentosus transversus* exists, but does not reach to separate it in two portions. The more caudal *extremita distalis* is narrow and it does not extend beyond the diaphysis. The *sulcus humerotricipitalis* and the *sulcus scapulotricipitis* are similar in size and extension.

Comments

Archaeospheniscus lopdelli has a medium-sized humerus (Table 1), which is larger than that of *A. lowei* from the early Late Oligocene of New Zealand (Marples 1952) and compared to that of *A. wimani* from the late Eocene of Seymour Island (Myrcha *et al.* 2002), which is the smallest species of the genus. *A. lopdelli* was previously known only from the early Late Oligocene of New Zealand and this is the first record of the species in Antarctica.

Spheniscidae Bonaparte, 1831 *incertae sedis*

It is not the scope of our study to discuss the subfamiliar diagnostic characters given by Simpson (1946). However, the following specimens could not be clearly assigned to any known subfamily because of the unique features that are present. These humeri differ from the Anthropornithinae by having a thin and straight shaft; from the Palaeeudyptinae by the presence of subequal proximal and distal width and a large *fossa tricipitalis*; from the Paraptenodytinae by having a shaft with subequal proximal and distal widths, and a smaller shaft–trochlear angle; and, finally, the humerus differs from those of Palaeospheniscinae because these humeri have a smaller shaft–trochlear angle and a *fossa tricipitalis* undivided instead of bipartite.

Tonniornis gen. nov

Type species. *Tonniornis mesetaensis* MLP 93-X-1-145

Included species

Tonniornis mesetaensis and *T. minimum*.

Etymology

After Eduardo Tonni, the Argentinian palaeontologist. Masculine in gender.

Diagnosis

Small-sized penguin bones characterized by their small and straight (without sigmoid curvature) humeri. Flattened *caput humeri*. Short and shallow *incisura capitis*. Relatively large and undivided *fossa tricipitalis*. *Sulcus ligamentosus transversus* divided in two subequal portions by a partition. Lacks a distinct point of insertion for the *musculi brachialis internus*. A low angulus preaxialis (*c.* 12°). *Facies musculi pectoralis* well marked. *Sulcus humerotricipitalis* deeper and wider than the *sulcus escapulotricipitis*. Shaft–trochlear angle greater than 31°.

Comments

The straight diaphysis, the *sulcus ligamentosus transversus* separated into two subequals portions, and the undivided *fossa tricipitalis* distinguish the new taxon from all the known fossil and living penguins. The genera *Mesetaornis*, *Marambiornis* and *Delphinornis* were originally defined on the basis of their tarsometatarsal features only. However, Kandefer (1994), in her unpublished thesis, assigned some humeri to these genera. *Mesetaornis polaris* has a humeral diaphysis that is wider proximally than in *Tonniornis*; the humeri of *Marambiornis* are smaller and more slender than those of *Tonniornis*, and the humeri of *Delphinornis* are conspicuously smaller and more slender than those of *Tonniornis*. Two penguin taxa registered outside of Antarctica, *Pachydyptes* and *Platydyptes*, as well as *Archaeospheniscus* discovered in Antarctica and New Zealand, have humeri bigger and stronger than the new genus.

Tonniornis mesetaensis sp. nov Fig. 6a, b

Holotype

MLP 93-X-1-145 (complete humerus).

Type horizon

Submeseta Allomember, La Meseta Formation (Late Eocene), Seymour Island, Antarctica.

Etymology

From the La Meseta Formation, in reference to the provenance of the material.

Measurements

Total length 115 mm; distal width 23.5 mm; proximal width 30.4 mm; shaft–trochlear angle 35°; *angulus preaxialis* 13°.

Diagnosis

Straight shaft without *angulus preaxialis*. Narrow *extremita distalis caudalis* that is extended a little beyond the shaft. The main difference to *Tonniornis minimum*, the other known species of the genus, is the last feature.

Occurrence

Seymour Island, La Meseta Formation, Submeseta Allomember (Late Eocene).

Tonniornis minimum sp. nov. Fig. 6c, d

Holotype

MLP 93-I-6-3 (complete humerus).

Referred material

MLP 93-X-1-22 (diaphysis and distal epiphysis).

Type horizon

Submeseta Allomember, La Meseta Formation (Late Eocene), Seymour Island, Antarctica.

Etymology

In reference to the small size of the specimen.

Measurements

Total length 95.4 mm; distal width 23.6 mm; shaft–trochlear angle 31°, *angulus preaxialis* 11° (Table 1).

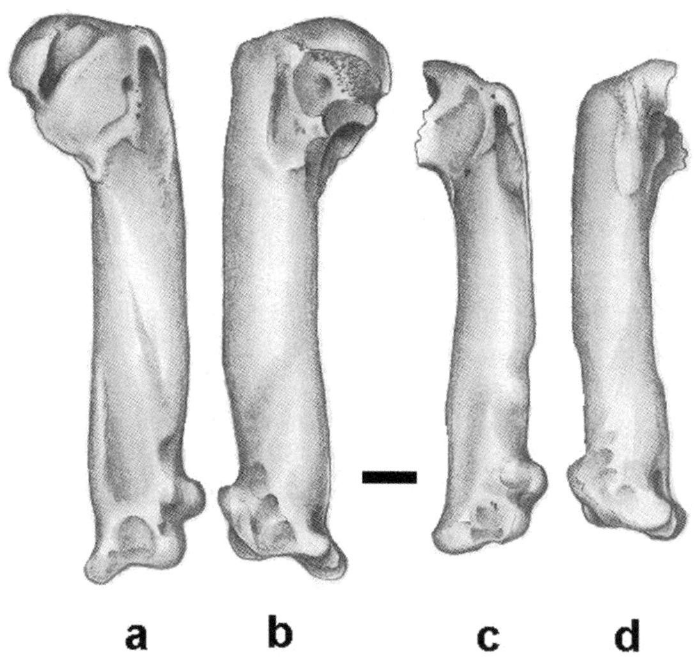

Fig. 6. *Tonniornis mesetaensis* gen. et sp. nov., MLP 93-X-1-145. (**a**) Cranial view. (**b**) Caudal view. *Tonniornis minimun* gen. et sp. nov., MLP 93-I-6-3. (**c**) Cranial view. (**d**) Caudal view. Scale bar represents 1 cm.

Diagnosis

Shorter than *Tonniornis mesetaensis*, the other species of the genus. Low *angulus preaxialis*. Width subequal proximally and distally. Width *extremita distalis caudalis* does not extend beyond the shaft. In this way, in caudal view, the caudal *extremita distalis humeri* is rounded, with a wide base that partially covers the base of the intermediate *extremita distalis humeri*.

Occurrence

Seymour Island, La Meseta Formation, Submeseta Allomember (Late Eocene).

Discussion

Systematic remarks

As the result of this study we recognize 10 species of penguins (Table 2). The distribution of three of them includes areas outside the James Ross Basin: *A. nordenskjoeldi* is known in sediments of the Late Eocene from Australia (Jenkins 1974; Fordyce & Jones 1990); *Palaeeudyptes antarcticus* is recorded in the Early Oligocene from New Zealand and Australia (Simpson 1970); and *Archaeospheniscus lopdelli* is recorded in the early Late Oligocene from New Zealand (Marples 1952). Until now, the presence of the two last species was restricted to New Zealand and Australia.

In addition, the discovery of a single partial skeleton of an undetermined Anthropornithinae from the Late Eocene of La Leticia Formation, Tierra del Fuego (J. Clarke pers. comm.) has an important palaeobiogeographical significance. Based on its marine invertebrates, Olivero & Malumián (1999) regarded this unit as equivalent to the upper part of the La Meseta Formation. Ongoing systematic study by Clarke and Olivero will shed new light on the relationships between Antarctic and Patagonian fossil penguins.

Finally, we have described a new genus, *Tonniornis*, with two small species, *T. mesetaensis* and *T. minimum*. Both new taxa are recorded exclusively in the uppermost level of the La Meseta Formation.

Delphinornis cf. *arctowski* is the smallest penguin recorded from the James Ross Basin. Very few cases of tiny penguins are known: *Eretiscus tonnii* from the Early Miocene of Patagonia and the 'Hakataramea bird' (previously mentioned as *Eretiscus* by Cozzuol *et al.* 1991) from the latest Oligocene–earliest Miocene of New Zealand.

The combination of diversity, abundance and occurrence of penguins in the Submeseta horizons makes this group very useful for biostratigraphical studies.

Anthropornis nordenskjoeldi *Biozone*

As we mentioned previously, although penguins remains are widespread in all the allomembers of the La Meseta Formation, it is in the uppermost 100 m of the Submeseta Allomember that the highest concentration of penguin bones occur and all the penguin species known for the Eocene of Antarctica. Fifteen species are known

Table 2. *Penguin species of the La Meseta Formation, Seymour Island, Antarctic Peninsula*

Species	Previous work *,†		This work	
	Cucullaea I A.	Submeseta A.	*Cucullaea* I A.	Submeseta A.
Anthropornis nordenskjoeldi		X		X
Anthropornis grandis	X	X		X
Palaeeudyptes antarcticus				X
Palaeeudyptes gunnari	X	X	X	X
Palaeeudyptes klekowsky		X	X	X
Delphinornis larseni	X	X	X	X
Delphinornis gracilis		X		
Delphinornis arctowski		X		
Mesetaornis polaris		X		
Marambiornis exilis		X		
Archaeospheniscus wimani		X		
Archaeospheniscus lopdelli				X
Tonniornis mesetaensis				X
Tonniornis minimum				X

Penguin species recorded in the La Meseta Formation, Seymour Island, Antarctic Peninsula after * Simpson, (1971), † Myrcha *et al.* (2002) and this paper.

from these horizons, including giant and tiny species. These strata belong to Facies Association III of Marenssi *et al.* (1998*a*) and document the highest morphological and taxonomical diversity of penguins in the world that lived sympatrically. Myrcha *et al.* (1990) had noted this relationship and, moreover, they delimited these high concentrations of penguin bones to horizons between two shell banks: a lower one bearing the gastropod *Turritella*, and a higher one bearing *Modiolus* sp. and *Lingula* sp.

As a result of this study we determined that *Anthropornis nordenskjoeldi*, *Delphinornis gracilis*, *D. arctowski*, *Archaeospheniscus lopdelli* and *Palaeeudyptes antarcticus* are species exclusively recorded in this interval of strata. The first and last appearances of these species have been found within these strata. The age of these horizons has been established between 34.2 and 36.13 Ma (late Late Eocene) (Dingle & Lavelle 1998; Dutton *et al.* 2002). On this basis, we define a new biostratigraphic unit, the *Anthropornis nordenskjoeldi* Biozone, easily distinguishable by the common occurrence of penguin bones and the phospatic brachiopod *Lingula*.

We selected *Anthropornis nordenskjoeldi* to identify the zone, as: (1) it is restricted to an interval of strata traceable through the Submeseta Allomember; and (2) it is numerically predominant over the other penguin species.

We designate the stratotype of this biozone to the section measured by Marenssi *et al.* (1998*b*) on the east flank of the plateau, facing the Weddell Sea (Fig. 1b). The interval of strata is located nearly 30–35 m below the top of the 145 m-thick Submeseta Allomember at this place (Fig. 3).

The thickness of these horizons, which crop out continuously around the uppermost flanks of the plateau, is approximately 100 m (Fig. 2b).

The marine and terrestrial macrofossil assemblage of the *Anthropornis nordenskjoeldi* Biozone is depicted in Tables 3 & 4. Gadiforms '*Mesetaichthys*', sharks (*Pristiophorus* and *Carcharias*) and a primitive mysticete whale (*Llanocetus denticrenatus*) are also present in this zone. Interestingly, reworked plant remains and isolated leaves of *Nothofagus* have been occasionally found within this zone (Vizcaíno *et al.* 1997).

In total, the faunistic content of this zone represents an important source of palaeontological information on the Late Eocene Antarctic biota.

Stillwell & Zinsmeister (1992) have defined, on the basis of gastropods, different biozones

Table 3. *Associated macroinvertebrate fauna of the* Anthropornis nordenskjoeldi *Biozone*

Suprageneric taxa	Species
Bivalves	
Hyatellidae	*Hiatella tenuis*
	Panopea philippii
Mytilidae	*Arcuatula sootryeni*
	Botula pirriei
	Modiolus thomsoni
Cucullaeidae	*Cucullaea donaldi*
Pinnidae	*Pinna sobrali*
Pectinidae	*Chlamys* sp.
Ostreidae	*Ostrea seymouriensis*
Lucinidae	*Saxolucina sharmani*
Veneridae	*Eurhomalea florentinoi*
Myidae	*Mya nucleoides*
Sportellidae	*Anisodonta subovata*
Pteriidae	*Electroma notiala*
Gastropods	
Patellidae	*Celliana feldmanni*
Vermetidae	*Serpulorbis hormathos*
Aporrhaidae	*Arrhoges (Antarctohoges) arcuacheilos*
	Arrhoges (Antarctohoges) diversicostata
Struthiolariidae	*Perissodonta laevis*
Naticidae	*Polinices (Polinices) cf. subtenuis*
Muricidae	*Lenitrophon suteri*
	Xymene lamesetaensis
Buccinidae	*Cyrtochetus (Cyrtochetus) bucciniformis*
	Cominella (Josepha) ottoi
	Aeneator lawsi
Nassariidae	*Sudanassarius antarctohimaleos*
Fasciolariidae	*Fusinus suraknisos*
Turridae	*Cosmasyrinx brychiosinus*
	Zemacies finlayi
	Zemacies canalomos
Bullidae	*Bulla glacialis*
	Scafander (Kaitoa) schmitti
Turritellidae	*Turritella*
Scaphopods	
Dentaliidae	*Dentalium (Dentalium) pulchrum*
Crinoids	*Metacrinus*
Echinoids	*Abatus*
Decapods	
Ophiuroids	*Ophiura*
Bryozoans	*Smittina*
Brachiopods	*Lingula antarctica*
	Bouchardia

Marine macroinvertebrates associated to the *Anthropornis nordenskjoeldi* Biozone, Submeseta Allomember, La Meseta Formation, Seymour Island, Antarctic Peninsula. Compiled from Stilwell & Zinsmeister (1992), Aronson & Blake (2001), and Myrcha *et al.* (2002).

Table 4. *Associated vertebrate fauna of the* Anthropornis nordenskjoeldi *Biozone*

Suprageneric taxa	Species
Fish	
Gadiformes Merluccidae	*'Mesetaichthys'*
Sharks	
Pristiophoridae	*Pristiophorus lanceolatus*
Cetaceans	
Archaeoceti	*Zeuglodon* sp. *Zygorhiza* sp.
Mysticeti Crenaticeti	*Llanocetus denticrenatus*
Terrestrial mammals	
Sparnotheriodontidae	
Birds	
Ratitae	
Phorusrhacidae	

Non-penguin vertebrates associated to the *Anthropornis nordenskjoeldi* Biozone, Submeseta Allomember, La Meseta Formation, Seymour Island, Antarctic Peninsula. Compiled from Reguero *et al.* (2002).

within the La Meseta Formation. One of them, the *Perissodonta laevis* zone, stratigraphically overlaps our *Anthropornis nordenskjoeldi* Biozone and includes both *Cucullaea* II and Submeseta allomembers. The local range of that zone is somewhat imprecise because its upper boundary is not recognizable in the sequence and temporally spans approximately 10 Ma (from 34.2 to 45 Ma). For these reasons, we prefer to characterize a new biozone with more accurate upper and lower boundaries, fossil content, regional extension and chronological span.

As pointed out previously, penguin bones of Facies Association III are frequently well preserved, and, although disarticulated and with diverse degrees of abrasion, they are not usually broken. Most, if not all, were transported at least for a short time before burial and concentrated in discrete horizons. We agree with Stilwell & Zinsmeister (1992) in that time-averaging is at a minimum and fossils are approximately contemporaneous with the entombing sediments. Schäfer (1972) states that whale and penguin bones usually accumulate in shallow-marine or strandline deposits. The brachiopod *Lingula* sp., which usually appears in these levels, is thought to have preferred shallow waters (less than 30 m deep) in low-energy subtidal environments (Heckel, 1972), further supporting this palaeoenvironmental interpretation.

Rocks from these horizons were interpreted as shallow-marine, low-energy (strand plain) deposits by Trautman (1976), and lately as moderate-energy deposits associated to the transition between delta-front and delta-plain environments by Pezzetti (1987). Stilwell & Zinsmeister (1992) proposed a shallow, protected lagoonal setting with open bays behind barrier islands with tidal channels affected by storms.

Most penguin bones were found on the surface of loose sand and only a few were found in conglomerate. Only a single partial articulated skeleton of penguin was found in this level. In addition, the good preservation of the bones suggests that the conditions of deposition were quiet and of low energy. This depositional condition is not seen in the underlying stratigraphic members of the sequence, where the bones are found reworked and with major breakage.

The reconstruction in Figure 7 shows the great diversity of penguins during the Late Eocene on the eastern shore of the Antarctic Peninsula. This diversity is mainly documented in the body size and other morphological traits. On the basis of the body size, *Anthropornis nordenskjoeldi* was evidently the largest penguin known. Its hydrodynamic constraints suggest that it was a rather slow swimmer with speeds perhaps in the order 7–8 km h^{-1} (Jenkins 1985). The angled shape of the flippers, not straightened as in modern penguins, also suggest that *A. nordenskjoeldi* did not have the specialization to dive. The long neck of *A. nordenskjoeldi* probably favoured the capture of motile prey (fish) rather than krill and small squid.

The current knowledge of the fossil Antarctic penguins is based on fragmentary, but very informative, evidence. Future prospecting and new collections will lead to a substantial increase in our understanding of the Eocene penguin evolution. From this understanding, it may be possible to decipher the role of the spheniscids in Eocene marine ecosystems and the relationships with penguins from other latitudes.

We especially thank D. Pirrie, J. Francis and A. Crame for the opportunity to participate in this volume; and J. Case and another anonymous referee, whose commentaries substantially improved the quality of this manuscript. We show gratitude to M. Cozzuol and to T. Ando for sharing information on Eocene whales and New Zealand penguins, respectively. J. González drew the line drawings. Fieldwork on Seymour Island has been supported by the Instituto Antártico Argentino. We also thank CONICET for permanent research support.

Fig. 7. Reconstruction of the environment of the Allomember Submeseta, La Meseta Formation, Seymour Island. Many species of small and giant penguins congregating on the eastern shore of the ancient Antarctic Peninsula 30 Ma ago. In the background *Nothofagus* forest, a ratite bird and a sparnotheriodontid mammal feed quietly.

References

ANDERTON, R. 1976. Tidal shelf sedimentation: an example from the Scottish Dalradian. *Sedimentology*, **23**, 429–458.

ARONSON, R. & BLAKE, D. 2001. Global climate change and the origin of modern benthic communities in Antarctica. *American Zoologist*, **41**, 27–39.

BAUMEL, J. & WITMER, L.M. 1993. Osteología. *In*: BAUMEL, J., KING, A., BREAZILE, J.E., EVANS, H. & VANDEN BERGUE, J.C. (eds) *Handbook of Avian Anatomy: Nomina Anatomica Avium*. Publications of the Nuttall Ornithological Club, No. 23. Cambridge, MA, 45–132.

BELLOSI, E.S. 1987. *Litoestratigrafía y sedimentación del 'Patagoniano' en la cuenca de San Jorge, Terciario de Chubut y Santa Cruz*. PhD Thesis, Universidad de Buenos Aires.

CASE, J.A. 1992. Evidence from fossil vertebrates for a rich Eocene Antarctic marine environment. *In*: KENNETT, J.H. & WARNKE, D.A. (eds) *The Antarctic Paleoenvironment: A Perspective on Global Change.* American Geophysical Union, Antarctic Research Series, **56**, 119–130.

COZZUOL, M.A., FORDYCE, R.E. & JONES, C.M. 1991. La presencia de *Eretiscus tonni* (Aves, Spheniscidae) en el Mioceno temprano de Nueva Zelanda y Patagonia. *Ameghiniana*, **28**, 406.

DEL VALLE, R.A., ELLIOT, D.H. & MACDONALD, D.I. 1992. Sedimentary basins on the east flank of the Antarctic Peninsula: proposed nomenclature. *Antarctic Science*, **4**, 477–478.

DINGLE, R.V. & LAVELLE, M. 1998. Late Cretaceous Cenozoic climatic variations of the northern Antarctic Peninsula: new geochemical evidence and review. *Palaeogeography, Palaeoclimatology, Palaeoecology*, **141**, 215–232.

DINGLE, R., MARENSSI, S. & LAVELLE, M. 1998. High latitude Eocene climate deterioration: evidence from northern Antarctic Peninsula. *Journal of South American Earth Sciences*, **11**, 571–579.

DUTTON, A.L., K.C., LOHMANN, K.C. & ZINSMEISTER, W.J. 2002. Stable isotope and minor element proxies for Eocene climate of Seymour Island, Antarctica, *Paleoceanography*, **17**, 1016–1029.

ELLIOT, D.H. & TRAUTMAN, T.A. 1982. Lower Tertiary strata on Seymour Island, Antarctic Peninsula. *In*: CRADDOCK, C. (ed.) *Antarctic Geoscience*. University of Wisconsin Press, Madison, WI, 287–297.

FORDYCE, R.M. & JONES, C. 1990. Penguin history and new fossil material from New Zealand. *In*: DAVIS, L.S. & DARBY, J.T. (eds) *Penguin Biology*. Academic Press, New York, 419–446.

HECKEL, P.H. 1972. Recognition of ancient shallow marine environments. *In*: RIGBY, J.K. & HAMBLIN, W.L. (eds) *Recognition of Ancient Shallow Marine Environment*, Society of Economic Paleontologists and Mineralogists, Special Publications, **16**, 226–286.

JADWISZCAK, P. 2001. Body size of Eocene Antarctic penguins. *Polish Polar Research*, **22**, 147–158.

JADWISZCZAK, P. 2003. Eocene penguins of Seymour Island: systematics, evolution and paleoecology. *In*: *Terra Nostra 2003/4: 9th International Symposium on Antarctic Earth Sciences (ISAES IX), 8–12 September 2003, Potsdam, Germany*, 169.

JENKINS, R. 1974. A new giant penguin from the Eocene of Australia. *Palaeontology*, **17**, 291–310.

JENKINS, R. 1985. *Anthropornis norddenskjoeldi* Wiman, 1905. *In*: RICH, P.V. & VAN TETS, G.F. (eds) *Kadimakara, Extinct Vertebrates of Australia.* Pioneer Design Studio, Victoria, Australia, 183–187.

KANDEFER, H.M. 1994. *Róznorodnosc fauny pingwinów kopalnych antarktycznej Wyspy Seymour w oparciu o analize humeri z kolekcji Instytutu Biologii Filii Uniwersytetu Warszawskiego w Bialymstoku.* Degree thesis, Warszawski University, Poland.

MARENSSI, S.A., SANTILLANA, S.N. & RINALDI, C.A. 1998*a*. *Paleoambientes sedimentarios de la Aloformación La Meseta (Eoceno), Isla Marambio (Seymour), Antártida.* Instituto Antártico Argentino, Contribución, **464**.

MARENSSI, S.A., SANTILLANA, S.N. & RINALDI, C.A. 1998*b*. Stratigraphy of La Meseta Formation (Eocene), Marambio Island, Antarctica. *In*: CASADÍO, S. (ed.) *Paleógeno de América del Sur y de la Península Antártica.* Revista de la Asociación Paleontológica Argentina, Publicación Especial, **5**, 137–146.

MARPLES, B.J. 1952. Early Tertiary penguins of New Zealand. *New Zealand Geological Survey, Palaeontological Bulletin*, **20**, 1–66.

MARPLES, B.J. 1953. *Fossil Penguins From the Mid-Tertiary of Seymour Island.* Falkland Islands Dependencies Survey Scientific Reports, **5**, 1–15.

MYRCHA, A., JADWISZCZAK, P., TAMBUSSI, C., NORIEGA, J., GAZDZICKI, A., TATUR, A. & DEL VALLE, R. 2002. Taxonomic revision of Eocene Antarctic penguins based on tarsometatarsal morphology. *Polish Polar Research*, **23**, 5–46.

MYRCHA, A., TATUR, A. & DEL VALLE, R. 1990. A new species of fossil penguin from Seymour Island, West Antarctica. *Alcheringa*, **14**, 195–205.

O'HARA, R. 1989. *Systematics and the study of natural history, with an estimate of the phylogeny of the living penguins (Aves: Spheniscidae).* Ph.D. thesis, Harvard University, Cambridge.

OLIVERO, E.B. & MALUMIÁN, N. 1999. Eocene stratigraphy of Southeastern Tierra del Fuego, Argentina. *AAPG Bulletin*, **83**, 295–313.

PEZZETTI, T.F. 1987. *The sedimentology and provenance of the Eocene La Meseta Formation, Seymour Island, Antarctica.* Ph. thesis, The Ohio State University, City of Columbus.

REGUERO, M.A., MARENSSI, A.M., & SANTILLANA, S.N. 2002. Antarctic Peninsula and South America (Patagonia) Paleogene terrestrial faunas and environments: biogeographic relationships. *Palaeogeography, Palaeoclimatology, Palaeoecology*, **179**, 189–210.

RINALDI, R.A., MASSABIE, A., MORELLI, J., ROSENMANN, L. & DEL VALLE, R. 1978. Geología de la Isla Vicecomodoro Marambio, Antártida. *Contribución Instituto Antártico Argentino*, **217**, 1–37

SIMPSON, G.G. 1946. Fossil penguins. *Bulletin of the American Museum of Natural History,* **87**, 1–100.

SIMPSON, G.G. 1970. Miocene penguins from Victoria, Australia, and Chubut, Argentina. *Memories Natural History Museum, Victoria*, **31**, 17–24.

SIMPSON, G.G. 1971. Review of fossil penguins from Seymour Island. *Proceedings of the Royal Society of London*, **B178**, 357–287.

SCHÄFER, W. 1972. *Ecology and Paleocology of Marine Environments.* University of Chicago Press, Chicago, IL.

STILWELL, J.D. & ZINSMEISTER, W.J. 1992. *Molluscan Systematics and Biostratigraphy, Lower Tertiary La Meseta Formation, Seymour Island, Antarctic Peninsula.* American Geophysical Union, Antarctic Research Series, **55**.

TRAUTMAN, T.A. 1976. *Stratigraphy and petrology of Tertiary clastic sediments, Seymour Island, Antarctica.* Ph.D. thesis, The Ohio State University, City of Columbus.

VIZCAÍNO, S., BOND, M., REGUERO, M. & PASCUAL, R. 1997. The youngest record of fossil land mammals from Antarctica: its significance for the evolution of the terrestrial environment of the Antarctic Peninsula during the Late Eocene. *Journal of Paleontology*, **71**, 348–350.

WIMAN, C. 1905. Vorläufige Mitteilung über die alttertiären Vertebraten der Seymourinsel. *Bulletin of the Geological Institute of Upsala*, **6**, 247–253.

A new 'South American ungulate' (Mammalia: Litopterna) from the Eocene of the Antarctic Peninsula

M. BOND[1], M. A. REGUERO[1], S. F. VIZCAÍNO[1] & S. A. MARENSSI[2]

[1]*División Paleontología Vertebrados, Museo de La Plata, Paseo del Bosque s/n, 1900 La Plata, Argentina (e-mail: regui@fcnym.unlp.edu.ar)*

[2]*Instituto Antártico Argentino, Cerrito 1248, 1010 Buenos Aires, Argentina*

Abstract: *Notolophus arquinotiensis*, a new genus and species of the family Sparnotheriodontidae (Mammalia, Litopterna), is represented by several isolated teeth from the shallow-marine sediments of the La Meseta Formation (late Early–Late Eocene) of Seymour Island, Antarctic Peninsula, which have also yielded the youngest known sudamericids and marsupials. The new taxon belongs to the extinct order of 'South American native ungulate' Litopterna characterized by the convergence of the later forms with the equids and camelids. *Notolophus arquinotiensis* shows closest relationships with *Victorlemoinea* from the Itaboraian (middle Palaeocene) of Brazil and Riochican–Vacan (late Palaeocene–early Eocene) of Patagonia, Argentina. Although still poorly documented, this new taxon shows that the early Palaeogene Antarctic faunas might provide key data concerning the problems of the origin, diversity and basal phylogeny of some of the 'South American ungulates' (Litopterna). This new taxon shows the importance of Antarctica in the early evolution of the ungulates and illustrates our poor state of knowledge.

Initial palaeontological work in early 1980 on Seymour Island produced a modest assemblage of terrestrial fossil mammals (marsupials and South American ungulates). During the 1989–1990 season, geologists of the Instituto Antártico Argentino, while mapping Eocene marine rocks in Seymour Island, discovered small- and medium-sized land mammals, including two representatives of the South American native ungulates, Litopterna and Astrapotheria (Marenssi *et al.* 1994 and see also Hooker 1992). The Antarctic litoptern was referred by Bond *et al.* (1990) to the eolitoptern sparnotheriodontid genus *Victorlemoinea*. Renewed field efforts on Seymour Island (1992–2000) greatly enhanced the original collection and the sites are known to contain a high number of sparnotheriodontids, as well as many other taxa previously unknown from the area (Reguero *et al.* 2002). This new material allows us to reinterpret the teeth initially attributed to *Victorlemoinea*.

Litopterna is considered a natural group of South American native ungulates. Miocene–Pleistocene forms show a notable convergence with equids (Proterotheriidae) and camelids (Macraucheniidae). One of the most unusual of the litopterns was the Pleistocene camel-like *Macrauchenia* with large size and proboscis. The early Palaeogene forms (Palaeocene–Eocene) show morphological resemblances with the 'ancestral ungulates', the 'condylarths'. Sparnotheriodontids were medium- to large-sized ungulates. The family is known in the middle Palaeocene Itaboraian South American Land Mammal Age (SALMA) of Brazil and the late Palaeocene Riochican SALMA of Patagonia, and survived through at least the Late Eocene (Divisaderan SALMA) of Mendoza, Argentina. The species of sparnotheriodontids are classified in three genera and are listed in the Table 1.

The fossil record of the family Sparnotheriodontidae in South America is rather sparse;

Table 1. *Sparnotheriodontid species formally recognized*

Species	Geographic location	Age	Source
Victorlemoinea labyrinthica	Cañadón Vaca, Chubut	Riochican	Ameghino 1901
Victorlemoinea prototypica	Itaboraí, Brazil	Itaboraian	Paula Couto 1952
Sparnotheriodon epsilonoides	Cañadón Vaca, Chubut	Vacan	Soria 1980*a*
Phoradiadus divortiensis	Divisadero Largo, Mendoza	Divisaderan	Simpson *et al.* 1962
Sparnotheriodontidae?			
Heteroglyphis dewoletzky	Cerro del Humo, Chubut	Mustersan	Roth 1899

Species of Sparnotheriodontidae known in South America (Argentina and Brazil). Data from Soria (2001).

From: Francis, J. E., Pirrie, D. & Crame, J. A. (eds) 2006. *Cretaceous–Tertiary High-Latitude Palaeoenvironments, James Ross Basin, Antarctica.* Geological Society, London, Special Publications, **258**, 163–176. 0305–8719/06/$15

sparnotheriodontids are seldom common in any given locality, even during the late Palaeocene (Riochican SALMA), when the group reached its climax.

The taxonomy of Sparnotheriodontidae remains contentious and is currently based solely on teeth. This family has long been the subject of discussion over its systematic position, generic content and nomenclatural priorities. The main causes of these problems are the uniformity of its dental morphology (taxonomic differences are often minor and easily confused with intraspecific variation) and the poor quality of type specimens. Originally considered by Ameghino (1901) as a member of the meniscotheriid condylarths, Simpson (1945, 1948) regarded *Victorlemoinea* as a primitive Macraucheniidae (Litopterna). Later, this genus was included in the enigmatic family Sparnotheriodontidae (Soria 1980*b*, 2001; Cifelli 1983*a*, *b*, 1993). Morphological evidence suggests that sparnotheriodontids are most closely related to other primitive litopterns such as the eolitoptern Anisolambdidae (Hoffstetter & Soria 1986; Soria 2001; Anisolambdinae of Cifelli 1983*b*). However, other morphological studies, based on tarsals, argue that the Sparnotheriodontidae belongs to the Didolodontoidea, a group included in the paraphyletic Condylarthra (Cifelli 1983*a*, *b*, 1993). As the association of tarsal and dental elements that supports this last statement is not clear we follow here Soria (2001), treating the Sparnotheriodontidae as eolitopterns closely related to the Anisolambdidae.

The new sparnotheriodontid sample is important for a number of reasons: (1) it is valuable for systematic evaluation of previously collected specimens; (2) it can be used to test previous hypotheses about the age of the terrestrial mammal-bearing horizons of La Meseta Formation; and (3) it provides for a more complete assessment of the biogeographic associations of the La Meseta terrestrial fauna.

Material and methods

Comparisons were made with specimens in the Vertebrate Palaeontology collections of the Museo de La Plata (MLP), Museo Argentino de Ciencias Naturales 'Bernardino Rivadavia' (MACN), Museo Nacional of Rio de Janeiro (MNRJ) and the American Museum of Natural History (AMNH). All Seymour Island specimens are housed in the Vertebrate Palaeontology collection of the MLP. All listed specimens were collected from Instituto Antártico Argentino and División Paleontología de Vertebrados, Museo de La Plata localities designated by 'IAA' and 'DPV', respectively. Dry sieving and surface crawling were the primary techniques for specimen collection.

All measurements are reported in mm (Table 2). Terminology and measurements for litoptern teeth follow Nessov *et al.* (1998) and Soria (2001).

Table 2. *Dimensions of sparnotheriodontid teeth from Seymour Island, Antarctica. See abbreviations in the text*

Specimen	L (mm)	W (mm)
MLP 90-I-20-1	20	*c*. 20
MLP 90-I-20-3	15.8	12.7
M LP 90-I-20-5	*c*. 10.2	*c*. 10
MLP 91-II-4-1	21.7	12.6
MLP 91-II-4-5	10.90	6.70
MLP 92-II-2-135	–	–
MLP 94-III-15-3	10.8	8
MLP 95-I-10-6	25.6	*c*. 25
MLP 96-I-5-9	12.45	10.40
MLP 96-I-5-10	17.20	13.80
MLP 01-I-1-1	31	16.8
MLP 04-III-3-1	17.50	13.50

Institutional abbreviations

AMNH, American Museum of Natural History, New York, USA; DGM, Divisao de Geologia e Mineralogia do Departamento Nacional da Producao Mineral, Rio do Janeiro, Brazil; MACN, Museo Argentino de Ciencias Naturales 'Bernardino Rivadavia', Buenos Aires, Argentina; MLP, Museo de La Plata, La Plata, Argentina; MNRJ, Museu Nacional do Rio de Janeiro, Brazil.

Material

Comparisons to other sparnotheriodontid taxa were made using the following specimens: *Victorlemoinea prototypica*, MNRJ 1470-V (holotype), right M3, MNRJ 1471-V (paratype), left M3, MNRJ 1472V, left M3, MNRJ 1477V, right M1 or M2 (DP4?), MNRJ 1481V, right p3, MNRJ 1484, left M1, MNRJ 1487V, left p3, MNRJ 1402V, right m2, MNRJ 1488-V, left p3, DGM 268-M, left dp3-m1?, AMNH 49816, left M3; *Victorlemoinea* sp., AMNH 28465, left m1 or m2; AMNH 28466, left M1 or M2, AMNH 28467, right m3; AMNH 28468, left M1; AMNH 28508, right p2?; AMNH 28515, right upper premolar; AMNH 27895, right M3; MLP 61-VIII-3-163, fragmentary right upper molar; *Victorlemoinea labyrinthica*, MACN A-10671 (type), left P4-M1?; *Victorlemoinea emarginata*, MACN A-10672 (type), right M1–M2; *?Victorlemoinea longidens*, MACN A-10670 (type), right m1–m2?; *Sparnotheriodon epsilonoides*, MACN 18225 (holotype), incomplete lower jaw with left and right i1–m3; *Victorlemoinea* sp., MLP 66-V-12-1, right M3, MLP 66-V-12-2, right DP4-M2; *Phoradiadus divortiensis*, MACN 18061 (type), right M2–M3; MLP 87-III-20-7, left p4?; MLP 87-III-20-16, fragmentary rostrum with left and right I3–P3; MLP 87-III-20-17, right P4-M3; MLP 87-III-20-39, left m3; MLP 87-III-20-71, right P1–P3; MLP 87-III-20-72, very damaged skull and lower jaw with right P2–M3 and m2–m3 preserved; *Heteroglyphis dewoletzky*, MLP 12-1462 (type) left upper molariform.

Systematic palaeontology

Class MAMMALIA Linnaeus, 1758
Grandorder UNGULATA Linnaeus, 1766
Order LITOPTERNA Ameghino, 1889
Suborder EOLITOPTERNA Soria, 2001
Family SPARNOTHERIODONTIDAE Soria, 1980*a*

Emended diagnosis (after Soria 1980*a*)

Medium-sized (e.g. *Phoradiadus*) to large-sized (e.g. *Sparnotheriodon*) litopterns. Complete and closed dental series, i3/3, c1/1, p4/4, m3/3; teeth brachyodont, lophobunoselenodont to lophoselenodonts; i1–i3 relatively robust, foliform or spatuliform, increasing in size posteriorly (i1<i2<i3). The i3 is non-caniniform (e.g. *Sparnotheriodon*, *Phoradiadus*), lingual and labial cingula variably developed. The c1 are enlarged and conical (similar to those of the notoungulate Isotemnidae), with anterior and posterior crests, normally obliterated by wear, especially the anterior one; with labial and lingual cingula. Lower cheek-teeth (p1–m3) with labial and lingual cingula variably developed, but normally the lingual cingulum is weaker than the labial. The p1 is not molariform, simple and single rooted, elongated anteroposteriorly with a single main cuspid, prolonged by an anterior and a posterior crest or very short talonid. The p2 is more complex, with trigonid crescentic and short talonid (*Phoradiadus*) or trigonid and talonid subequal and bicrescentic (*Sparnotheriodon*).

The p3–p4 are molarized and, with the m1–m3, all are morphologically very similar, lophoselenodont and bicrescentic, dp4 fully molarized, with talonid and trigonid subequal (*Sparnotheriodon*) or trigonid somewhat smaller than the talonid (*Notolophus*), with well-developed labial (ectoflexid) and lingual (meta and entoflexid) flexids. Trigonid with a very well developed paralophid, its lingual end with a cuspid (paraconid? or ?neoparaconid) rapidly coalescent with wear. Metaconid high, but especially conspicuous on m1–m3 (e.g. *Sparnotheriodon*); lingual wall of the metaconid flattened, with descending crest enclosing part of the talonid basin that is more conspicuous on p3–p4. Talonid with cristid obliqua connected to the lingual end of the metalophid (metaconid). Entoconid very small (*Sparnotheriodon*) to well developed (*Phoradiadus*) and coalescent at the base with the hypoconulid (e.g. *Sparnotheriodon*). The m3 is larger than the m1 and m2, with talonid of m3 subequal to the trigonid or longer and narrower than the trigonid with a posteriorly projecting hypoconulid (e.g. *Notolophus*).

I1–I3 with lingual cingulum well developed. I3 equal or larger than the I1–I2. C1 very well developed, robust, similar to those of the Isotemnidae notoungulates, with sharp anterior and posterior crests (*Phoradiadus*). P1–P4 with labial and lingual cingula, variably developed, continuous or not. P1 simple, enlarged anteroposteriorly, with a single labial cusp, single rooted but bilobed lingually. P2–P4 increasingly complex and expanded transversally. P2–P3 with a labial (paracone) cusp showing no or very little differentiation of the metacone and a well-developed anterior parastyle. P2 with a very small protocone with anterior and posterior crests enclosing a basined trigon. The P3 is more complex, with a well-developed protocone, high and enclosing with the paraloph and metaloph a central fossette in the trigon basin. Paraloph connected to the ectoloph, with one or two cuspules trending lingually to the trigon basin from the ectoloph. Protostyle variably developed in the anterolingual cingulum. The P4 is molariform, with a metacone well differentiated, protocone very well developed and a crescentic metaconule. Lingual cingulum continuous or interrupted, always with well-developed pre- and post-cingulum, sometimes with a double cingulum. The M1–M2 with a strongly lophoselenodont ectoloph. Parastyle and mesostyle very well developed, with strong labial columns projected labially or anterolabially. Metastyle fairly to little developed. Paracone and metacone selenodont, with labial columns little developed or absent; projecting lingually into the trigon and variably developed there are cuspules, forming one or two short crests. Protocone bunoid, connected by a short crest to the paraconule and to the hypocone, closing the internal valley, but with a shallow lingual sulcus. Hypocone smaller than the protocone and connected by a short crest to the metaconule (e.g. *Victorlemoinea*, *Phoradiadus*) or directly to it (*Notolophus*). Paraconule and metaconule subcrescentic, sometimes connected to the ectoloph by very short and low crests. In some cases (*Notolophus*) the paraconule is no longer recognizable as an independent cusp, present as a short paraloph connected to the anterior cingulum. Post-metaconule crista present but variably developed. Labial cingulum not very strong, sometimes restricted to the posterior portion; lingual cingulum variably developed. Precingulum, with a very well developed protostyle, sometimes connected to the paraloph (*Notolophus*). Post-cingulum encloses a low fossette. Pre- and post-cingulum present as a low extra cingulum, forming a 'double cingulum' that occurs also in the Anisolambdidae litopterns. The M3 is similar to the M1–M2, but with the hypocone absent. Of the deciduous molars known, the DP4 is fully molarized, with prominent mesostyle, hypocone, paraloph, metaloph, postcingulum fossette, accessory cusps projecting lingually from the ectoloph as a 'double post-cingulum'. As so far known, the recognized taxa in this family posses enamel with vertically oriented Hunter–Schreger bands.

Comments

Soria (1980*a*) established the Sparnotheriodontidae as an undetermined notoungulate monotypic family based on *Sparnotheriodon epsilonoides* from the Vacan subage (late Palaeocene–early Eocene) of Patagonia. Subsequently, Soria (2001) characterized the family and included with it the Anisolambdidae (regarded by Cifelli as a subfamily of Proterotheriidae) in a new suborder, Eolitopterna.

Cifelli (1993) defined the Sparnotheriodontidae by several advanced characters, including a lophoid metaconule and an expanded post-cingulum, but included in this family the Indaleciinae, a group of very small ungulates traditionally considered as Adianthidae litopterns (Cifelli & Soria 1983) or as a family, Indaleciidae, of the Order Notopterna (Soria 1989). Bonaparte & Morales (1997) followed Cifelli (1993) in the grouping of *Victorlemoinea* and *Indalecia*, but considered them all litopterns. Here, we exclude the indaleciids from the Sparnotheriodontidae and, as stated earlier, follow Soria (2001) in his use of the Sparnotheriodontidae.

Notolophus gen. nov.

Type species

Notolophus arquinotiensis, sp. nov.

Diagnosis

Same as for the type species.

Etymology

Notos, is derived from the greek νοτοσ, south, in reference to the geographical area where the taxon was found; and λοφοσ, lophs, crests.

Notolophus arquinotiensis sp. nov. (Figs 2a, b, 4a, c & 5a, b)

Holotype

MLP 95-I-10-6, left M3 incomplete (the buccal part of paracone and metacone is missing) (Fig. 2a). La Meseta Formation, Submeseta Member (TELM 7), DPV 16/84 locality. This molar was briefly described and figured by Vizcaíno *et al.* (1997).

Hypodigm

Holotype plus MLP 90-I-20-1, left upper molariform (M1 or M2?), *Cucullaea* I Member (TELM 5), IAA 1/90. MLP 91-II-4-1, right p4, *Cucullaea* I Member (TELM 4), DPV 2/84 locality. MLP 95-I-10-7, fragmentary left upper molariform, *Cucullaea* I Member (TELM 5), MLP 01-I-1-1, right m3, *Cucullaea* I Member (TELM 5), IAA 1/90 locality. MLP 04-III-3-1, incomplete right p4, *Cucullaea* I Member (TELM 5), IAA 1/95.

Referred specimens

MLP 90-I-20-3, right I3?, *Cucullaea* I Member (TELM 5), IAA 1/90 locality. MLP 90-I-20-5, left upper premolar incomplete (P2 or P3?), *Cucullaea* I Member (TELM 5), IAA 1/90 locality. MLP 91-II-4-5, right upper premolar (P1), *Cucullaea* I Member (TELM 5), IAA 1/90 locality. MLP 92-II-2–135, fragment of a molariform (lower?), Campamento Member (TELM 3), IAA 1/92. MLP 94-III-15-3, left lower incisive, *Cucullaea* I Member (TELM 5), IAA 1/90 locality. MLP 96-I-5-5, left upper incisiviform (I1?), *Cucullaea* I Member (TELM 5),

IAA 2/95 locality. MLP 96-I-5-9, left lower incisiviform or first premolar?, *Cucullaea* I Member (TELM 5), IAA 3/96 locality.

Type locality

Museo de La Plata locality DPV 16/84, Seymour Island, Antarctic Peninsula (Fig. 1). GPS data: 64°14′04.672″S and 56°39′56.378″W. Sr isotope dating from this horizon yields an age of approximately 34.2 Ma (Dingle & Lavelle 1998). Additional specimens are known from other localities (Fig. 1) in lower levels (*Cucullaea* I and Campamento Members) of the La Meseta Formation.

Stratigraphy and age

La Meseta Formation (late Early Eocene–Late Eocene), Campamento (Early Eocene), *Cucullaea* I (Middle Eocene) and Submeseta (Late Eocene) members.

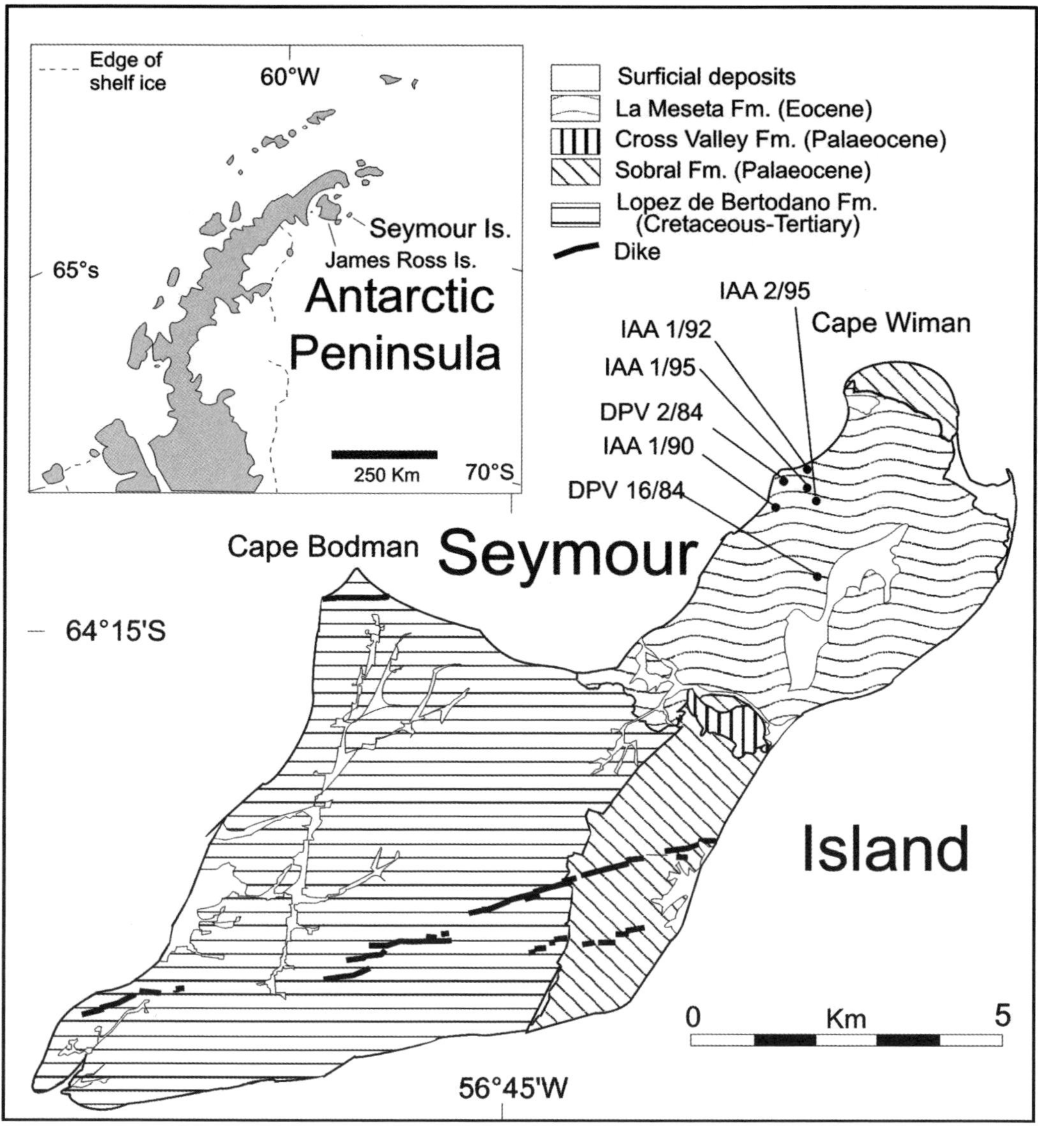

Fig. 1. Map of Seymour (Marambio) Island (Antarctic Peninsula) showing the IAA and DPV localities mentioned in the text.

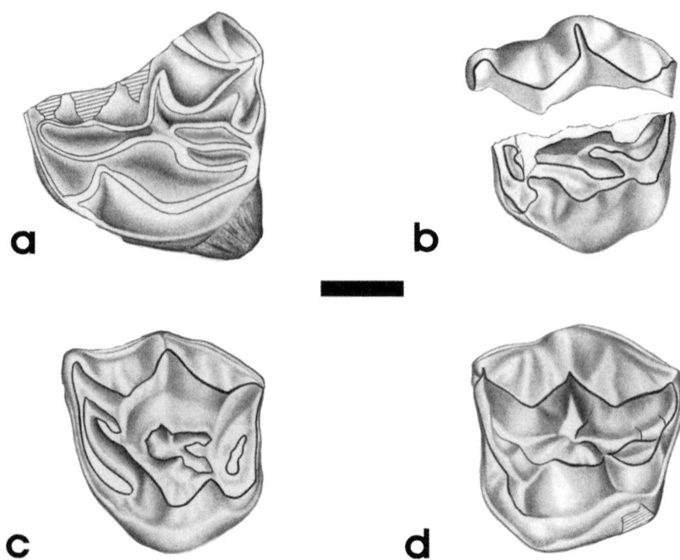

Fig. 2. Occlusal views of upper molars of sparnotheriodontids from Antarctica and Patagonia. (**a**) *Notolophus arquinotiensis*, gen. et sp. nov. MLP 95-I-10-6, left M3, holotype; (**b**) *Notolophus arquinotiensis*, gen. et sp. nov. MLP 90-I-20-1, left M1 or M2; (**c**) *Victorlemoinea labyrinthica*, MACN A-10871, left M1 (based also on M2 of the same individual); and (**d**) *Victorlemoinea* sp. MLP 66-V-12-2, right M1 (reversed). The scale bar equals 5 mm (drawing by A. Viñas).

Etymology

The specific epithet, *arquinotiensis*, is in reference to the Ihering's (1927) Archinotis continent.

Short diagnosis

A sparnotheriodontid larger than *Phoradiadus* and nearly equal as *Sparnotheriodon*. Differs from the other known taxa in having the upper molars with a protocone projected anteriorly by a short paraloph which connected to the protostyle in the second anterior cingulum (precingulum). Metaconule lophoid anteriorly extended and connected directly to the poorly developed hypocone, without intermediate crest as in the other taxa known. In the M3 the hypocone is very weak or absent, also the paraloph and the protostyle connects directly to the protocone. Lower molariforms with the trigonid smaller than the talonid. The m3 has a well-developed bunoid entoconid and a posteriorly projecting hypoconulid.

Differential diagnosis

Sparnotheriodontid much larger than *Phoradiadus divortiensis*, and nearly as large as *Sparnotheriodon epsilonoides*. Upper molars with a very strong and well-developed ectoloph, labial cingulum very weak and restricted to the posterior portion of the ectoloph, between the metacone and the short metastyle. Short but strong lingual projections, one from the posterior part of the paracone and other from the anterior part of the metacone. Protocone elongated and projected anteriorly by a paraloph in which no paraconule is visible as separate cusp. Hypocone little developed connected by a short crest to the protocone; the hypocone is vestigial or absent on M3. Metaconule lophoid, nearly straight and very anteriorly extended, directly connected to the hypocone, without the short intermediate crest connecting these cusps as in *Victorlemoinea* or *Phoradiadus*. Post-metaconule crista low and post-cingulum enclosing a small basin that is not so developed as in *Victorlemoinea* or *Phoradiadus*. In the M3, the metaconule is connected with the first post-cingulum. The second anterior cingulum (or precingulum) possess a prominent protostyle, which is connected by a short posterolabial isthmus to the paraloph. In the M3 the protostyle directly connects to the anterior portion of the protocone, the paraloph being vestigial or absent.

Lower molariforms with trigonid somewhat smaller than the talonid. Trigonid and talonid basins not so narrow and *Phoradiadus* and *Sparnotheriodon*. The trigonid exhibits a very well developed paralophid, with an engrossed lingual end that could represents a paraconid or ?neoparaconid. Posterior premolars with a conspicuous descending crest posterior to the metaconid; small entoconid connected with the hypoconulid. The m3 with a talonid larger and more elongated than the trigonid, with a projecting hypoconulid in a rudimentary 'third lobe'; well-developed bunoid entoconid antero-posteriorly enlarged and connected to the hypoconulid. Labial cingulum variably developed; lingual cingulum low and continuous to absent.

Description

As stated earlier, our knowledge of previously known Sparnotheriodontidae is meager. Taking this in account, the unassociated and, sometimes, fragmentary nature of the Antarctic sparnotheriodontid specimens precludes an adequate interpretation. Most of the Antarctic ungulate teeth undoubtedly can be assigned to Sparnotheriodontidae, and with a high degree of confidence to *Notolophus arquinotiensis*. Notwithstanding, some of them are difficult to interpret, not in taxonomical reference but in its proper position in the dental series given the aforementioned scanty knowledge of the complete dental anatomy of this group.

The molariform teeth, upper and lower, can be referred with a high degree of confidence to this new taxon because all upper molars known have the same derived features. The lower molariform teeth match well in size with the upper ones and are therefore referred to the same taxon.

MLP 90-I-20-3 (Fig. 5) and MLP 96-I-5-10 are incisiviforms rather than caniniforms and match in size with the other teeth assigned to *Notolophus arquinotiensis*. They are robust, with a straight root and a labial single cusp with a convex labial wall (in MLP 90-I-20-3 the wear has obliterated this cusp). There are very well developed labial and lingual cingula. The enamel is thick with strongly marked alternating bands of the vertically oriented Hunter–Schreger bands. The morphology of these teeth is very different from the incisors known of Pyrotheria and Astrapotheria (see Simpson 1967) (astrapotheres do not have upper incisors), which also have vertically oriented Hunter–Schregger enamel bands (see Fortelius 1985) as in the Sparnotheriodontidae. Also, these teeth resemble the I3 of some notoungulate families such as the Isotemnidae, but the enamel structure of the isotemnids is completely different lacking the vertically oriented Hunter–Schreger bands (see Fortelius 1985). By comparison with the upper and lower incisors known in *Phoradiadus* and *Sparnotheriodon* we refer tentatively these specimens as probable upper incisors (I3?) of *Notolophus arquinotiensis*, but recognize that they are more robust than the I3 of these species. It cannot be ruled out that the teeth aforementioned could represent upper canines, but since in sparnotheriodontids, like *Phoradiadus divortiensis*, the canines (upper and lower) are pointed and with sharp edges, we therefore, identify these teeth tentatively as I3. MLP 96-I-5-5 is a very worn compressed mesiodistal incisiviform with an ellipsoid coronal figure, with no trace of a labial cingulum; it is very probably an anterior incisor, perhaps the I1. By comparison with the anterior lower dentition known in *Sparnotheriodon epsilonoides*, MLP 94-III-15-3 is considered as a probable lower incisiviform (right i3?); it is a simple tooth, very worn, with a principal labial cusp and a short anterior crest; there is also a lingual cingulum connected with the anterior crest and it has a middle lingual cuspule. The specimen MLP 96-I-5-9 very probably represents a first lower premolar (left p?1); this tooth, although very worn occlusally, shows an antero-posteriorly enlarged and wide trigonid, with a principal labial cusp area and a very short talonid, somewhat different then to the more elongated p1 of *Sparnotheriodon epsilonoides*. This tooth is single rooted with a very oblique root. No teeth were found that could be referred confidently as the canines, upper or lower, of *Notolophus arquinotiensis*.

MLP 91-II-4-5 is a very simple tooth, single rooted, with a flattened crown by wear. It has a principal labial cusp (paracone) with a short anterolabial crest interpreted as a parastyle, and a shorter posterior crest (metastyle?). Labially, the principal cusp has a convex surface and an anterior shallow fold which delimitates the parastyle from the paracone. A strong lingual cingulum is connected to the parastyle and metastyle; this lingual cingulum has a well defined cuspule which is connected to the paracone by a short posterolabially directed crest. This tooth is interpreted here as a P1.

MLP 90-I-20-5, by comparison with the upper premolars of *Phoradiadus*, represents an upper premolar, possibly a left P3. The specimen is not complete, but has a well-developed protostyle, a bunoid protocone, apparently lacks the hypocone and short lingual crests project from the ectoloph, and the posterior fossette formed by the metaloph and posterior cingulum has been obliterated by wear. Posterior to the cingulum there is an extra cingulum.

Sparnotheriodontid molars are quite uniform in form, and those of *N. arquinotiensis* share the same general pattern with other Patagonian sparnotheriodontids, but their proportions and especially the morphology of the upper molars is quite characteristic.

The holotype, MLP 95-I-10-6 (Fig. 2a), is of roughly rectangular outline, and the anterior and medium part of the ectoloph is missing. The preserved ectoloph shows a lophoselenoid metacone with a flattened labial area and a short metastyle which descends posteriorly; there is also preserved part of a low labial cingulum, but which may or may not have been continuous. The shallow internal basin or principal valley is formed between the protocone and the ectoloph, and exhibits two short, low crests projecting from the ectoloph. The protocone is large, anteroposteriorly elongated and connected to a very well developed anterolingual cusp. This lingually displaced cusp is interpreted here as an enlarged protostyle cingular cusp, although we do not rule out the possibility that it could also be a displaced paraconule fused with the protostylar cusp. Nevertheless, its position and the relationships with the second precingulum are more indicative of an enlarged protostyle. The protocone possess a posterior crest that connects to the post-cingulum. The metaconule is strongly lophoid and projected mesiodistally to the internal valley, post-metaconular crista well developed and directed labially connecting the metaconule to the metacone area. No hypocone exists, and the metaconule connects directly with the posterior projection of the protocone. The post-cingulum, connected to the protocone and metaconule, is expanded and encloses a small fossette; this basined post-cingulum is proportionally more developed in *Victorlemoinea* (Fig. 3) and *Phoradiadus* than in *Notolophus* (Fig. 2). Pre- and post-cingula with a very low extra cingulum. The lingual cingulum is very low and restricted to the anterior part of the protocone.

MLP 90-I-20-1 is very probably a left M1 or M2 (Fig. 2b), although it could represent a molariform DP4. It is very similar to the above described M3, but has a complete ectoloph. No labial columns are present on the paracone and metacone, and, except in the middle, which is slightly convex, the walls of the paracone and metacone are flattened to slightly concave. The parastyle is conspicuous, but the mesostyle represents the strongest element of the ectoloph with a very wide base. The labial cingulum is restricted to the posterior part of the ectoloph. The hypocone is small and connected to the protocone by a short crest, with a very shallow sulcus between the protocone and hypocone. The metaconule is lophoid, strongly projected mesiodistally as in the M3, and is connected directly to the hypocone without the short intermediate lingually projected crest that connects the metaconule and hypocone in *Victorlemoinea* (e.g. *V. labyrinthica*) and *Phoradiadus*, but which is very short and nearly absent in MLP 66-V-12-2 (Fig. 2d) identified as *Victorlemoinea* sp. from the Vacan (early Casamayoran) of Patagonia. Post-metaconular crista is similar in form and direction as in the M3, although it is lower and not so well developed. Lingual cingulum apparently restricted to the anteriormost part of the protocone. Anterior and posterior cingula with low extracingula, conforming the double cingulum of the Sparnotheriodontidae.

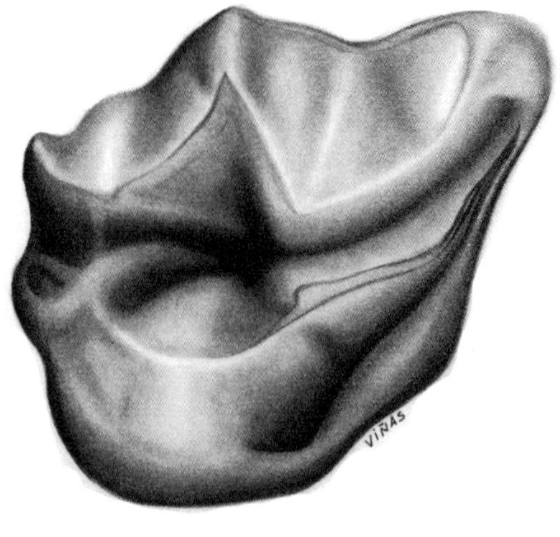

Fig. 3. Occlusal view of right M3 (MLP 66-V-12-1) of *Victorlemoinea* sp. The scale bar equals 5 mm (drawing by A. Viñas).

Two lower molariforms, MLP 91-II-4-1 (Fig. 4c) and MLP 04-III-3-1, are tentatively assigned to the 'molarized' premolars of this species, and they probably represent two right p4, or a p4 and a p3, respectively. They are fully molariform with the trigonid crescent relatively shorter than that of the talonid and not so labially projected. The paralophid is very well developed and lingually projected as a small

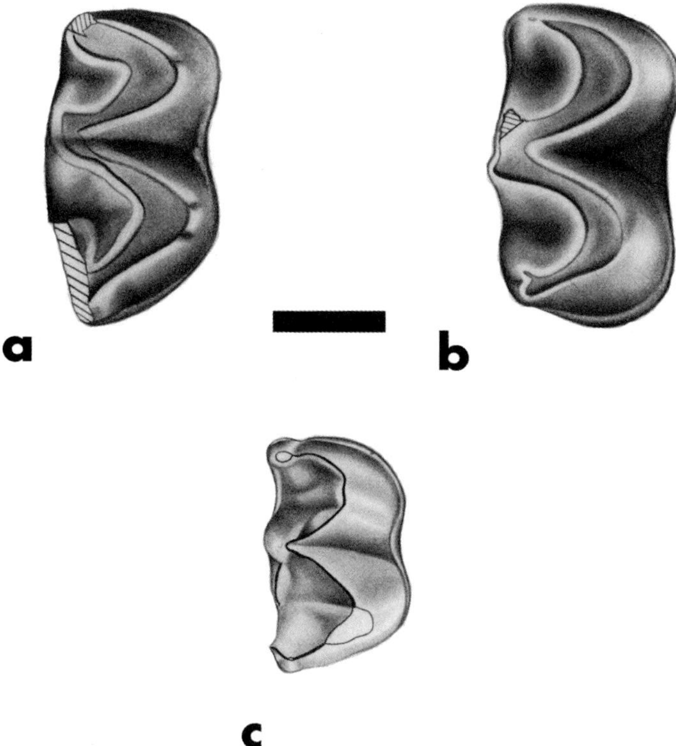

Fig. 4. Occlusal views of lower molars of sparnotheriodontids from Antarctica and Patagonia. (**a**) *Notolophus arquinotiensis*, gen. et sp. nov. MLP 01-I-1-1, right m3; (**b**) *Sparnotheriodon epsilonoides*, MACN 18225, right m3. The scale bar equals 5 mm (drawing by A. Viñas). (**c**) *Notolophus arquinotiensis*, gen. et sp. nov. MLP 91-II-4-1, occlusal view of right p4. The scale bar equals 5 mm (drawing by A. Viñas).

cuspid (paraconid or ?neoparaconid). Metaconid with a very sharp descending crest, similar to that observed in the p3–p4 of *Sparnotheriodon elipsonoides* and *Phoradiadus divortiensis.* The entoconid is reduced and coalescent with a very short hypoconulid. Labial fold (ectoflexid) and lingual folds (meta and entoflexid) very well developed. The ectoflexid is deeper and more penetrating than the lingual flexids, with the entoflexid more open than the metaflexid. Well-developed anterior and posterior cingula extend labiolingually and may or not be connected to the labial and lingual cingula. The labial cingulum is present in these two specimens, but it is continuous (MLP 04-III-3-1) or is restricted to the base of the labial fold (ectoflexid) (MLP 91-II-4-1). The lingual cingulum is low but continuous (MLP 04-III-3-1) or absent (MLP 91-II-4-1).

A nearly complete right m3, MLP 01-I-1-1 (Fig. 4a), has a trigonid shorter than the more elongated talonid. The trigonid shows the lingual portion of the paralophid engrossed (paraconid or ?neoparaconid) as in the premolars described above, but (at least in this state of wear) with no trace of an independent cusp. The metaconid is the highest cusp and has a relatively wide descending posterior crest. The talonid is more elongated anteroposteriorly than the trigonid, with a posteriorly projected hypoconulid separated by a labial fold forming a short and rudimentary 'third lobe'. The entoconid is bunoid, projects anteriorly and is connected to the hypoconulid; it is more developed and inflated than in *Sparnotheriodon epsilonoides* and similar to ?*V. longidens*, but the entoconid is not so differentiated from the hypolophid as in *Phoradiadus divortiensis*. The ectoflexid is more open, deep and penetrating than the lingual folds, which are relatively shallow. The anterior cingulum is well developed and extends transversely with the lingual portion higher and directed to the paralophid; it is not connected to the labial cingulum that extends from the hypoconulid lobe to the posterior part of the protoconid column. Some cuspules occur in the ectoflexid valley. The lingual cingulum is apparently restricted to the trigonid, extending from the paralophid to the anterior portion of the metaconid.

Discussion

Notolophus arquinotiensis is one of the most abundant taxa among the terrestrial mammals from the La Meseta Formation. *N. arquinotiensis* is currently represented by a small number of specimens collected at six localities in Seymour Island (Fig. 1). Its tooth anatomy, as described above, is distinctive and allows a clear differentiation from other Palaeocene and Eocene sparnotheriodontids.

Only three sparnotheriodontid genera are so far known in South America (Table 1). A fourth genus, *Heteroglyphis*, from the Mustersan Age (late Eocene) was included tentatively within the family by Soria (2001), although restudy of the type and only known specimen suggests that *Heteroglyphis dewoletzky*, Roth 1899 belongs to the Anisolambdinae or Anisolambdidae eolitopterns. The specimens discussed here were initially referred to *Victorlemoinea* (Bond *et al.* 1990). The genus *Victorlemoinea* was erected by Ameghino (1901), who recognized two species: *V. labyrinthica*, the genotypical one (Fig. 2c) and *V. emarginata*, both based on upper molariform teeth (see Simpson 1948) from the Casamayoran SALMA (possibly Vacan 'subage') of Patagonia. From the same area and age, Simpson (1948) doubtfully referred *Victorlemoinea* to the species *Anisolambda longidens* Ameghino, 1901, based on lower teeth. Later, Paula Couto (1952) referred a fourth species to *Victorlemoinea*: *V. prototypica* from the Itaboraian SALMA (middle Palaeocene) of Brazil and based on upper and lower teeth.

Notolophus arquinotiensis (Fig. 2) is different from *V. labyrinthica*: *V. emarginata* and *V. prototypica* being somewhat larger than *V. labyrinthica*, and definitely larger than *V. emarginata* and *V. prototypica*. The peculiar connection of the protocone–paraloph with the enlarged protostyle is clearly distinct from the morphology observed in the species of *Victorlemoinea*. It is interesting to note that upper molars from the Early Casamayoran SALMA (Vacan subage), referred here as *Victorlemoinea* sp., MLP 66-V-12-2, have a similar size to those of the type of *Victorlemoinea labyrinthica*, but differ in having a smaller hypocone and a shorter crest connecting the metaconule with this cusp. These molars, similar to those figured by Simpson (1948) (i.e. AMNH 28466), also

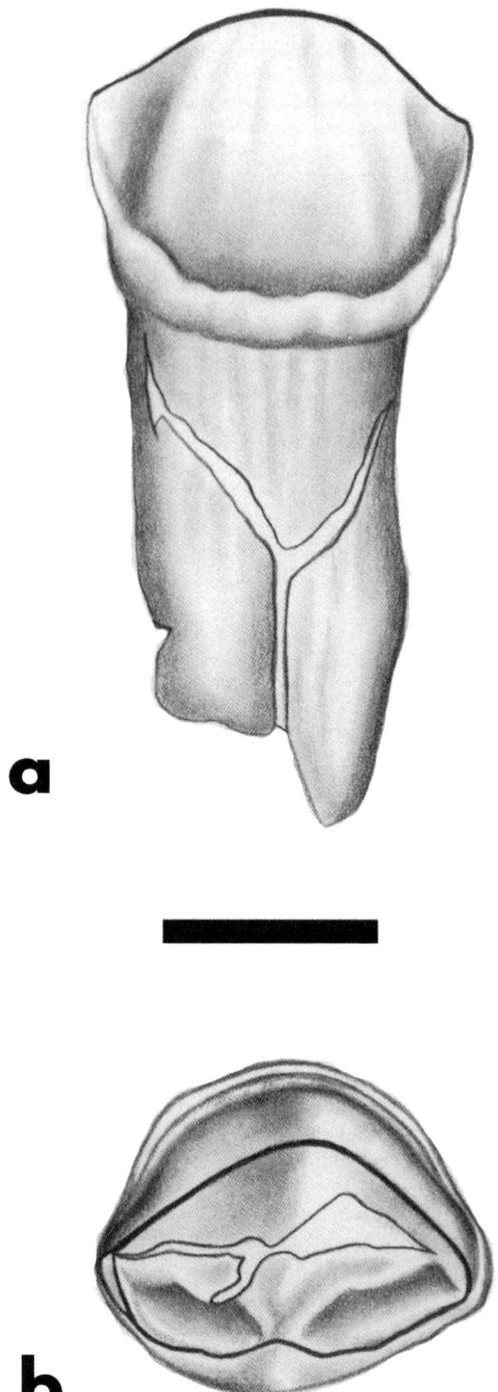

Fig. 5. *Notolophus arquinotiensis*, gen. et sp. nov. MLP 90-I-20-3, right I3?. (**a**) Labial view and (**b**) occlusal view. The scale bar equals 5 mm (drawing by A. Viñas).

from the Vacan subage (Casamayoran age), approach the condition observed in *Notolophus arquinotiensis*, but clearly differ by the paraloph which in MLP 66-V-12-2 is not united to the protostyle as in *V. labyrinthica*.

?*Victorlemoinea longidens* is based on lower premolars and molars not clearly associated. The lower premolars are different from other known Sparnotheriodontidae, and do not have vertically oriented Hunter–Schreger bands; its morphology is more reminiscent of a notoungulate Isotemnidae than a litoptern, and we do not consider this premolar as those of a sparnotheriodontid. The incomplete right lower molars (m1–m2), although of smaller size than those of *Notolophus,* have an enlarged entoconid and a weak lingual cingulum, which are characters also observed in the lower molars of *Notolophus*, but they differ in the more narrow and penetrating meta and entoflexids of ?*V. longidens*. Also, it is very possible that *?V. longidens* could represent the lower teeth of *Victorlemoinea labyrinthica*.

Sparnotheriodon epsilonoides is only known from its lower teeth and mandible (Soria 1980*a*), so no direct comparison can be made between it and MLP 95-I-10-6. However, the lower molars of the hypodigm of *Notolophus* (MLP 91-II-4-1 and MLP 01-I-1-1) are clearly lophoselenodonts and match very well in size and general anatomy with those of *Sparnotheriodon*.

Recent work on the faunal similarities of the La Meseta fauna indicate a strong biogeographical connection with the southern tip of South America (Patagonia) (Goin *et al.* 1999; Reguero *et al.* 2002), and the identification of archaic marsupial prepidolopids and derorhynchids at Seymour Island reinforces that link. Similarly, the recovery of sudamericid gondwanatheres from Seymour Island and the recognition of strong morphological correspondence between the Seymour gondwanathere and *Sudamerica ameghinoi* also demonstrate a late Palaeocene connection with Patagonia (Reguero *et al.* 2002).

The rare occurrences of sparnotheriodontids in an otherwise very well recorded faunal context of the Palaeocene of Patagonia and Brazil leads to the assumption that they could be extreme ecological specialists. They show a number of dental characteristics that may be adaptations to forested habitats, and the striking dental features of the Antarctic taxon are brachyodonty and the particular structure of the enamel (vertically oriented Hunter–Schreger bands) (Reguero *et al.* 2002). Janis (1984) pointed out that brachyodonty is associated with browsing herbivores that are adapted to forest habitats. In particular, *Notolophus* could browse, stripping off twigs and saplings from evergreen trees even during winter months (Vizcaíno *et al.* 1998*b*). No post-cranial information is available for the Antarctic ungulates, but information from the nearest relatives (all of them fossils) can be used to infer the locomotor adaptation to the cursoriality. Cifelli (1983*a*, *b*) associated teeth and astragalus and calcaneum to the Itaboraian (Palaeocene) species *Victorlemoinea prototypica* of Brazil.

The faunal evidence, mainly that provided by the marine invertebrates (Stilwell & Zinsmeister 1992), indicates the deposition of the Submeseta Member, where the holotype was recovered, was in cool-temperate conditions, unlike the underlying *Cucullaea* I Member. A sharp decrease of diversity near the contact between the upper members of La Meseta Formation (*Cucullaea* II and Submeseta) may be correlated with the climatic cooling event which culminated at the time of deposition of the uppermost part of the La Meseta Formation (Gazdzicki *et al.* 1992). The presence of *Notolophus*, together with a ground-dwelling bird (ratite) and *Nothofagus* leaves from the same horizon, suggest that the terrestrial environment during the time of deposition of at least part of the Submeseta Member was apparently not dissimilar to that reconstructed by Reguero *et al.* (2002) for the *Cucullaea* I Member with *Nothofagus* forests and mountainous cordillera.

Notolophus had a more bilophodont than bunodont dentition, and their molariforms teeth had strong enamel ridges extending between the cusps. These enamel ridges serve as shearing surfaces, and the formation of dentine 'lakes' along the ridges produce double-edged shearing blades. These mainly performed a shearing action, slicing leaves into quite large pieces like a modern tapir that feeds almost entirely on leaves of forest trees. The body size of the Antarctic sparnotheriodontid (395–400 kg) indicates that it was the largest terrestrial herbivore living in Antarctica at this time (Vizcaíno *et al.* 1998*b*). Evidently, the large size of this herbivore favoured the exploitation of leaves because a longer time in the gut for bacterial fermentation is required to obtain sufficient nutrients from leaves. Based on dental morphology, sparnotheriodontids were probably hindgut fermenters like non-ruminant artiodactyls and perissodactyls (Fortelius 1985; Rensberger & Pfretzschner 1992). Astrapotheres and sparnotheriodontids also have teeth with vertical Hunter–Schreger bands. Fortelius (1985) indicated that a number of lophodont ungulates have evolved vertically oriented Hunter–Schreger bands, a modification that

Fig. 6. Reconstruction of the archaic litoptern *Notolophus* and the opossum-like marsupial *Antarctodolops* on the eastern shore of the Antarctic Peninsula during the Middle Eocene. In the background *Nothofagus* forest and mountains (drawing by A. Viñas).

involves the mode of prism decussation and three-dimensional arrangement of the bands. This condition has been interpreted as an adaptation to resist cracking when the enamel edges are loaded in a direction away from the supporting dentine (Boyde & Fortelius 1986). In *Notolophus*, as in the rest of the representatives of the family, the ectoloph forms a thin, vertical, blade-like ectoloph with a strong mesostyle.

Notolophus arquinotiensis is a large sparnotheriodontid (Fig. 6), larger and different in morphology than the last ones of the Divisaderan SALMA (late Eocene), and more similar in size to some remains of the Vacan Subage (Casamayoran Age; early Eocene). The material of *Notolophus* from Seymour Island exhibits no change of size through the Campamento Member (TELM 3) to the Submeseta Member (TELM 7) of the La Meseta Formation, indicating that the individuals of *Notolophus arquinotiensis* were of very large size existing over a large timespan. Also, related forms in the Vacan Subage (early Eocene) may tempt one to propose an immigration event for the sparnotheriodontids in Antarctica near the Vacan Subage or Riochican Age (late Paleocene). However, other taxa (e.g. the marsupials) could indicate an earlier migration datum, but additional taxa from the La Meseta Formation are required to demonstrate either an impoverished fauna of a previous, single immigration event or a cluster of taxa arriving on the Antarctic Peninsula at different times by chance routes.

A more precise reconstruction of the palaeoecology of *Notolophus* would be possible if cranial and post-cranial remains were known. Clearly, much remains to be learned about this rare Antarctic litoptern, questions that only future discoveries of additional material can answer.

Conclusion

The new taxon reported here is the first well-documented Antarctic 'South American ungulate', and it belongs to an archaic and uncommon lineage whose ultimate ancestry may be Laurasiatic 'condylarths'. *Notolophus arquinotiensis* definitively confirms the occurrence of an archaic ungulate population in Antarctica and supports the role of the continent as a probable centre of eutherian evolution (Vizcaíno *et al.* 1998*a*). *Notolophus arquinotiensis* has close affinities with *Victorlemoinea*, indicating at least a very close common ancestor, probably a 'condylarth' despite its strikingly molariform P3–4/p3–4.

We express our gratitude to Dr J. E. Martin and an anonymous reviewer for critical review of the manuscript. We thank the personnel and authorities of the

Instituto Antártico Argentino, especially S. Santillana and E. Yermolin for their logistic support during fieldwork at locality IAA 1/90 in the Antarctic Peninsula; and Mr. J. J. Moly for his fieldwork in Antarctica. We also acknowledge the following people for access to fossils housed in their respective institutions: M. Norell and M. Novacek (AMNH), and J. Bonaparte and A. Kramarz (MACN). Fieldwork at Seymour Island and museum research was supported by the Instituto Antártico Argentino (IAA), Consejo Nacional de Investigaciones Científico y Técnicas (CONICET) and the National Geographic (grant to S. A. Marenssi). We gratefully acknowledge A. Viñas for his fine artwork in Figures 2–5.

References

AMEGHINO, F. 1901. Notices préliminaires sur des ongulés nouveaux des terraines crétacés de Patagonie. *Boletín de la Academia Nacional de Ciencias de Córdoba*, **16**, 349–426.

BONAPARTE, J.F. & MORALES, J. 1997. Un primitivo Notonychopidae (Litopterna) del Paleoceno inferior de Punta Peligro, Chubut, Argentina. *Estudios Geológicos*, **53**, 263–274.

BOND, M., PASCUAL, R., REGUERO, M.A., SANTILLANA, S.N. & MARENSSI, S.A. 1990. Los primeros ungulados extinguidos sudamericanos de la Antártida. *Ameghiniana*, **16**, 240.

BOYDE, A. & FORTELIUS, M. 1986. Development, structure and function of rhinoceros enamel. *Zoological Journal of the Linnean Society*, **87**, 181–214.

CIFELLI, R.L. 1983*a*. Eutherian tarsals from the late Paleocene of Brazil. *American Museum Novitates*, **1761**, 1–31.

CIFELLI, R.L. 1983*b*. The origin and affinities of the South American Condylarthra and early Tertiary Litopterna (Mammalia). *American Museum Novitates*, **2772**, 1–49.

CIFELLI, R.L. 1993. The phylogeny of the native South American ungulates. *In*: SZALAY, F.S., NOVACEK, M.J. & MCKENNA, M.C. (eds) *Mammal Phylogeny, Volume 2. Placentals*. Springer, New York, 195–216.

CIFELLI, R.L. & SORIA, M.F. 1983. Systematics of the Adianthidae (Litopterna, Mammalia). *American Museum Novitates*, **2771**, 1–25.

DINGLE, R. & LAVELLE, M. 1998. Late Cretaceous–Cenozoic climatic variations of the northern Antarctic Peninsula: new geochemical evidence and review. *Palaeogeography, Palaeoclimatology, Palaeoecology*, **107**, 79–101.

FORTELIUS, M. 1985. Ungulate cheek teeth: developmental, functional and evolutionary interrelations. *Acta Zoologica Fennica*, **180**, 1–76.

GAZDZICKI, A.J., GRUSZCZYNSKI, M., HOFFMAN, A., MALKOWSKI, K., MARENSSI, S.A., HALAS, S. & TATUR, A. 1992. Stable carbon and oxygen isotope record in the Paleogene La Meseta Formation, Seymour Island, Antarctica. *Antarctic Science*, **4**, 461–468.

GOIN, F.J., CASE, J.A., WOODBURNE, M.O., VIZCAÍNO, S. F. & REGUERO, M.A. 1999. New discoveries of 'oppossum-like' marsupials from Antarctica (Seymour Island, Medial Eocene). *Journal of Mammalian Evolution*, **6**, 335–365.

HOFFSTETTER, R. & SORIA, M.F. 1986. *Neodolodus colombianus* gen. et sp. nov., un nouveau Condylarthre (Mammalia) dans le Miocene de Colombie. *Comptes Rendus de l'Académie des Sciences, Paris, série II*, **17**, 1619–1622.

HOOKER, J.J. 1992. An additional record of a placental mammal (Order Astrapotheria) from the Eocene of Western Antarctica. *Antarctic Science*, **4**, 107–108.

IHERING, H.V. 1927. *Die Geschichte des Atlantischen Ozeans*. Gustav Fischer, Jena.

JANIS, C.M. 1984. The use of fossil ungulate communities as indicators of climate and environment. *In*: BRENCHLEY, P. (ed.) *Fossils and Climates*. Wiley, Chichester, 85–104.

MARENSSI, S.A., REGUERO, M.A., SANTILLANA, S.N. & VIZCAÍNO, S.F. 1994. Eocene land mammals from Seymour Island, Antarctica: Palaeobiogeographical implications. *Antarctic Science*, **6**, 3–15.

NESSOV, L.A., ARCHIBALD, J.D. & KIELAN-JAWOROSKA, Z. 1998. Ungulate-like mammals from the late Cretaceous of Uzbekistan and a phylogenetic analysis of Ungulatomorpha. *Bulletin of Carnegie Museum of Natural History*, **34**, 40–88.

PAULA COUTO, C. DE. 1952. Fossil mammals from the beginning of the Cenozoic in Brazil. Condylarthra, Litopterna, Xenungulata and Astrapotheria. *Bulletin of the American Museum of Natural History*, **99**, 355–394.

REGUERO, M.A., MARENSSI, S.A. & SANTILLANA, S.N. 2002. Antarctic Peninsula and South America (Patagonia) Paleogene terrestrial faunas and environments: biogeographic relationships. *Palaeogeography, Palaeoclimatology, Palaeoecology*, **179**, 189–210.

RENSBERGER, J.M. & PFRETZSCHNER, H.U. 1992. Enamel structure in Astrapotheres and its functional implications. *Scanning Microscopy*, **6**, 495–510.

ROTH, S. 1899. Aviso preliminar sobre mamíferos mesozoicos encontrados en Patagonia. *Revista del Museo de La Plata*, **9**, 381–388.

SIMPSON, G.G. 1945. The principles of classification and a classification of mammals. *Bulletin of the American Museum of Natural History*, **85**, 1–350.

SIMPSON, G.G. 1948. The beginning of the Age of the Mammals in South America. Part I. *Bulletin of the American Museum of Natural History*, **91**, 1–232.

SIMPSON, G.G. 1967. The beginning of the Age of the Mammals in South America. Part II. *Bulletin of the American Museum of Natural History*, **137**, 1–259.

SIMPSON, G.G., HINOPRIO, J.L. & PATTERSON, B. 1962. The mammalian fauna of the Divisadero Largo Formation, Mendoza, Argentina. *Bulletin of the Museum of Comparative Zoology*, **127**, 239–293.

SORIA, M.F. 1980*a*. Una nueva y problemática forma de ungulado del Casamayorense. *II Congreso Argentino de Paleontología y Bioestratigrafía y I Congreso Latinoamericano de Paleontología*, Buenos Aires, 1978, **2**, 193–203.

SORIA, M.F. 1980*b*. Las afinidades de *Phoradiadus divortiensis* Simpson, Minoprio and Patterson,

1962. *Circular Informativa Asociación Paleontológica Argentina*, **4**, 20.

SORIA, M.F. 1989. Notopterna: un nuevo orden de mamíferos ungulados eógenos de América del Sur. Parte I: Los Amilnedwardsiidae. *Ameghiniana*, **25**, 245–258.

SORIA, M.F. 2001. *Los Proterotheriidae (Litopterna, Mammalia), sistemática, origen y filogenia*. Monografías del Museo Argentino de Ciencias Naturales, **1**, 1–167.

STILWELL, J.D. & ZINSMEISTER, W.J. 1992. *Molluscan Systematics and Biostratigraphy, Lower Tertiary La Meseta Formation, Seymour Island, Antarctic Peninsula*. American Geophysical Union, Antarctic Research Series, **55**.

VIZCAÍNO, S.F., BOND, M., REGUERO, M.A. & PASCUAL, R. 1997. The youngest record of fossil land mammals from Antarctica, its significance on the evolution of the terrestrial environment of the Antarctic Peninsula during the late Eocene. *Journal of Paleontology*, **71**, 348–350.

VIZCAÍNO, S.F., PASCUAL, R., REGUERO, M.A. & GOIN, F.J. 1998*a*. Antarctica as a background for mammalian evolution. *In*: CASADÍO, S. (ed.) *Paleógeno de América del Sur y de la Península Antártica*. Asociación Paleontológica Argentina, Publicación Especial, **5**, 199–209.

VIZCAÍNO, S.F., REGUERO, M.A., GOIN, F.J., TAMBUSSI, C.P. & NORIEGA, J.I. 1998*b*. An approach to the structure of the Eocene terrestrial vertebrate community from Antarctic Peninsula. *In*: CASADÍO, S. (ed.) *Paleógeno de América del Sur y de la Península Antártica*. Asociación Paleontológica Argentina, Publicación Especial, **5**, 177–183.

The late Middle Eocene terrestrial vertebrate fauna from Seymour Island: the tails of the Eocene Patagonian size distribution

JUDD A. CASE

Department of Biology, Saint Mary's College of California, Moraga, CA 94575, USA
(e-mail: jcase@stmarys-ca.edu)

Abstract: The Middle Eocene Antarctic terrestrial vertebrate palaeofauna from the La Meseta Formation on Seymour Island (= Isla Marambio), Antarctic Peninsula has a U-shaped, bimodal distribution of body sizes. This palaeofauna includes a wide range of body sizes from small insectivorous, omnivorous and granivorous marsupials, a rodent-like non-therian gondwanathere, large-sized ungulates, a sloth and cursorial birds (a ratite and a phororachoid). Medium-sized, homeothermic animals in the size range represented by rabbit to small ungulate-sized animals have not been found.

For comparison, the Early Eocene Casamayoran (Vacan 'subage') mammalian palaeofauna from Patagonia has a reasonably normal distribution of body sizes, with the modal class represented by medium-sized mammals, a distribution that is the direct opposite of the Antarctic palaeofauna. A comparison of the Middle Eocene Antarctic palaeofauna from the La Meseta Formation to the early Eocene Vacan-aged mammal palaeofauna is appropriate, due to the taxonomic affinities of the Antarctic palaeofauna to Riochican (latest Palaeocene) and Vacan-aged palaeofaunas of Patagonia. If these Patagonian mammalian palaeofaunas (PMP) were the source for the La Meseta palaeofauna (LMP), then a similar normal distribution with less taxonomic diversity would be expected. However, the LMP does not meet this expectation or even a distribution where all size classes are equally represented. Thus, the pattern of size distribution is quite different from the PMPs.

Floral data for the Early Eocene of Patagonia indicate subtropical conditions with mean annual temperatures (MAT) of 15.6 °C and equable winter temperatures (>10 °C) generating high taxonomic diversity at the species level. Floral data from the La Meseta Formation of equivalent age to the mammalian fauna indicate a cooler MAT of 11–13 °C with a highly seasonal climate, where the mean winter temperature could have ranged from –3 to 2 °C. There is also a significant drop in floral taxonomic diversity, which is dominated by *Nothofagus*.

A bimodal body size distribution pattern is not an unusual pattern for higher latitude mammalian faunas. Modern boreal mammalian faunas of North America have a low frequency of species in the medium body size range in response to cold winter temperatures in these higher latitudes. The smaller-sized mammals have adapted their physiology to the cold winter temperatures. The larger animals have adapted to the cold winter conditions by conserving heat through small surface-area-to-volume ratios as a result of their greater bulk. The low frequency of medium-sized animals is due to the fact that neither of these thermal strategies is available to them and thus they are at a selective disadvantage.

The only discovery of fossil mammals from Antarctica has been recorded from the NW portion of Seymour Island, Antarctic Peninsula (Fig. 1). The record began with the initial finding of a polydolopid marsupial, *Antarctodolops dailyi*, reported by Woodburne & Zinsmeister (1982, 1984; now *Polydolops dailyi* – see Candela & Goin 1995). Since the beginning, the affinity of the Antarctic marsupials to counterparts in South America has been examined with the goal being to establish a time of dispersal to Antarctica. The interpretation as to when the dispersal may have occurred has changed as more taxa have been added to the Antarctic palaeofauna, and with the increase in the number of new South American localities and new taxa (Bond *et al.* 1990; Goin *et al.* 1999; Reguero *et al.* 2002). This topic will be also discussed here. Also examined here is the size composition of the Antarctic mammalian palaeofauna and its relationship to Antarctic climate in the Middle Eocene.

Subsequent to the initial mammal findings, there were additions of a second species of polydolopid (Case *et al.* 1988), two ungulate (hoofed mammal) species, a litoptern (Bond *et al.* 1990) and an astrapothere (Bond *et al.* 1990; Hooker 1992), and a tardigrade sloth

From: Francis, J. E., Pirrie, D. & Crame, J. A. (eds) 2006. *Cretaceous–Tertiary High-Latitude Palaeoenvironments, James Ross Basin, Antarctica*. Geological Society, London, Special Publications, **258**, 177–186. 0305–8719/06/$15

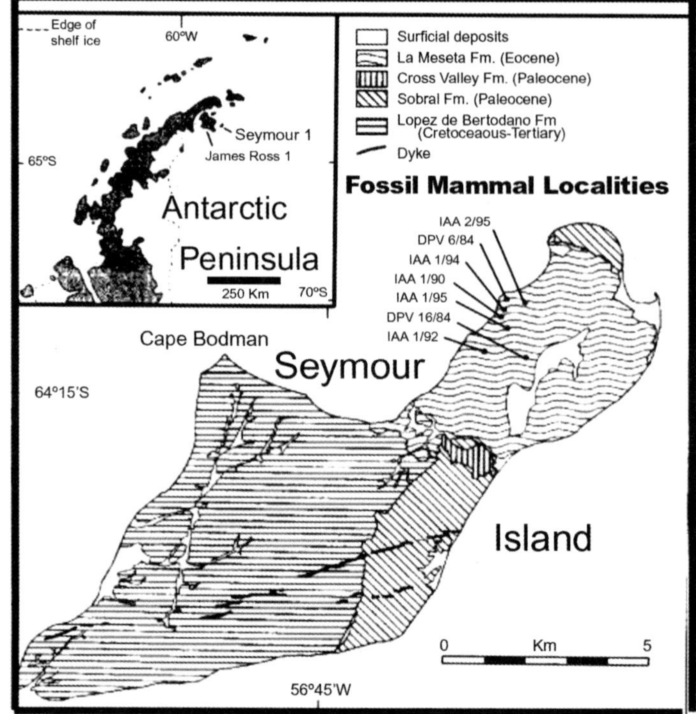

Fig. 1. Geological map of Seymour Island, Antarctic Peninsula, illustrating the geological formations and the locations of the Middle Eocene mammal-bearing localities in the La Meseta Formation. Inset shows the location of Seymour Island in the James Ross Basin on the NE side of the Antarctic Peninsula (after Goin *et al.* 1999).

(Vizcaíno & Scillato-Yane 1995). The palaeomammalian fauna from the La Meseta Formation now consists of seven species of marsupials from four different families (Goin *et al.* 1999), that not only includes the two species of polydolopid, but also three species of opossum-like derorhynchids, a prepidolopid and a microbiotheriid (Table 1). In addition, a non-therian mammal, a sudamericid gondwanathere (Reguero *et al.* 2002), has been recovered from the La Meseta Formation.

A single stratigraphic interval within the La Meseta Formation, referred to by Sadler (1988) as Telm 5 and by Marenssi & Santillana (1994) as *Cucullaea* I Allomember (= Telm 4 and 5), has produced all of these mammalian species recovered to date (Fig. 2, Table 1). The base of the *Cucullaea* I Allomember has produced a $^{87}Sr/^{86}Sr$ date of 49.5 Ma (Marenssi 2006) and the mammal-bearing beds are considered to be early Middle Eocene in age (*c.* 45 Ma; Reguero *et al.* 2002).

There are two different taxa of large flightless, running birds – a ratite (ostrich-like birds: Tambussi *et al.* 1994) and a phororhacoid (a cursorial carnivorous bird: Case *et al.* 1987) – that have been recovered from the near-shore deposits of the La Meseta Formation and can be added to the mammalian palaeofauna from these Eocene deposits. The two bird taxa and a second occurrence of the litoptern ungulate species are found at the very top of the Submeseta Allomember (or Telm 7) and are Late Eocene in age, as the top of unit has a date of 34.2 Ma (Reguero *et al.* 2002).

Affinities of the La Meseta mammalian palaeofauna

The conjectures as to the origin of the La Meseta palaeofauna (LMP), and which South American palaeofauna(s) may have the greatest similarity to this early Middle Eocene palaeofauna from the La Meseta Formation, has evolved over time. After the initial recovery of two species of polydolopid marsupials and the two ungulates, the LMP was considered to be a

Table 1. *Mammalian taxa from the La Meseta Formation*

Marsupialia
Didelphimorphia
Derorhynchidae
Derorhynchus minutus
Pauladelphys juanjoi
Xenostylus peninsularis
Microbiotheria
Microbiotheriidae
Marambiotherium glacialis
Polydolopimorphia
Polydolopididae
Polydolops dailyi
P. seymouriensis
Prepidolopidae
Perrodelphys coquinense
Placentalia
Astrapotheria
Trigonostylopidae
Trigonostylops sp.
Litopterna
Sparnotheriodontidae
Victorlemoinea sp.
Xenarthra
Tardigradidae
Tardigradidae indet.
Gondwanathria
Sudamericidae
Sudamericidae indet.

Data from Woodburne & Zinsmeister (1982, 1984); Case *et al.* (1988); Bond *et al.* (1990); Hooker (1992); Vizcaíno & Scillato-Yane (1995); Goin *et al.* (1999); Reguero *et al.* (2002).

Late Eocene representative of a Casamayoran-aged mammalian palaeofauna from Patagonia (Marenssi *et al.* 1994). The Casamayoran is an Early Eocene South American Land Mammal Age (SALMA, which is a chronology based on land mammal assemblages: Flynn & Swisher 1995). However, more recent data provide us with a different interpretation as to the origins of this now Middle Eocene Antarctic palaeofauna.

The recovery of the variety of small marsupials from the La Meseta Formation (Goin *et al.* 1999) that included three species of derorhynchids, a microbiotheriid and a prepidolopid, most of which are unknown from Casamayoran palaeofaunas, begins to suggest where the possible affinities of the LMP may lie. Continued work in Patagonia in pre-Casamayoran-aged deposits has revealed a host of small marsupials from the same families as those represented in the LMP (Goin *et al.* 1999; Reguero *et al.* 2002 and references therein). Derorhynchid marsupials are limited to Palaeocene palaeofaunas in Patagonia, where they are the most abundant family (Goin *et al.* 1997). Microbiotheres are first known from Middle Palaeocene (=Itaborian-age) palaeofaunas in both the northern and southern regions of South America (Marshall 1987; Goin *et al.* 1997), and the family continues to survive until the present day in the Patagonian region in both Chile and Argentina (Woodburne & Case 1996). Prepidolopid and polydolopid marsupials both make their first appearances in the Riochican-age (latest Palaeocene) palaeofaunas of Patagonia and continue on through the end of the Eocene (Goin *et al.* 1999). The ungulate taxa, the sparnotheriodontid, *Victorlemoinea* and the trigonostylopid, *Trigonostylops,* occur first in Riochican-age palaeofaunas of Patagonia and continue to the late Eocene (Casamayoran) Barrancan palaeofaunas.

The Casamayoran South American Land Mammal Age (Fig. 3) had been represented by palaeofaunas from Grand Barranca and elsewhere from Patagonia, and was considered as Early Eocene in age. However, there were no radiometric dates from these units, which contain a variety of marsupial and ungulate taxa. Subsequent work and radiometric dating ($^{40}Ar/^{39}Ar$) by Kay *et al.* (1999) was able to demonstrate that the Grand Barrancan palaeofauna is Late Eocene in age and the palaeofaunas of this age are assigned to the Barrancan 'subage' of the Casamayoran. The Casamayoran palaeofaunas from elsewhere in Patagonia remain Early Eocene in age and these palaeofaunas are now referred to the Vacan 'subage' of the Casamayoran (Cifelli 1985). There is a substantial time gap between the Vacan and Barrancan subages, as there are no palaeofaunas representing the Middle Eocene time range in Patagonia.

Kay *et al.* (1999), based on the dating of the Barrancan 'subage', considered the LMP as being Late Eocene in age, correlating ungulate taxa between the Barrancan and La Meseta palaeofaunas, the latter of which had been previously stated as being Late Eocene in age (based on the invertebrate taxa: Case *et al.* 1988). Temporally, the LMP was originally considered as a post-Mustersan palaeofauna, yet similarities of the Antarctic and Patagonian palaeofaunas are most comparable to those of Casamayoran (Vacan subage) palaeofaunas (Woodburne & Zinsmeister 1984; Case *et al.* 1988). New radiometric dates from the La Meseta Formation place the LMP firmly in the Middle Eocene (44.5–47.4 Ma; Dutton *et al.* 2002), precluding a correlation with the Barracan-aged palaeofaunas of Patagonia as suggested by Kay *et al.* (1999).

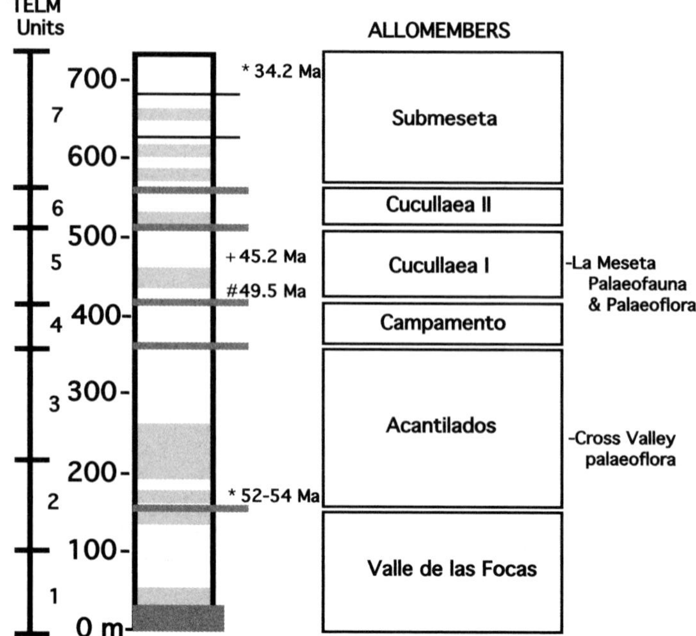

Fig. 2. Generalized stratigraphy for the Eocene La Meseta Formation, Seymour Island after Marenssi & Santillana (1994) including stratigraphic occurrences of the La Meseta palaeofauna and palaeoflora, plus the Cross Valley palaeoflora. Radiometric dates are based on $^{87}Sr/^{86}Sr$ isotopic ratios after * Dutton *et al.* (2002), + Reguero *et al.* (2002) and # Marenssi (2006). TELM units are from Sadler (1988).

Therefore, the origin for the Antarctic fauna, based on the biostratigraphic ranges of the La Meseta taxa in South America, is best correlated to Riochican or Vacan subage palaeofaunas from Patagonia. The early Late Palaeocene Itaborian-age palaeofaunas or Late Palaeocene Riochican-age palaeofaunas are possible sources for the Antarctic marsupials. Range extensions of the ungulate taxa into pre-Casamayoran palaeofaunas in Patagonia now indicate a possibility for a Riochican-age source for the La Meseta ungulates as well.

Differences between the Vacan and the La Meseta palaeofaunas

The La Meseta palaeofauna (LMP) and the Vacan-aged palaeofauna from Patagonia (PVP) differ with regard to major taxonomic groups (at the family level), taxonomic diversity and distribution of body size among the taxa present in each palaeofauna.

The Vacan PVP is a very diverse fauna with 12 different taxonomic orders of mammals represented, totalling 25 families and 43 genera. The majority of taxa are placental mammals (46 species), most of which are ungulates with some xenarthrans. Marsupial species make up the remainder of the palaeofauna (21 species). Conversely, the LMP appears to be a depauperate fauna relative to the PVP. Although the ordinal diversity of the LMP is at a reasonable level, with seven orders present, the familial (eight families), generic (10 genera) and species (11 species) diversity levels are much lower than for the PVP.

The PVP and the LMP share four families and three genera in common, but clearly have very different levels of diversity. In total, the LMP is missing seven orders and 20 families of mammals that are present in the early Eocene PVP. The LMP completely lacks the marsupial carnivores and medium-sized ungulates, the latter of which are the most abundant taxa in the PVP. In addition, the LMP is under represented in large-sized ungulate taxa as well, although all of the ungulates in the LMP are in the upper size range. However, the small-sized marsupials in the insectivore and frugivore niches are nearly equal in representation in both the LMP and the PVP.

A third major difference between these two palaeofaunas is the distribution of body sizes

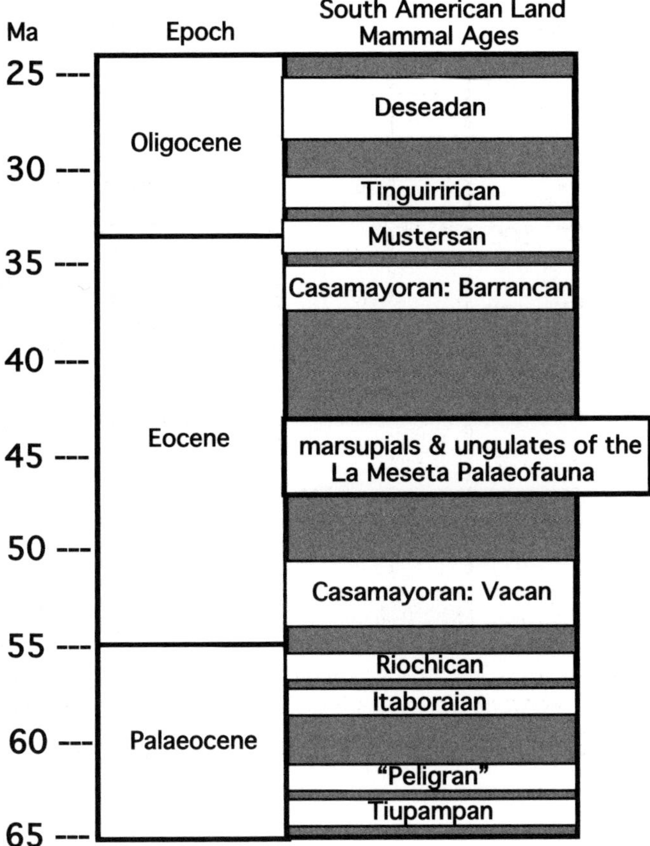

Fig. 3. Chronostratigraphic representation of South American Land Mammal Ages (SALMA based on distinct land mammal assemblages) for the Palaeogene. Note the separation of the Casamayoran SALMA into two 'subages', as indicated by Cifelli (1985) and Kay *et al.* (1999). The box inset represents the temporal position of the La Meseta Palaeofauna at a time span of approximately 45 Ma.

within each fauna. The PVP pattern of body size distributions is represented by a normal distribution (i.e. bell-shaped distribution; Fig. 4), where the modal class is in the middle of the range of body sizes present in the fauna. The LMP body size pattern is a quite different pattern with a bimodal or U-shaped distribution (Fig. 5), where the 'tails' of the body size range are the size classes with the highest frequencies, while those in the middle portion of the range exhibit the lowest frequencies.

Why should there be differences between the Patagonian Vacan palaeofauna and the La Meseta palaeofauna in regard to the distribution of body sizes? If the LMP represents a subset (i.e. fewer taxa), and thus was a representative sample of the PVP, then it should have a normal distribution pattern. The only difference would be that the LMP would have reduced frequencies for each of the classes, as it is much less diverse and abundant in its number of species. Yet, analysing the expected frequencies from a sample normal distribution vs the observed frequencies from the actual bimodal distribution using chi-square analysis ($P << 0.05$), there is a significant difference between these two patterns. Alternatively, the LMP could hypothetically be a sample from the PVP where there was an equal number species in each size class (at 1.25 taxa/size class), resulting in an intermediate pattern between the two known distributions that would be flat or platykurtotic in shape. The hypothetical flat distribution vs the actual bimodal distribution of the LMP has also a statistically significant difference ($P << 0.05$) between the hypothetical and observed frequencies. This further illustrates that the PVP and LMP distributions are quite distinct from each other.

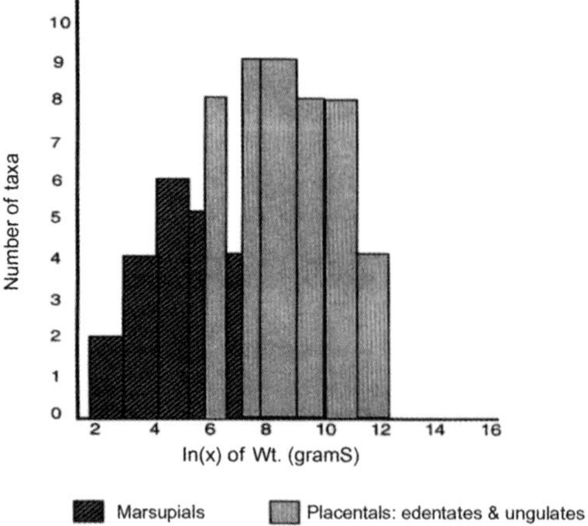

Fig. 4. The Casamayoran Vacan 'subage' taxa arranged by the log of estimated body size for the species present in the palaeofaunas of that age (data sources are Bond *et al.* 1995 and Reguero *et al.* 2002). The marsupial taxa are represented by the hatched vertical bars, while the placental taxa are represented by the plain vertical bars. The counts or *Y*-axis are the number of taxa in each size category. Note the distribution of body sizes is that of a normal distribution (bell-shaped).

Sample bias?

Should then the La Meseta palaeofauna be considered as a biased sample of the Patagonian Vacan palaeofauna and, if so, what might the casual reason be for such a biased sample?

Could there have been a collecting or a preservational bias to generate the bimodal distribution of the LMP? At least seven of the eight sites, which have produced terrestrial mammal material in the form of teeth and jaws, have been initially surface collected and then extensively sieved. This recovered large teeth, very small teeth and jaw fragments, which would be in the size range of teeth of medium-sized taxa. Thus, what was found by these methods was found with no apparent bias as to size.

This now brings up a possible preservational bias, where material of a size that would represent the medium-sized mammals from the PVP was differentially not preserved. Case (1992) demonstrated that sharks from the La Meseta Formation (particular from sites in TELM 4 and 5) produced a diversity of shark taxa with a range of sizes from the small dogfish with tiny teeth (2–3 mm wide) to relatives of modern great white sharks with very large teeth (3–4 cm wide). The majority of the taxa had teeth that were between these extremes in size. In addition, fossil penguin material collected from the La Meseta Formation also exhibits a complete range of body sizes from the largest penguins known, some 1.5 m in height, to those that match the smallest of the modern penguins, 0.3 m in height (Case 1992). The majority of penguins in the distribution are again between these two extremes in size, representing a normal distribution for body size, in fact there are several localities that have the complete range of body sizes in the penguin palaeofauna (Case 1992). Thus, data from other vertebrate taxa indicate that there appears to be no noticeable bias with regard to preservation of fossil material of particular body sizes within the La Meseta Formation.

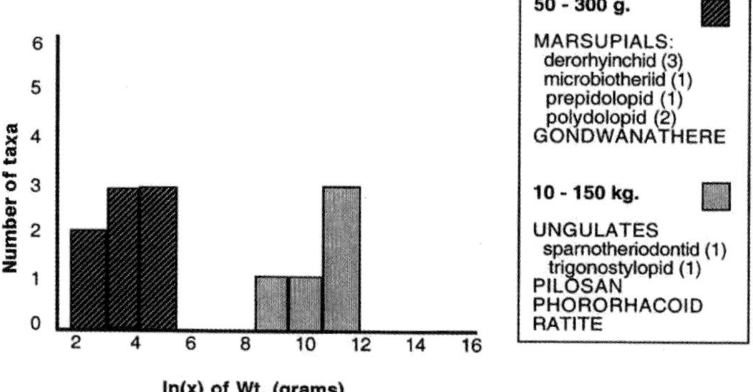

Fig. 5. The taxa of La Meseta palaeofauna, Seymour Island, Antarctic Peninsula are arranged by the natural log of estimated body size for the species present in that palaeofauna (data source is Goin *et al.* 1999). The marsupial taxa are represented by the hatched vertical bars, while the placental and avian taxa are represented by the plain vertical bars. The counts on the *Y*-axis are the number of taxa in each size category. Note the distribution of body sizes is that of a bimodal distribution (U-shaped).

Floral and faunal diversity

Could changes in taxonomic diversity result in the bimodal pattern of body sizes seen in the LMP? To help examine this question it is useful to compare the palaeofloras of Seymour Island with those of Patagonia and see what information can be gleaned from such a comparison. Case (1989) was able to successfully correlate levels of floral diversity with levels of faunal diversity through the early and mid-Cenozoic of Australia to explain patterns of origination and diversification among marsupial taxa. So there is a connection that can be determined between floral diversity and faunal diversity.

An Early Eocene flora from Patagonia, near Gran Barranca, has recently been described (Wilf *et al.* 2003) and shows some remarkable parallels regarding diversity to the palaeofauna from the same region and time. This is a highly diverse palaeoflora, with more than 100 species of plants having been identified. Dicot species are the most abundant, contributing 88 species to the palaeoflora with the remainder being represented by monocot angiosperms, ferns, conifers, cycads and ginkgoes. The species-level morphotypes and the floral composition suggest a rich subtropical forest (Wilf *et al.* 2003).

When the leaf morphology of many of the plant species are examined using CLAMP analysis, a warm mean annual temperature (MAT) of 13.6–17.6 °C is predicted with an equable climate and winter-time temperatures of above 10 °C (Wilf *et al.* 2003). The Early Eocene southern Patagonia palaeofloras have comparable floral diversity to lower latitude floras in subtropical regions today, suggesting that high plant diversity appears to have been a phenomenon of long duration.

The Late Palaeocene (Nordenskjöld) palaeoflora from Seymour Island is dominated by angiosperm taxa with some 36 species present, 13 of which have entire margins and 23 with toothed margins (Francis *et al.* 2003). In addition, there are both araucarian and podocarp conifers and at least three pterophyte (fern) species (Case 1988). CLAMP analysis of the angiosperm leaves suggest that the MAT was 13.5 °C for this time period in the Antarctic Peninsula (Francis *et al.* 2003).

The Middle Eocene palaeofloras of the La Meseta Formation (where the mammal palaeofauna has been recovered) show a distinct change from those of the Late Palaeocene. Diversity is lower with only 19 angiosperm species in the La Meseta palaeoflora, with most of the subtropical components of the earlier Nordenskjöld palaeoflora being lost (Francis *et al.* 2003). This Eocene flora is dominated by *Nothofagus* species (Case 1988; Francis *et al.* 2003) as a response to the cooler temperatures (MAT 10.8 °C) and lower precipitation (2110 mm in the Late Palaeocene to 1534 mm in the Middle Eocene) (Francis *et al.* 2003).

There is also a distinct difference between the Early Eocene, Cross Valley Formation palaeoflora from the Acantilados Allomember and the Middle Eocene, La Meseta palaeoflora from the *Cucullaea* I Allomember. Leaf size decreases from the Acantilados palaeoflora to the *Cucullaea* I palaeoflora, suggesting a decrease in temperature (Case 1988). CLAMP analysis on the leaf morphology from *Cucullaea* I suggests a MAT of 11–13 °C, with a mean winter temperature (MWT) of –3 to 2 °C (Gandolfo *et al.* 1998*a*, *b*; Fig. 2).

The La Meseta palaeoflora is as equally diverse as the Cross Valley palaeoflora at the family level, with eight angiosperm families present (Gandolfo *et al.* 1998*a*, *b*). However, the La Meseta palaeoflora is lower in species diversity within each family, except for the genus *Nothofagus*, which is the dominant plant genus. This cool temperate rainforest includes podocarp, araucarian and Cupressaceae-type gymnosperms, and at least three species of fern are present (Case 1988). Torres *et al.* (1994) indicate the presence of at least six fossil wood taxa with affinities to extant taxa living today in the cold-temperate rainforests of southern South America. Modern counterparts of the Middle Eocene mammal palaeofauna from the La Meseta Formation live in these same cold-temperate rainforests of southern Chile and Argentina (Woodburne & Case 1996) with a climate comparable to that interpreted from the La Meseta palaeoflora with cold wet winters.

Body size and thermal strategies

One explanation for the lower diversity of the Antarctic Peninsula palaeofauna compared to those from Patagonia in the Early–Middle Eocene is likely to have been tied to climatic differences, perhaps linked to the difference in the palaeolatitudes between the two areas. Mean annual temperatures were lower and seasonality was increased with cold winter temperatures in the Middle Eocene on the Antarctic Peninsula. The climatic changes have also manifested themselves in vegetation changes; that is, lower species diversity and species abundance difference between the regions through time. As with the Middle Eocene La Meseta palaeoflora, the La Meseta palaeofauna shows a substantial decrease in species-level diversity. The decreased diversity of LMP compared to the PVP could still predict a normal distribution of size classes only with a decreased number of taxa; however, that does not match the bimodal body size pattern for the LMP. It is proposed here that the lower temperatures and more seasonal climatic conditions created circumstances for changes in faunal composition from Patagonia (normal distribution of body sizes) to that seen in the Antarctic Peninsula (bimodal pattern of body size distribution) by the Middle Eocene.

Are there other examples of bimodal body size distributions and are these patterns correlated with climatic factors, especially temperature? An example of a modern mammalian fauna with a similar bimodal pattern in the distribution of body sizes to the Antarctic LMP comes from boreal Northern America (*c.* 45°N latitude; Fig. 6). In this distribution we again see a decrease in expected numbers of animals in the 0.5–10 kg range of weights. There are only three species here compared to five species each in the adjacent size categories. The expectation would be for a similar number of species (five each) or more in each of the depressed size units in the range. Thus, there must be some correlation with this pattern of body size distribution and climatic conditions, specifically cold, wintertime temperatures where winter mean temperatures (WMT) have a range that includes a lower end below 0 °C.

This bimodal pattern of body size distribution represents a series of strategies that mammals employ as a response to cold winter climatic conditions. The small mammals (<0.5 kg) have large surface-area-to-volume ratios that are highly prone to heat loss and thus they have high metabolic rates to maintain their high body temperatures. In order to maintain their high metabolic rates these animals must eat a high percentage of their body weight in food each day. However, their food supply becomes scarce or non-existent during the winter. Consequently, these animals hibernate, reducing body metabolism, lowering body temperatures and slowly utilizing body fat stores while 'sleeping' through the winter months (Campbell & Reece 2004). At the opposite end of the distribution are the large mammals, which employ a different adaptive strategy. These mammals, mostly large ungulates and some large carnivores, have sufficient body mass and small surface-area-to-volume ratios that allow these animals to more efficiently retain body heat even during cold temperatures. These animals are capable of being active during the winter as they are able to find sufficient food, and, with adequate fat stores, they can survive until spring and food supplies are increased again. These higher latitude animals tend to be larger in body size than their sister species living at lower latitudes, which is defined as 'Bergmann's rule' (Campbell & Reece 2004).

Unfortunately, the mammals in the middle of the body size distribution are unable to employ either of these adaptive strategies. The middle-sized mammals are too small to be able to adopt the winter-active strategy as their surface-area-to-volume ratios are too large to hold in heat

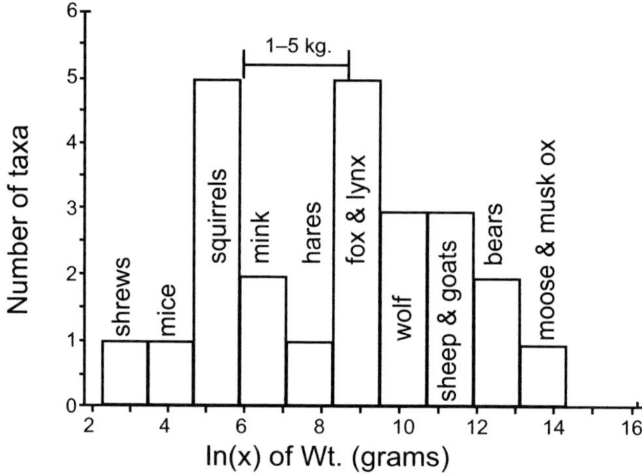

Fig. 6. The distribution of the natural log of average body weights for the species present in the modern boreal (≥ 45° latitude) mammalian fauna of North America. The counts or *Y*-axis are the number of taxa in each size category. Note the distribution of body sizes is that of a bimodal distribution (U-shaped). The pattern is probably based on thermal strategies (see text) employed by different sized taxa for adaptation to cold winter temperatures.

efficiently. Further, these mammals are too large to hibernate by storing adequate body fat and to undergo sustained torpor and an early spring arousal. Thus, these animals are at a selective disadvantage in this environment and hence their vastly lower representation in the high-latitude faunas compared to their abundance in lower temperate–tropical latitudes.

The bimodal size distribution of the La Meseta palaeofauna is consistent with physiological adaptation to cold environmental conditions. The bimodal distribution pattern itself is indicative of cold winter temperatures, just as serrated leaf-margin morphology is also indicative of cold winter temperatures.

Summary

The Middle Eocene Antarctic terrestrial vertebrate palaeofauna from the La Meseta Formation on Seymour Island (Marambio), Antarctic Peninsula, exhibits a wide, but bimodal, range of body sizes from small marsupials, as well as a rodent-like gondwanathere, to large-sized ungulates, along with a large-bodied sloth and cursorial birds including ratites and phororachoids. The palaeofauna lacks medium-sized mammals and thus has a U-shaped, bimodal distribution of body sizes.

An Early Eocene or Casamayoran (Vacan 'subage') mammalian palaeofauna from Patagonia has a reasonably normal distribution of body sizes, with the modal class represented by medium-sized mammals. This body size distribution is directly opposite to that of the Antarctic palaeofauna. The taxonomy of the Vacan-aged Patagonian mammal palaeofaunas compares well with the Middle Eocene La Meseta palaeofauna, suggesting that the early Palaeogene Patagonian faunas are a probable source for the La Meseta palaeofauna, despite differences in body size distribution.

Floral data from the Early Eocene of Patagonia indicate subtropical conditions with moderate mean annual temperatures and an equable climate, and exhibits a high taxonomic diversity, as does the mammalian palaeofauna. Floral data from the La Meseta Formation of an equivalent age to the mammalian palaeofauna are less diverse and indicate a cooler mean annual temperature with a highly seasonal climate, with winter temperatures averaging from –3 to 2 °C. Palaeomammalian diversity is also lower in the La Meseta palaeofauna.

The bimodal body size distribution pattern in higher latitude extant mammalian faunas appears to be a response to cold winter temperatures in higher latitudes. The smaller-sized mammals have adapted to the cold winter temperatures through the physiological strategies of torpor and hibernation. The larger animals have adapted to the winter conditions by having larger body sizes, creating small surface-area-to-volume ratios as a result of their greater bulk and thus can better conserve heat. The low frequency (or absence in the La Meseta palaeofauna) of medium-sized animals probably indicates that neither thermal adaptive strategies are available to them, resulting in a selective disadvantage in these cold winter environments.

This project was partially funded by the US National Science Foundation, Office of Polar Programs grant # 0003844 to the author. I acknowledge the time and generosity of the following colleagues and institutions for their fieldwork and research on the La Meseta palaeofauna and who have advanced these scientific investigations: M. O. Woodburne and his students formally from the University of California, Riverside, including D. Chaney; M. Reguero, F. Goin, S. Vizcaíno and J. J. Moly from the Museo de La Plata; S. Marenssi and S. Santillana from the Instituto Antártico Argentino; A. Crame and J. Hooker from the British Antarctic Survey. I thank Saint Mary's College of California in helping to fund and providing time to participate in the symposium. Finally, I would like to thank J. Francis for her patience and her editorial advice, and two anonymous reviewers for their helpful suggestions for improving the quality of this study.

References

BOND, M., CARLINI, A.A., GOIN, F.J., LEGARRETA, L. ORTIZ-JUAREGUIZAR, E., PASCUAL, R. & ULIANA, M.A. 1995. Episodes in South American land mammal evolution and sedimentation: testing their apparent concurrence in a Palaeocene succession from central Patagonia. *In*: *Congreso Argentino de Paleontología y Bioestratigrafia*. Asociación Paleontológica Argentina, Trelew, 47–58.

BOND, M., PASCUAL, R. REGUERO, M.A., SANTILLANA, S.N. & MARENSSI, S.A. 1990. Los primeros ungulados extinguidos sudamericanos de la Antártida. *Ameghiniana*, **16**, 240.

CAMPBELL, S. & REECE, M. 2004. *Biology*, 7th edn. Benjiman Cummings, Menlo Park, CA.

CANDELA, A.M. & GOIN, F.J. 1995. Revisión de las especies antárticas de marsupials Polidolopinos (Polydolopimorphia, Polydolopidae). *In*: *III Jornadas de Comunicaciones Sobre Investigaciones Antárticas*. Instituto Antártico Argentino, Buenos Aires, 55–56.

CASE, J.A. 1988. Paleogene floras from Seymour Island, Antarctic Peninsula. *In*: FELDMANN, R.M. & WOODBURNE, M.O. (eds) *Geology and Paleontology of Seymour Island*. Geological Society of America, Memoirs, **169**, 523–530.

CASE, J.A. 1989. Antarctica: the effect of high latitude heterochroneity on the origin of the Australian marsupials. *In*: CRAME, J.A. (ed.) *Origins and Evolution of the Antarctic Biota*. Geolological Society, London, Special Publications, **47**, 217–226.

CASE, J.A. 1992. Evidence from fossil vertebrates for a rich Eocene, Antarctic marine environment. *In*: KENNETT, J. & WARNKE, D. (eds) *Paleoenvironment Evolution of Antarctica and the Southern Oceans*. American Geophysical Union, Antarctic Research Series, **56**, 119–130.

CASE, J.A., WOODBURNE, M.O. & CHANEY, D.S. 1987. A gigantic phororhacoid(?) bird from Antarctica. *Journal of Paleontology*, **6**, 1280–1284.

CASE, J.A., WOODBURNE, M.O. & CHANEY, D.S. 1988. A new genus and species of polydolopid marsupial from the La Meseta Formation, late Eocene, Seymour Island Antarctic Peninsula. *In*: FELDMANN, R.M. & WOODBURNE, M.O. (eds) *Geology and Paleontology of Seymour Island*. Geological Society of America, Memoir, **169**, 505–521.

CIFELLI, R.L. 1985. Biostratigraphy of the Casamayoran, early Eocene, of Patagonia. *American Museum Novitates*, **2820**, 1–26.

DUTTON, A.L., LOHMANN, K.C. & ZINSMEISTER, W.J. 2002. Stable isotope and minor element proxies for Eocene climate of Seymour Island, Antarctica. *Paleoceanography*, **17**, 1016.

FLYNN, J.J. & SWISHER, C.C. 1995. Cenozoic South American Land Mammal Ages: correlation to global geochronologies. *In*: BERGGREN, W.A., KENT, D.V., AUBRY, M.P. & HARDENBOL, J. (eds) *Geochronology, Time Scales and Global Stratigraphic Correlation*. Society of Sedimentary Geology, Special Publications, **54**, 317–333.

FRANCIS, J., TOSOLINI, A-M. & CANTRILL, D.J. 2003. Biodiversity and climatic change in Antarctic Palaeogene floras. *In*: *Antarctic Contributions to Global Earth Sciences. 9th International Symposium on Antarctic Earth Sciences (ISAES IX), Potsdam, Germany*, Alfred Wegener Institute for Polar and Marine Research, Berlin, 107.

GANDOLFO, M.A., HOC, P., SANTILLANA, S.N. & MARENSSI, S.A. 1998*a*. Una flór fosil morfologicamente afín a las Grossulariaceae (Orden Rosales) de la Formación La Meseta (Eoceno medio), Isla Marambio, Antártida. *In*: CASADIO, S. (ed.) *Paleógeno de América del Sur y de la Península Antártica*. Asociación Paleontológica Argentina, Publicación Especial, **5**, 147–153.

GANDOLFO, M.A., MARENSSI, S.A. & SANTILLANA, S.N. 1998*b*. Flora y paleoclíma de la Formación La Meseta (Eoceno medio), Isla Marambio, Antártida. *In*: CASADIO, S. (ed.) *Paleógeno de América del Sur y de la Península Antártica*. Asociación Paleontológica Argentina, Publicación Especial, **5**, 155–162.

GOIN, F.J., CANDELA, A.M. & FORASIEPI, A. 1997. New middle Palaeocene marsupials from Central Patagonia. *Journal of Vertebrate Paleontology*, **17**, (Suppl.), 49A.

GOIN, F.J., CASE, J.A., WOODBURNE, M.O., VISCAÍNO, S.F. & REGUERO, M.A. 1999. New discoveries of 'opposum-like' marsupials from Antarctica (Seymour Island, Middle Eocene). *Journal of Mammalian Evolution*, **6**, 335–655.

HOOKER, J.J. 1992. An additional record of a placental mammal (Order Astrapotheria) from the Eocene of West Antarctica. *Antarctic Science*, **4**, 107–108.

KAY, R.F. & MADDEN, R.H. *ET AL.* 1999. Revised geochronology of the Casamayoran South American Land Mammal Age: climatic and biotic implications. *Proceedings of the National Academy of Sciences*, **96**, 13 235–13 240.

MARENSSI, S.A. 2006. Eustatically controlled sedimentation recorded by Eocene strata of the James Ross Basin, Antarctica. *In*: FRANCIS, J.E., PIRRIE, D. & CRAME, J.A. (eds) *Cretaceous–Tertiary High-Latitude Palaeoenvironments: James Ross Basin, Antarctica*. Geological Society, London, Special Publications, **258**, 125–133.

MARENSSI, S.A. & SANTILLANA, S.N. 1994. Unconformity bounded units within the La Meseta Formation, Seymour Island, Antarctica: a preliminary approach. *In*: ZALEWSKI, M. (ed.) *XXI Polar Symposium*, Institute of Geophysics of Polish Academy of Sciences, Warsaw, 33–37.

MARENSSI, S.A., REGUERO, M.A., SANTILLANA, S.N. & VIZCAÍNO, S.F. 1994. Eocene land mammals from Seymour Island, Antarctica: palaeobiogeographical implications. *Antarctic Science*, **6**, 3–15.

MARSHALL, L.G. 1987. Systematics of Itaboraian (middle Palaeocene) age 'opossum-like' marsupials from the limestone quarry at Sao Jose de Itaborai, Brazil. *In*: ARCHER, M. (ed.) *Possums and Opossums: Studies in Evolution*. Surrey Beatty, Royal Zoological Society of New South Wales, Sydney, 91–160.

REGUERO, M.A., MARENSSI, S.A. & SANTILLANA, S.N. 2002. Antarctic Peninsula and South America (Patagonia) Paleogene terrestrial faunas and environments: biogeographic relationships. *Palaeogeography, Palaeoclimatology, Palaeoecology*, **179**, 189–210.

SADLER, P.M. 1988. Geometry and stratification of the uppermost Cretaceous and Paleogene units on Seymour Island, northern Antarctic Peninsula. *In*: FELDMANN, R.M. & WOODBURNE, M.O. (eds) *Geology and Paleontology of Seymour Island*. Geological Society of America, Memoirs, **169**, 303–320.

TAMBUSSI, C.P., NORIEGA, J.I., GAZDZICKI, A., TATUR, A., REGUERO, M.A. & VIZCAÍNO, S.F. 1994. Ratite bird from the Paleogene La Meseta Formation, Seymour Island, Antarctica. *Polish Polar Research*, **15**, 15–20.

TORRES, M.T., MARENSSI, S.A. & SANTILLANA, S.N. 1994. Maderas fósiles de la Isla Seymour, Formación La Meseta. *Serie Científica Instituto Antártico Chileno*, **44**, 17–38.

VIZCAÍNO, S.F. & SCILLATO-YANE, G.J. 1995. An Eocene tardigrade (Mammalia, Xenarthra) from Seymour Island, West Antarctica. *Antarctic Science*, **7**, 407–408.

WILF, P., CUNEO, N.R., JOHNSON, K.R., HICKS, J.F., WING, S.L. & OBRADOVICH, J.D. 2003. High plant diversity in Eocene South America: evidence from Patagonia. *Science*, **300**, 122–125.

WOODBURNE, M.O. & CASE, J.A. 1996. Dispersal, vicariance, and the Late Cretaceous to Early Tertiary land mammal biogeography from South America to Australia. *Journal of Mammalian Evolution*, **3**, 121–161.

WOODBURNE, M.O. & ZINSMEISTER, W.J. 1982. Fossil land mammal from Antarctic. *Science*, **218**, 284–286.

WOODBURNE, M.O. & ZINSMEISTER, W.J. 1984. The first land mammal from Antarctica and its biogeographic implications. *Journal of Paleontology*, **58**, 913–948.

Distribution, lithofacies and environmental context of Neogene glacial sequences on James Ross and Vega islands, Antarctic Peninsula

MICHAEL J. HAMBREY[1] & JOHN L. SMELLIE[2]

[1]*Centre for Glaciology, Institute of Geography & Earth Sciences, University of Wales, Aberystwyth SY23 3DB, UK (e-mail: mjh@aber.ac.uk)*

[2]*British Antarctic Survey, Natural Environment Research Council, High Cross, Madingley Road, Cambridge CB3 0ET, UK (e-mail: jlsm@bas.ac.uk)*

Abstract: Considerable controversy exists concerning the stability of the Antarctic Ice Sheet during the Neogene Period. The northern Antarctic Peninsula is in a critical position in this debate as it represents a peripheral area of the ice sheet and is therefore likely to have been sensitive to climatic changes. This paper is concerned with Neogene glacial deposits that occur on James Ross and Vega islands. They occur between a thick volcanic sequence, the James Ross Island Volcanic Group, and Upper Cretaceous sedimentary rocks; they also occur within the volcanic sequence itself. The glacial deposits, where dated, give a series of snapshots of glacial conditions in Neogene time. The deposits are characterized by diamictite and sandy mudstone. Published $^{87}Sr/^{86}Sr$ ages on shelly fossils in some deposits range from 9.9 to 4.7 Ma, although additional $^{40}Ar/^{39}Ar$ ages on interbedded volcanic rocks suggest that younger sedimentary deposits are also present. On James Ross Island the basal diamictite is interpreted as glaciomarine sediment that has undergone subaqueous mass movement, and on Vega Island as basal till originating from the west. Provenance studies indicate that the Antarctic Peninsula Ice Sheet expanded sufficiently to deposit these sediments. These diamictites, in places, are overlain by waterlain tuffaceous rocks that include a minor ice-rafted component. Complex deformation of sedimentary and volcanic deposits and contact-metamorphism relationships confirm that volcanism was contemporaneous with glaciation. Later glacial events (within the volcanic sequence) are characterized by glacial erosion of basalt followed by basal till and, possibly, glaciofluvial deposition. The clasts in the latter are almost exclusively local, hence later glaciation was as a small ice cap constructed on the growing volcanic complex of James Ross Island.

The aim of this paper is to present representative stratigraphic and sedimentological data from an investigation concerning Neogene glacigenic and associated volcanic strata in the James Ross Island Volcanic Group, near the northern tip of the Antarctic Peninsula. Lying near the northern limit of ice-sheet development in Antarctica, the James Ross Island area is likely to show more clearly than most parts of the continent the nature of ice-sheet fluctuations in response to climate. Sedimentological and volcanological investigations allow us to define the scale and thermal characteristics of the glaciers, and hence palaeoclimate, whilst the volcanic rocks also provide a means of dating events radiometrically.

Neogene glacial climates in Antarctica are particularly important for several reasons. First, in a global context, IPCC (2001) announced that the latest climate models predict a global temperature rise as a result of CO_2 emissions of between 1.4 and 5.8 °C by the end of the 21st century. If emissions continue to be unrestricted, the models are projected to give by 2100 AD an atmosphere on Earth not seen since approximately 12–13 Ma. Therefore understanding the scale of Antarctic ice masses in pre-Quaternary times is relevant to predicting their response to future climatic change, and their impact on sea-level change (for discussions of the importance of the history of the Antarctic Ice Sheet in a global context see: Webb & Harwood 1991; Wilson 1995; Barrett 1996; Sugden 1996). Neogene and Palaeogene ice sheets in Antarctica predate all Northern Hemisphere glaciations, and therefore represent the only significant eustatic influence on sea level at that time.

Secondly, a wide range of evidence from sites on the continental shelf of Antarctica, such as Prydz Bay (Barron *et al.* 1991; O'Brien *et al.* 2001; Hambrey & McKelvey 2000*a*) and the western Ross Sea (Hambrey & Wise 1998; Barrett & Ricci 2000, 2001), suggests that the pre-Quaternary East Antarctic Ice Sheet was

From: FRANCIS, J. E., PIRRIE, D. & CRAME, J. A. (eds) 2006. *Cretaceous–Tertiary High-Latitude Palaeoenvironments, James Ross Basin, Antarctica.* Geological Society, London, Special Publications, **258**, 187–200. 0305–8719/06/$15

polythermal or even temperate in character, and may have behaved quite differently from the present-day ice sheet, possibly strongly fluctuating in association with major changes in sea level. Similarly, sedimentological evidence from onshore areas, notably the Prince Charles Mountains (Hambrey & McKelvey 2000*b*; Whitehead *et al.* 2003) and the Transantarctic Mountains (Webb *et al.* 1996; Wilson *et al.* 1998; Hambrey *et al.* 2003), indicates that the East Antarctic Ice Sheet was subject to major fluctuations in pre-Quaternary time.

Extensive latest Oligocene and earliest Miocene glacial outcrops (Polonez Cove and Cape Melville formations) are preserved on King George Island, in the Antarctic Peninsula region (Troedson & Riding 2002; Troedson & Smellie 2002). The thermal regimes for associated glacier ice were interpreted to be subpolar (i.e polythermal) and temperate, respectively. Diverse reworked and ice-rafted clasts are similar within both formations, and originated from both local and distal sources, possibly as far as the Transantarctic and Ellsworth mountains. King George Island also contains outcrops of a glacial sedimentary unit known as the Vauréal Peak Formation that originated entirely on King George Island (Birkenmajer 1996). Although Birkenmajer (1996) thought it was broadly related in time to an underlying volcanic sequence, said to be 30–26 Ma in age, the field relations indicate that the glacial unit is younger. Its age is thus very poorly constrained and it may be substantially younger than the volcanic sequence. Elsewhere in the region, glacial sedimentary deposits are present in the James Ross Island area, on James Ross, Vega and Cockburn islands (Jonkers & Kelley 1988; Pirrie *et al.* 1997*b*; Jonkers *et al.* 2002). They occur within and at the base of the Neogene James Ross Island Volcanic Group, and are the subject of this paper. Two formations have previously been defined. The Miocene Hobbs Glacier Formation, ascribed an origin by glaciomarine sedimentation close to a grounding line of either a floating ice shelf or grounded tidewater glacier (Pirrie *et al.* 1997*b*); and the Pliocene Cockburn Island Formation, composed of sandstone and very coarse conglomerate that is thought to have been deposited under ice-free, fully-marine interglacial conditions (Jonkers 1998).

Up to now, the interest in Cenozoic glacial history has focused most strongly on the switch from a strongly fluctuating ice sheet over East Antarctica to the present-day stable ice sheet. The timing of this crucial switch is highly controversial. There is the 'stabilist' view that argues that the East Antarctic Ice Sheet has remained stable for at least 15 Ma (Denton *et al.* 1993; Marchant *et al.* 1993; Sugden *et al.* 1993; Sugden 1996; Stroeven & Kleman 1999), which contrasts with the 'dynamicists' view that stability was only achieved as late as 3–4 Ma (Webb & Harwood 1991; Wilson 1995; Webb *et al.* 1996; Harwood & Webb 1998; Hambrey & McKelvey 2000*a*). Resolution of this question requires investigation of other parts of the Antarctic Ice Sheet, especially those areas most sensitive to climatic fluctuations, such as the Antarctic Peninsula.

The significance of the Northern Antarctic Peninsula region is that:

- it provides snapshots of glacial conditions from latest Palaeogene time and through the Neogene Period;
- it allows evaluation of glacier thermal regime (principally from sediment facies associations) and, hence, contributes to our understanding of the palaeoclimate;
- being interbedded with volcanic rocks provides insight into the understanding of glacier–volcano interactions and can also be radiometrically dated;
- it facilitates reconstruction of the former ice-flow direction that enables the location of the ice mass to be determined, which in turn can contribute to the understanding of regional palaeotopography and, hence, tectonic controls.

The fieldwork for this project included sites on James Ross Island that had already been investigated in some detail (Pirrie *et al.* 1997*b*). Other sites were also visited on James Ross Island and Vega Island, some of which were mentioned briefly by Smellie (1999), but most of which were discovered during the present investigation.

Stratigraphic setting

Miocene glacigenic sequences occur as lenses typically only a few metres thick between the Cretaceous Marambio Group, a shallow-marine sequence (Crame *et al.* 1991; Pirrie 1991; Pirrie *et al.* 1991, 1997*a*), and the Neogene James Ross Island Volcanic Group (Nelson 1975; Smellie 1999). Exceptionally, the sedimentary deposits form 27 to 90 m-thick successions. The glacigenic sequence was formally defined as the Hobbs Glacier Formation on SE James Ross Island, and includes diamictite overlain by waterlain tuff (Pirrie *et al.* 1997*b*). The formation lies unconformably on the Cretaceous strata and is transitional with the overlying hyaloclastites of the volcanic group. Prior to the current investigation, the Hobbs Glacier Formation was documented from about 15 localities, but some 30 more outcrops were discovered in 2002 and 2003 (Fig. 1). The true extent of the

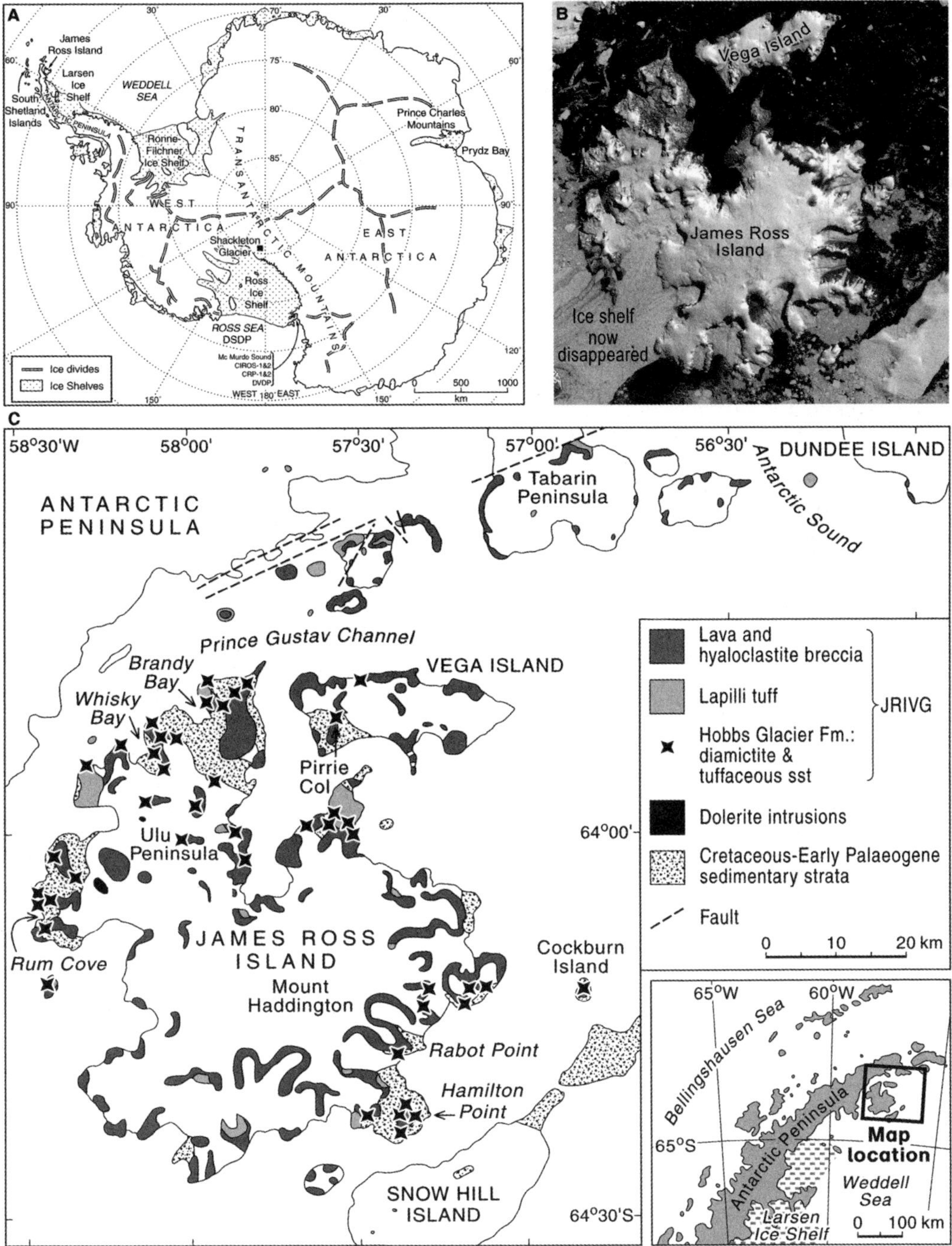

Fig. 1. Geographical context of James Ross Island glacial sediments. (**A**) Location map of key Cenozoic glacial sites in Antarctica. (**B**) Satellite image of James Ross Island (part of mosaic by Bundesamt für Kartographie und Geodäsie, Germany. Landsat Thematic Mapper, Path 215 Row 105 (29 February 1988) and Path 217 Row 105 (1 March 1986)). (**C**) Summary geological map of James Ross Island (Cretaceous strata after Crame *et al.* 1991; volcanic strata and positions (stars) of Neogene glacial outcrops after Smellie 1999, and this paper), showing locations of stratigraphic sections described in this paper and position of James Ross Island in the Antarctic Peninsula region; the number and locations of glacial sedimentary outcrops on James Ross Island is highly simplified. JRIVG, James Ross Island Volcanic Group.

formation is probably greater than is currently known. The relatively thin sedimentary outcrops typically occur at the base of volcanic cliffs and are commonly obscured by scree.

Field sites and methods

The principal sites investigated were in the vicinity of Rabot Point and Hamilton Point on SE James Ross Island, Whisky Bay on the NW side of James Ross Island, and Pirrie Col on Vega Island.

To elucidate the environmental context of the glacigenic sediments and associated volcanic rocks, lithofacies logging and textural analyses were undertaken. This paper focuses on presenting representative logs from each of the three main sites in order to demonstrate the variety of facies and their potential for interpreting palaeoenvironments. Logging was undertaken through the Hobbs Glacier Formation diamictite and tuffaceous rocks, which generally total less than 6 m in thickness, and continued below into uppermost Cretaceous sediments, and above into hyaloclastite and tuff. Facies analysis was undertaken in the field using informal lithofacies designations previously developed for Antarctic glacigenic sediments (e.g. Hambrey & McKelvey 2000*b*). Lithofacies were documented in measured sections in terms of texture (proportions of gravel, sand and mud, estimated visually), sedimentary structures, boundary relations and fossil content. Lithofacies were classified according to a scheme for poorly sorted sediments (Moncrieff 1989; modified by Hambrey & Glasser 2003).

Stratigraphy and sedimentology

The range of facies and their interpretation are summarized in Table 1 and illustrated in Figure 2. Interpretations are based on data from modern glacial environments as summarized, for example, in Hambrey (1994), Bennett & Glasser (1996), Benn & Evans (1998) and Evans (2003). $^{87}Sr/^{86}Sr$ dating of barnacle plates from

Table 1. *Summary of lithofacies, their respective associations and interpretation of depositional setting*

Facies association	Lithofacies	Interpretation	Figure
Volcanic (James Ross Island Volcanic Group)	Hyaloclastite breccia	Formed by rapid chilling in water and deposition by avalanching and as mass flows. Form main (lower) part of lava-fed delta, analogous to foreset beds	2a
	Lithic breccia	Lava or sill fragmented by rapid chilling beneath ice or water, or against water-rich sediments	
	Basalt pillow lavas	Isolated clasts and protrusions from far-travelled lava flows on delta foresets; subaqueous extrusion	2b
	Pahoehoe lava	Compound lava extruded subaerially; forms horizontal lava caprock on lava-fed delta, analogous to sedimentary topset beds	
Transitional	Sandy or silty hyaloclastite breccia	Tuff, diamict and hyaloclastite breccia, mixed by loading or slumping	2b lower
Volcaniclastic rocks with some glacial influence (Hobbs Glacier Formation; upper member)	Coarse tuffaceous sandstone Laminated siltstone Sandstone/siltstone rhythmites Interlaminated grey siltstone and sandstone with disseminated clasts Gritstone Laminated sandstone Chert	Tuff deposited in subaqueous setting; subject to some reworking by slumping; minor ice-rafted component	2b, c, e 2b
Glacigenic (Hobbs Glacier Formation; lower member)	Diamictite, weakly stratified Diamictite, massive Diamicton, massive Mudstone with lonestones	Proximal glaciomarine; slumped or debris flow Contact metamorphosed basal till Basal till Distal glaciomarine	2d, e 2d

Fig. 2. Representative volcanogenic and sedimentary facies on James Ross Island as described in Table 1. (**A**) Hyaloclastite breccia overlying basalt pillow lava at Rabot Point; the hammer is 30 cm long. (**B**) Laminated volcaniclastic siltstone (dark grey) and sandstone (buff), overlying hyaloclastite breccia at Hamilton Point; slumped and disrupted laminae are evident; the pen used for scale is on the left. (**C**) Rhythmically laminated fine and medium sandstone affected by slump-folding at Hamilton Point; the pen is for scale. (**D**) Weakly stratified diamictite at Rabot Point interbedded with discontinuous laminated siltstone containing sparse clasts; the pen at the bottom is for scale. (**E**) Weakly stratified diamictite with both buff (light) and brown (dark) matrix, affected by soft-sediment deformation; overlain in turn by laminated tuffaceous sandstone and, above overhang, hyaloclastite breccia; underlain by Cretaceous sandstone below the person's knees; this represents a complete section through the Hobbs Glacier Formation near the type locality at Rabot Point.

the type section at Rabot Point gave an age of 9.9 ± 0.97 Ma (Dingle & Lavelle 1998). That age probably also applies to the formation at nearby Hamilton Point, where it occupies a similar stratigraphical position at the base of the volcanic sequence. The Pirrie Col exposure is not fossiliferous and remains undated. However, volcanic units elsewhere on Vega

Island have yielded ages of 2.7 and 2.03 Ma ($^{40}Ar/^{39}Ar$ dating: J. L. Smellie & W. C. McIntosh unpublished data), and provide a late Pliocene minimum age for the Pirrie Col exposure. The exposure at Whisky Bay underlies a volcanic unit with a relatively imprecise $^{40}Ar/^{39}Ar$ age of 5.64 ± 0.25 Ma. The sedimentary and volcanic units at Whisky Bay show synsedimentary deformation and were probably coeval.

Rabot Point

This locality, on the SE side of James Ross Island, has one of the thickest sections of the formation so far discovered (*c.* 5 m), and represents the type locality (Pirrie *et al.* 1997*b*). It comprises a lower diamictite member overlain by a tuffaceous member, before passing up via a sharp contact into hyaloclastite breccia (Fig. 3). The diamictite is clast-rich and ranges from massive to well stratified. A wide range of igneous and metamorphic clasts have been described by Pirrie *et al.* (1997*b*) from this site. An intraformational breccia containing fragments of the underlying Cretaceous siltstone occurs at the base. Lithofacies in the tuffaceous member include sandstone, siltstone, minor breccia and laminite with dispersed pebble-sized clasts (lonestones). The hyaloclastite breccia is part of a thick (>100 m) unit showing large-scale, steep-dipping (c. 30°) homoclinal cross-bedding and basalt lava lenses. It is overlain by a thinner (30–40 m) horizontal capping unit of basaltic pahoehoe lava (Smellie 1999).

Whisky Bay

A quite different succession is visible on the east side of Whisky Bay in NW James Ross Island (Fig. 4). The clast-rich sandy diamictite is less than 0.5 m thick, and ranges from highly weathered to well indurated. It is overlain or partly interbedded with volcanic (hyaloclastite) breccia similar to that at Rabot Point, but with its original capping lava sequence removed by erosion. In addition, diamictite (along with sandstone) occurs as large intraclasts (>1 m diameter) in the breccia. In places dyke-like features of diamictite extend upwards into the breccia.

Pirrie Col, Vega Island

The contact between the volcanic strata and the underlying Upper Cretaceous strata at Pirrie Col is mostly scree-covered, but in a few places after some excavation a transitional unit is observed (Fig. 5). This unit at its base has a boulder pavement consisting of exotic clasts (mostly quartzite with some amphibolite and granite) with flat, smooth upper surfaces. Then follows a few tens of centimetres of soft clast-rich muddy diamicton, overlain by veined, possibly contact-metamorphosed, diamictite. A tabular basalt flow or sill, a few centimetres thick, with chilled lower contact rests on top, and is succeeded by crumbly basaltic lithic breccia, and then strongly palagonite-altered coarse-grained tuff with dispersed granite pebbles.

Other glacigenic sedimentary successions

In addition to the successions described above that occur at the base of the James Ross Island Volcanic Group, several localities reveal diamictite units, each overlying an unconformity, within the volcanic succession. The unconformities typically cut out parts or (commonly) all of the underlying lava units that in turn cap the coeval hyaloclastite breccia foresets of each lava-fed delta sequence. The unconformities are smooth ('polished'), striated and ice-moulded, signifying the passage of wet-based ice; hence, they relate to a time interval of erosion prior to the eruption of the succeeding lava-fed delta. Most of these higher-level deposits only include local clasts, notably crystalline (subaerial) lavas, glassy lava blocks derived from hyaloclastite breccias, hyaloclastite breccia itself and occasional red-oxidized scoria. In addition, there are angular–subangular clasts of Cretaceous siltstone and sandstone, as well as a variety of Antarctic Peninsula lithologies derived from the Cretaceous conglomerates on James Ross Island. The latter are invariably rounded–well rounded, unlike the glacially derived rocks of the Antarctic Peninsula that occur within the diamicton of the Hobbs Glacier Formation (most of which are subangular and subrounded). Several clasts at each locality reveal faceting and rare striations.

Elsewhere, at the base of the volcanic succession, facies are typically composed solely of clast-rich muddy (silty) diamicton or diamictite with pebble- to boulder-sized clasts, and are massive or faintly planar bedded. A few exposures show a closely spaced, anastomosing, low-angle joint fabric. Lower surfaces are sharp but upper surfaces are more transitional, with overlying deposits (generally coarse hyaloclastite breccia, locally with pillows) commonly loaded from several decimetres to 1 or 2 m into the diamictite (cf. Pirrie *et al.* 1997*b*, fig. 11). The diamictite often forms squeeze-up structures into the breccia. Abraded pebbles–boulders, clearly derived from the diamictites, are commonly dispersed within the basal few metres of the volcanic breccias, which may also have patchy admixed silty matrix. These deposits are rarely fossiliferous (e.g. Jonkers 1998; Jonkers *et al.* 2002).

Interpretation of lithofacies

Diamicton and diamictite of the lower member of the Hobbs Glacier Formation show evidence of transport in the zone of traction at the base of a sliding glacier, including a wide range of clasts in terms of size and lithology, a clay- or silt-rich matrix, clasts bearing facets and striations, and a pavement of boulders with striated tops. Furthermore, exotic clasts (such as foliated metasedimentary rocks and granitoids) indicate an Antarctic Peninsula provenance. Most sections studied show crudely stratified diamictite with abundant signs of soft-sediment deformation. As also noted by Pirrie *et al.* (1997*b*), many clasts in the basal deposits have coatings of bryozoa, and broken shells are present, suggesting a marine setting. Pirrie *et al.* (1997*b*) inferred glaciomarine sedimentation for the Hamilton Point and Rabot Point sections. Deformation structures, such as synsedimentary folding and crude grading, point to remobilization of the glaciomarine facies as slumps and debris flows. At one locality at least (Pirrie Col on Vega Island), there is unequivocal evidence of basal glacial deposition. There a massive diamicton (and the overlying diamictite) not only has a strong clast-orientation fabric (visible on the upper bedding surface), but also overlies a discontinuous boulder pavement with striations indicating ice flow from the west.

Volcanic (tuffaceous) facies of the upper Hobbs Glacier Formation are not widely distributed away from eastern James Ross Island. They are either the product of ash falling directly into a water body (Pirrie *et al.* 1997*b*) or are reworked with a considerable amount of slumping. Alternatively, these facies are mainly relatively far-travelled, fine-grained delta-bottomset beds related to an overlying lava-fed delta or deltas, a suggestion also made by Pirrie & Sykes (1987). This unit is not present at the Pirrie Col and Whisky Bay localities, but stratified sandy beds are present at a few other localities on James Ross Island (cf. Jonkers *et al.* 2002). At most localities diamictite is followed abruptly, but with indications of contemporary deposition, by volcanic rocks. At Pirrie Col a thin basalt unit, showing apparent contact metamorphism with the diamictite, is overlain by a basaltic lithic breccia. The relationships suggest that either a basalt flow was erupted in contact with water or ice, or else it is a sill that intruded along the diamictite–tuff interface and interacted with water-saturated tuff, thus generating the lithic breccia. Bedding, clast and alteration characteristics of the overlying tuffs at Pirrie Col suggest that they were erupted explosively, probably forming a subaqueous tuff cone, and were emplaced as a series of sediment gravity flows (cf. Smellie 2001). At Whisky Bay, and most localities elsewhere, hyaloclastite breccia overlies thin diamictite, and the latter shows soft-sediment 'intrusion' into the volcanic rocks. Together, the hyaloclastite breccia and capping subaerial lava form a cogenetic lava-fed delta (*sensu* Skilling 2002), which is the characteristic volcanic unit of the James Ross Island Volcanic Group (Nelson 1975; Smellie 1999). With the abundant signs of slumping in the sedimentary rocks, and the incorporation of blocks of diamictite into the lowermost volcanic rocks, it is envisaged that the hyaloclastite deltas on James Ross Island were emplaced rapidly onto the then soft diamicton. The volcanic processes were contemporaneous with regional-scale glaciation. The higher level diamictites, within the volcanic succession, only very rarely contain exotic clasts (i.e. Antarctic Peninsula-derived), so glaciation was subsequently local. James Ross and Vega islands were probably covered by ice caps, and, between eruptions, were subject to glacial erosion (as indicated by striated surfaces). The eruptive activity, particularly heat released by advancing lava-fed deltas, may have been responsible for the melting out of the diamictite from basal debris layers within the ice cap.

Discussion

Within the diamictites the degree of clast-rounding and abundant fine-grained matrix, together with striations, indicates the presence of wet-based glaciers with little supraglacial input. The dominance of diamictite within the overall association is similar to the facies association found at polythermal glaciers in the high-Arctic, such as Svalbard (Bennett *et al.* 1999; Hambrey & Glasser 2002; Glasser & Hambrey 2003). There is little evidence of substantial meltwater deposition, as would have been expected in a temperate glacier association (Benn & Evans 1998). The facies associated with modern local glaciers, which were also examined in some detail, are quite different from the Miocene deposits described above. Those glaciers terminating on land are cold-based, and, apart from supraglacial meltwater

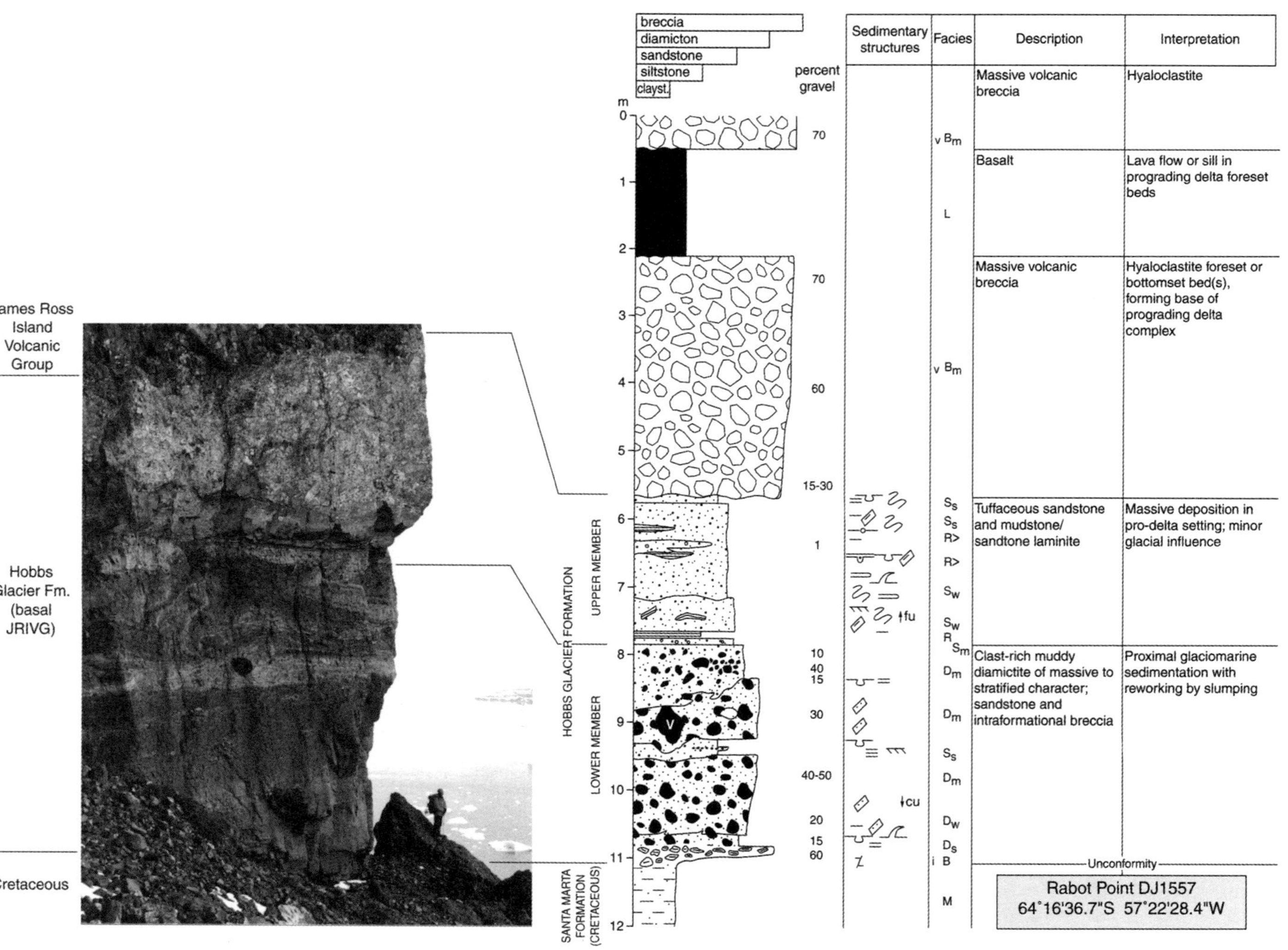

James Ross Island Volcanic Group
Hobbs Glacier Fm. (basal JRIVG)
Cretaceous
HOBBS GLACIER FORMATION
UPPER MEMBER
LOWER MEMBER
SANTA MARTA FORMATION (CRETACEOUS)
breccia
diamicton
sandstone
siltstone
clayst.
m
0
1
2
3
4
5
6
7
8
9
10
11
12
percent gravel
70
70
60
15-30
1
10
40
15
30
40-50
20
15
60
Sedimentary structures
Facies
Description
Interpretation
Massive volcanic breccia
Hyaloclastite
Basalt
Lava flow or sill in prograding delta foreset beds
Massive volcanic breccia
Hyaloclastite foreset or bottomset bed(s), forming base of prograding delta complex
Tuffaceous sandstone and mudstone/ sandtone laminite
Massive deposition in pro-delta setting; minor glacial influence
Clast-rich muddy diamictite of massive to stratified character; sandstone and intraformational breccia
Proximal glaciomarine sedimentation with reworking by slumping
Unconformity
Rabot Point DJ1557
64°16'36.7"S 57°22'28.4"W

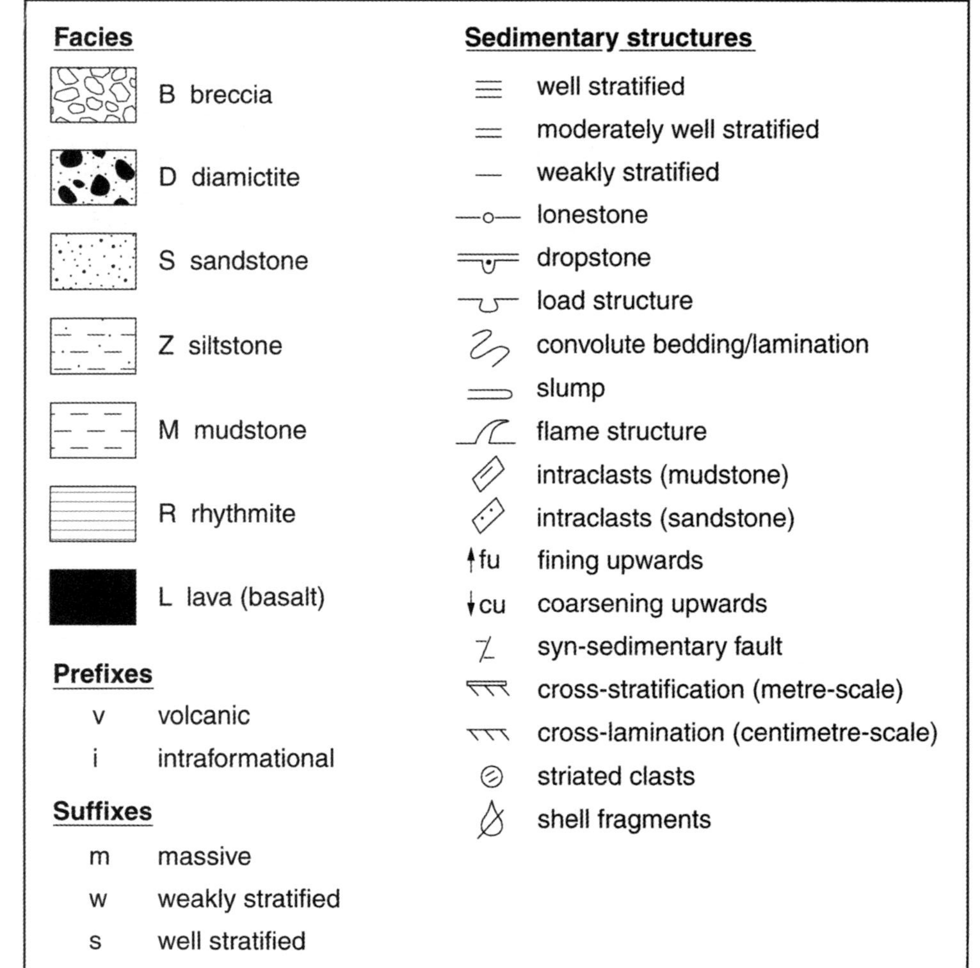

Fig. 3. Lithostratigraphy and interpretation of depositional environments of the Hobbs Glacier Formation and adjacent strata, Rabot Point, SE James Ross Island.

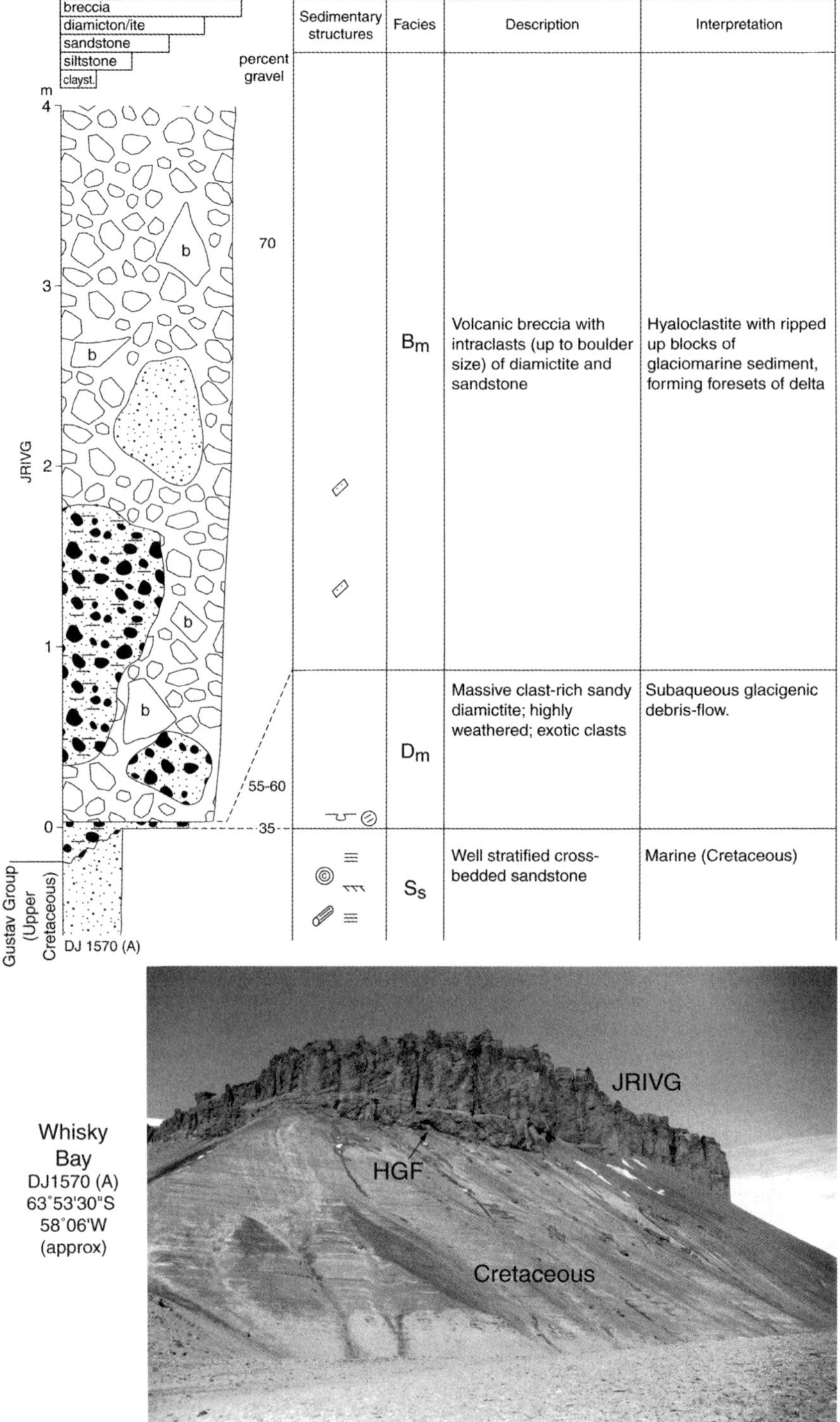

Fig. 4. Lithostratigraphy and interpretation of depositional environments of the glacigenic and adjacent strata at the base of the James Ross Island Volcanic Group, Whisky Bay, NW James Ross Island. The volcanic cliffs above the Hobbs Glacier Formation (HGF) outcrop are about 130 m high. JRIVG, James Ross Island Volcanic Group. See Fig. 3 for key.

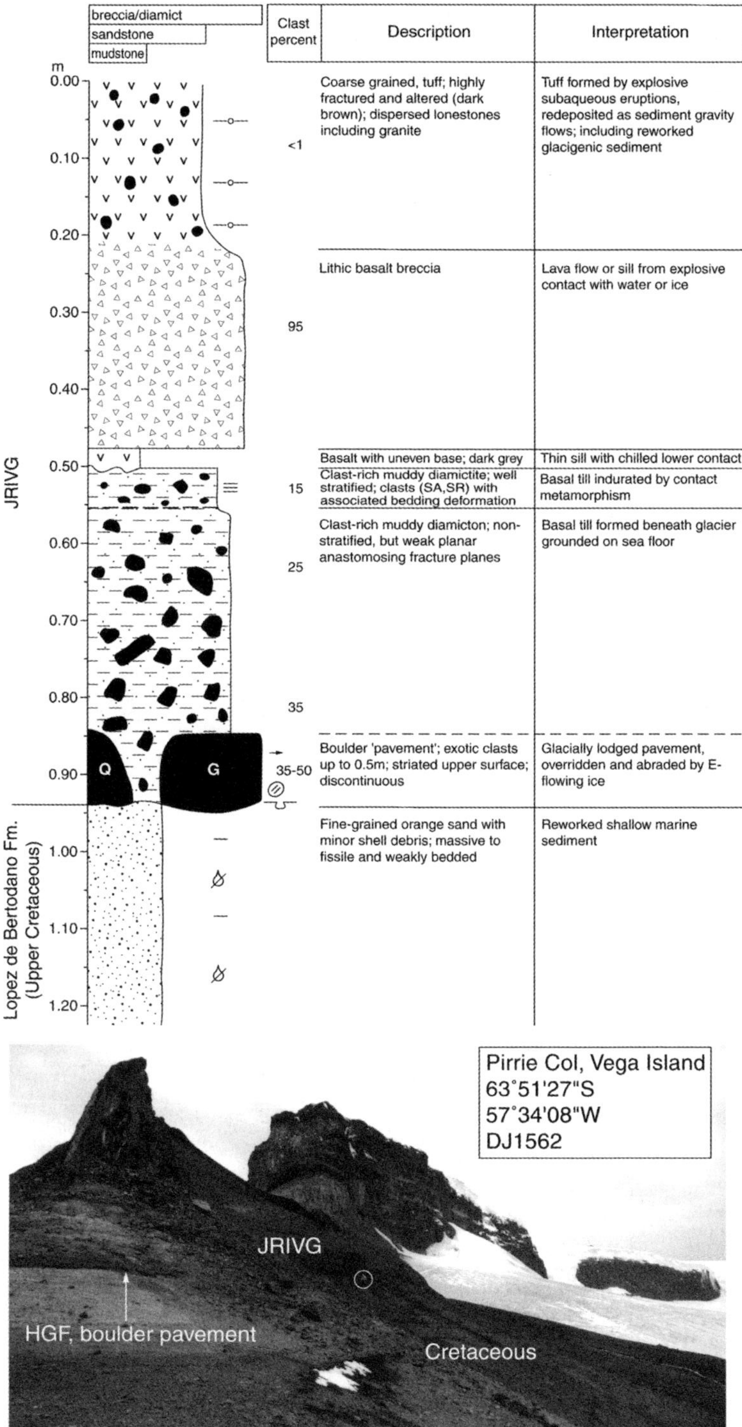

Fig. 5. Lithostratigraphy and interpretation of depositional environments of the glacigenic and adjacent strata at the base of the James Ross Island Volcanic Group, Pirrie Col, Vega Island. Person (ringed) gives scale. JRIVG, James Ross Island Volcanic Group. See Fig. 3 for key.

runoff causing gullying in the soft Cretaceous rocks, have little impact on the landscape (cf. Chinn & Dillon 1987). They carry little debris at their base, and the bulk of the supraglacial debris is angular. There are no signs of glacial abrasion of clasts, as would be expected at the base of a wet-based glacier. Tidewater glaciers in the area generally appear to be floating. During the period of investigation, sediment plumes associated with subglacial meltwater discharge were small in comparison with tidewater glaciers outside Antarctica, such as those in Svalbard (Boulton 1990) or Alaska (Powell & Alley 1997; Powell 2003). Therefore to use modern cold Antarctic glaciers as an analogue for the Hobbs Glacier Formation is inappropriate. Furthermore, the absence of any obvious supraglacial debris in the formation suggests that few rock surfaces were exposed; this in turn implies that the ice mass supplying the sediment was an ice cap or part of an ice sheet rather than a series of valley glaciers.

The provenance of many of the clasts in the basal exposures of the Hobbs Glacier Formation is clearly not local. The lithologically distinctive Trinity Peninsula Group of the Antarctic Peninsula is an obvious major source, as stated by Pirrie *et al.* (1997*b*) from a detailed petrographic study. Additional evidence for a provenance in the west is the striated boulder pavement at Pirrie Col. However, other sites imply deposition in a water body, with glacial sedimentation abruptly terminated by the onset of deposition of hyaloclastite breccias in lava-fed deltas. Glacial deposits also occur within the James Ross Island Volcanic Group at several higher levels, commonly in association with a striated unconformity. In these cases the clasts are almost exclusively of local provenance and bedrock striations are crudely radial from Mount Haddington, so the glaciation style is one of a local ice cap similar to, but somewhat larger than, that of the present day. Linking together the glacial processes to the onset of volcanism and placing them within the overall framework of the glacial history of Antarctica are challenges to overcome in the future.

Conclusions

The James Ross Island region lies near the northern limit of ice-sheet development in Antarctica, where fluctuations in ice-sheet extent and thermal regime should be most clearly shown. Our studies on James Ross and Vega islands indicate that glacial sedimentary sequences are much more common there than previously reported. The ages of many of the deposits are not yet well constrained, but published and ongoing new dating already indicate that a range of Miocene–Pliocene ages is present, and that the Hobbs Glacier Formation is diachronous. In particular, basal outcrops containing 'exotic' clasts (derived from the Antarctic Peninsula) have ages of 9.9, 5.64 and more than 2.7 Ma. This indicates that the Antarctic Peninsula Ice Sheet overrode parts of the region on at least three occasions, assuming that the dated fossils were not substantially reworked up-section by successive ice advances. In contrast, sedimentary units interbedded with the volcanic sequence are typically unfossiliferous and have clasts almost exclusively derived by erosion of their volcanic substrate, indicating that a small ice cap persisted on top of the volcanic sequence during most of its history. The sedimentary lithofacies present, in particular the dominance of diamictite at all levels in the sequence, suggest that the glacier thermal regime in the region was mainly subpolar. This is quite different from glaciers found on James Ross Island today, which are cold-based (polar). Relationships between the volcanic and sedimentary lithofacies indicate that eruptions and sedimentation were coeval, and that the sedimentary units were probably melted out of glacier ice by thermal effects of advancing lava-fed deltas. Finally, the unique association of glacial sedimentary deposits, several fossiliferous, and interbedded fresh volcanic rocks should render the region of vital importance in obtaining a chronologically well-constrained history of ice-sheet dynamics and palaeoenvironmental change for the Neogene Period.

This paper is a contribution to the British Antarctic Survey core project Late Cenozoic History of the Antarctic Ice sheet (LCHAIS). The authors thank Captain Moncrieff and the officers, crew and (particularly) helicopter pilots of HMS *Endurance*, which supported our field campaigns in January–February 2002 and January 2003, and our field assistants T. O'Donovan, A. Taylor and C. Day. The participation of M. J. Hambrey in the field programme was facilitated by the Antarctic Funding Initiative Collaborative Gearing Scheme (CGS90/01). Reviews by M. Bennett, D. Pirrie and D. Vaughan have helped clarify aspects of this paper.

References

BARRETT, P.J. 1996. Antarctic palaeoenvironment through Cenozoic times – a review. *Terra Antartica*, **3**, 103–119.

BARRETT, P.J. & RICCI, C.A. (eds). 2000. Studies from the Cape Roberts Project, Ross Sea, Antarctica: Scientific Report of CRP-2/2A. *Terra Antartica*, **7**, 213–412, 413–665.

BARRETT, P.J. & RICCI, C.A. (eds). 2001. Studies from the Cape Roberts Project, Ross Sea, Antarctica: Scientific Report of CRP-3. *Terra Antartica,* **7**, 123–308.

BARRON, J. & LARSEN, B. ET AL. 1991. *Kerguelen Plateau and Prydz Bay, Antarctica. Proceedings of the Ocean Drilling Program, Scientific Results*, **119**. Ocean Drilling Program, College Station, TX.

BENN, D.I. & EVANS, D.J.A. 1998. *Glaciers and Glaciation*. Arnold, London.

BENNETT, M.R. & GLASSER, N.F. 1996. *Glacial Geology: Ice Sheets and Landforms*. Wiley, Chichester.

BENNETT, M.R., HAMBREY, M.J., HUDDART, D., GLASSER, N.F. & CRAWFORD, K. 1999. The landform and sediment assemblage produced by a tidewater glacier surge in Kongsfjorden, Svalbard. *Quaternary Science Reviews*, **18**, 1213–1246.

BIRKENMAJER, K. 1996. Tertiary glacial/interglacial palaeoenvironments and sea-level changes, King George Island, West Antarctica. An overview. *Bulletin of the Polish Academy of Sciences, Earth Sciences*, **44**, 157–181.

BOULTON, G.S. 1990. Sedimentary and sea level changes during glacial cycles and their control on glacimarine facies architecture. *In*: DOWDESWELL, J.A. & SCOURSE, J.C. (eds) *Glacimarine Environments: Processes and Sediments*. Geological Society, London, Special Publications, **53**, 15–52.

CHINN, T.J.H. & DILLON, A. 1987. Observations on a debris-covered polar glacier 'Whisky Glacier', James Ross Island, Antarctic Peninsula, Antarctica. *Journal of Glaciology*, **33**, 300–310.

CRAME, J.A., PIRRIE, D., RIDING, J.B. & THOMSON, M.R.A. 1991. Campanian-Maastrichtian (Cretaceous) stratigraphy of the James Ross Island area, Antarctica. *Journal of the Geological Society, London*, **148**, 1125–1140.

DENTON, G.H., SUGDEN, D.E., MARCHANT, D.R., HALL, B.L. & WILCH, T.I. 1993. East Antarctic ice sheet sensitivity to Pliocene climate change from a Dry Valleys perspective. *Geografiska Annaler*, **75A**, 155–204.

DINGLE, R.V. & LAVELLE, M. 1998. Antarctic Peninsular cryosphere: Early Oligocene (*c.* 30 Ma) initiation and a revised glacial chronology. *Journal of the Geological Society, London*, **155**, 433–437.

EVANS, D.J.A. (ed.). 2003. *Glacial Landsystems.* Arnold, London.

GLASSER, N.F. & HAMBREY, M.J. 2003. Ice-marginal terrestrial landsystems: Svalbard polythermal glaciers. *In*: EVANS, D.J.A. (ed.) *Glacial Landsystems*. Arnold, London, 65–88.

HAMBREY, M.J. 1994. *Glacial Environments*. UCL Press, London.

HAMBREY, M.J. & GLASSER, N.F. 2002. Development of landform and sediment assemblages at maritime High-Arctic glaciers. *In:* HEWITT, K., BYRNE, M.-L., ENGLISH, M. & YOUNG, G. (eds) *Landscapes of Transition*. Kluwer, Dordrecht, 11–42.

HAMBREY, M.J. & GLASSER, N.F. 2003. The role of folding and foliation development in the genesis of medial moraines: examples from Svalbard glaciers. *Journal of Geology*, **111**, 471–485.

HAMBREY, M.J. & MCKELVEY, B. 2000*a*. Major Neogene fluctuations of the East Antarctic ice sheet: Stratigraphic evidence from the Lambert Glacier region. *Geology*, **28**, 887–891.

HAMBREY, M.J. & MCKELVEY, B. 2000*b*. Neogene fjordal sedimentation in the Prince Charles Mountains, East Antarctica. *Sedimentology*, **47**, 577–607.

HAMBREY, M.J. & WISE, S.W. (eds). 1998. Studies from the Cape Roberts Project, Ross Sea, Antarctica: Scientific Report of CRP-1. *Terra Antartica*, **5**, 255–713.

HAMBREY, M.J., WEBB, P.-N., HARWOOD, D.M. & KRISSEK, L.A. 2003. Neogene glacial record from the Sirius Group of the Shackleton Glacier area, central Transantarctic Mountains, Antarctica. *Geological Society of America Bulletin*, **115**, 994–1015.

HARWOOD, D.M. & WEBB, P.-N. 1998. Glacial transport of diatoms in the Antarctic Sirius Group: Pliocene refrigerator. *GSA Today*, **8**, (4), 1–8.

IPCC. 2001. HOUGHTON, J.T., DING, Y., GRIGGS, D.J., NOGUER, M., VAN DER LINDEN, P.J. & XIAOSU, D. (eds) The Scientific Basis Contribution of Working Group I to the Third Assessment Report of the Intergovernmental Panel on Climate Change (IPCC). Cambridge University Press, Cambridge.

JONKERS, H.A. 1998. The Cockburn Island Formation; Late Pliocene interglacial sedimentation in the James Ross Basin, northern Antarctic Peninsula. *Newsletters on Stratigraphy*, **36**, 63–76.

JONKERS, H.A. & KELLEY, S.P. 1998. A reassessment of the age of the Cockburn Island Formation, northern Antarctic Peninsula, and its palaeoclimatic implications. *Journal of the Geological Society, London*, **155**, 737–740.

JONKERS, H.A., LIRIO, J.M., DEL VALLE, R.A. & KELLEY, S.P. 2002. Age and depositional environment of the Miocene–Pliocene glaciomarine deposits, James Ross Island, Antarctica. *Geological Magazine*, **139**, 577–594.

MARCHANT, D.R., DENTON, G.H. & SWISHER, C.C., III. 1993. Miocene–Pliocene–Pleistocene glacial history of Arena Valley, Quartermain Mountains, Antarctica. *Geografiska Annaler*, **75A**, 269–302.

MONCRIEFF, A.C.M. 1989. Classification of poorly sorted sedimentary rocks. *Sedimentary Geology*, **65**, 191–194.

NELSON, P.H.H. 1975. *The James Ross Island Volcanic Group of North-east Graham Land*. British Antarctic Survey Scientific Report, **54**.

O'BRIEN, P.E., COOPER, A.K. ET AL. 2001. *Proceedings of the Ocean Drilling Program, Initial Reports, Leg 188*. Texas A&M University, College Station, TX.

PIRRIE, D. 1991. Controls on the petrographic evolution of an active margin sedimentary sequence: the Larsen Basin, Antarctica. *In*: MORTON, A.C., TODD, S.P. & HAUGHTON, P.D.W. (eds) *Developments in Sedimentary Provenance Studies*. Geological Society, London, Special Publications, **57**, 231–249.

PIRRIE, D. & SYKES, M.A. 1987. Regional significance of proglacial delta-front reworked tuffs, James Ross Island area. *British Antarctic Survey Bulletin*, **77**, 1–12.

PIRRIE, D., CRAME, J.A., LOMAS, S.A. & RIDING, J.B. 1997*a*. Late Cretaceous stratigraphy of the Admiralty Sound region, James Ross Basin, Antarctica. *Cretaceous Research*, **18**, 109–137.

PIRRIE, D., CRAME, J.A. & RIDING, J.B. 1991. Late Cretaceous stratigraphy and sedimentology of Cape Lamb, Vega Island, Antarctica. *Cretaceous Research*, **12**, 227–258.

PIRRIE, D., CRAME, J.A., RIDING, J.B., BUTCHER, A.R. & TAYLOR, P.D. 1997*b*. Miocene glaciomarine sedimentation in the northern Antarctic Peninsula region: the stratigraphy and sedimentology of the Hobbs Glacier Formation, James Ross Island. *Geological Magazine*, **136**, 745–762.

POWELL, R.D. 2003. Subaquatic landsystems: fjords. *In:* EVANS, D.J.A. (ed.) *Glacial Landsystems*. Arnold, London, 313–347.

POWELL, R.D. & ALLEY, R.B. 1997. Grounding-line systems: process, glaciological inferences, and the stratigraphic record. *In*: BARKER, P.F. & COOPER, A.C. (eds) *Geology and Seismic Stratigraphy of the Antarctic Margin*, Volume 2. American Geophysical Union, Antarctic Research Series, 169–187.

SKILLING, I.P. 2002. Basaltic pahoehoe lava-fed deltas: large-scale characteristics, clast generation, emplacement processes and environmental discrimination. *In*: SMELLIE, J.L. & CHAPMAN, M.G. (eds) *Volcano–Ice Interaction on Earth and Mars*. Geological Society, London, Special Publications, **202**, 91–113.

SMELLIE, J.L. 1999. Lithostratigraphy of Miocene–recent, alkaline volcanic field in the Antarctic Peninsula and eastern Ellsworth Land. *Antarctic Science*, **11**, 362–378.

SMELLIE, J.L. 2001. Lithofacies architecture and construction of volcanoes in englacial lakes: Icefall Nunatak, Mount Murphy, eastern Marie Byrd Land, Antarctica. *In*: WHITE, J.D.L. & RIGGS, N. (eds) *Lacustrine Volcaniclastic Sedimentation*. International Association of Sedimentologists, Special Publications, **30**, 73–98.

STROEVEN, A.P. & KLEMAN, J. 1999. Age of Sirius Group on Mount Feather, McMurdo Dry Valleys, Antarctica, based on glaciological inferences from the overridden mountain range of Scandinavia. *Global and Planetary Change*, **23**, 231–247.

SUGDEN, D.E. 1996. The East Antarctic ice sheet: unstable ice or unstable ideas? *Transactions of the Institute of British Geographers, New Series*, **21**, 443–454.

SUGDEN, D.E., MARCHANT, D.R. & DENTON, G.H. (eds). 1993. The case for a stable east Antarctic ice sheet. *Geografiska Annaler*, **75A**, 151–351.

TROEDSON, A.L. & RIDING, J.B. 2002. Upper Oligocene to lowermost Miocene strata of King George Island, South Shetland Islands, Antarctica: stratigraphy, facies analysis, and implications for the glacial history of the Antarctic Peninsula. *Journal of Sedimentary Research*, **72**, 510–523.

TROEDSON, A.L. & SMELLIE, J.L. 2002. The Polonez Cove Formation of King George Island, Antarctica: stratigraphy, facies and implications for mid-Cenozoic cryosphere development. *Sedimentology*, **49**, 277–301.

WEBB, P.-N. & HARWOOD, D.M. 1991. Late Cenozoic history of the Ross Embayment, Antarctica. *Quaternary Science Reviews*, **10**, 215–223.

WEBB, P.-N., HARWOOD, D.M., MABIN, M.C.G. & MCKELVEY, B.C. 1996. A marine and terrestrial Sirius Group succession, middle Beardmore Glacier–Queen Alexandra Range, Transantarctic Mountains, Antarctica. *Marine Micropaleontology*, **27**, 273–297.

WHITEHEAD, J.M., HARWOOD, D.M. & MCMINN, A. 2003. Ice-distal Upper Miocene strata from inland Antarctica. *Sedimentology*, **50**, 531–552.

WILSON, G.S. 1995. The Neogene East Antarctic ice sheet: a dynamic or stable feature? *Quaternary Science Reviews*, **14**, 101–123.

WILSON, G.S., HARWOOD, D.M., ASKIN, R.A. & LEVY, R.H. 1998. Late Neogene Sirius Group strata in Reedy Valley, Antarctica: a multiple-resolution record of climate, ice-sheet and sea-level events. *Journal of Glaciology*, **44**, 437–447.

Index

Page numbers in *italics* refer to Figures; page numbers in **bold** refer to Tables.

From: Francis, J. E., Pirrie, D. & Crame, J. A. (eds) 2006. *Cretaceous–Tertiary High-Latitude Palaeoenvironments, James Ross Basin, Antarctica.* Geological Society, London, Special Publications, **258**, 201–206. 0305–8719/06/$15 © The Geological Society of London 2006.